A Future for Public Service Television

A Future for Public Service Television

Edited by Des Freedman and Vana Goblot

Goldsmiths
Press

© 2018 Goldsmiths Press
Published in 2018 by Goldsmiths Press
Goldsmiths, University of London, New Cross
London SE14 6NW

Printed and bound by Clays Ltd, St Ives plc.
Distribution by the MIT Press
Cambridge, Massachusetts, and London, England

A CIP record for this book is available from the British Library.

Library of Congress Cataloging-in-Publication Data
Names: Freedman, Des, 1962- editor. | Goblot, Vana, editor.
Title: A future for public service television / edited by Des Freedman and Vana Goblot.
Description: London, England: Goldsmiths Press, [2018] | Includes bibliographical references and index.
Identifiers: LCCN 2017039952 | ISBN 9781906897710 (hardcover: alk. paper)
Subjects: LCSH: Public television–Great Britain. | Public service television programs–Great Britain. |
  British Broadcasting Corporation.
Classification: LCC HE8700.79.G7 F88 2018 | DDC 384.55/40941–dc23
LC record available at https://lccn.loc.gov/2017039952

ISBN 978-1-906897-71-0 (hbk)
ISBN 978-1-906897-76-5 (ebk)

www.gold.ac.uk/goldsmiths-press

Goldsmiths
UNIVERSITY OF LONDON

Contents

Part One

Introduction

Foreword

Lord Puttnam

To instruct democracy, if possible to reanimate its beliefs ... such is the first duty imposed on those who would guide society.

Alexis de Tocqueville (1863)

If the past few years have taught me anything, it's that our need for trusted sources of information, comprised of tolerant balanced opinion, based on the very best available evidence, has never been greater. In an era of fake news, alternative facts and online trolls our public service broadcasters (PSBs) stand as guarantors of accurate, informed and impartial information.

If only the same could be said elsewhere.

Our democracy is increasingly distorted by mendacious axe-grinding on the part of the tabloid press.

In his book, *Enough Said*, the former director general of the BBC Mark Thompson writes that:

Intolerance and illiberalism are on the rise almost everywhere. Lies go unchecked. At home, boundaries – of political responsibility, mutual respect, basic civility – which seemed secure a mere decade ago, are broken by the week.[1]

Our Inquiry set out to discover if the concept of public service broadcasting could survive in the hyper-commercial, market-dominated media environment of the 21st century.

In the pages that follow I believe that we have made the case that, not only do the public believe it should survive, but that our evolved PSB ecology functions as the most reliable bulwark available to a truly plural and informed democracy in its battle against ill-informed populism and market totalitarianism.

The truly successful societies of the 21st century are likely to be those in which the provision of news and information is rapid, accurate and trusted.

'Rapidity' is now a given, 'accuracy' remains a challenge, but 'trust' is proving increasingly elusive.

It's a commonplace to believe that trust lies at the heart of a sustainable democracy, yet, as Mark Thompson suggests, it is evaporating on a daily basis and, once shredded, could prove all but impossible to rebuild.

Clearly this is a battle we are losing, as the public has made it clear they no longer have any faith in the press and are developing increasing reservations about television.

I think most people accept that knowledge and understanding play a vital role in our ability to navigate the complexities and opportunities of our times. So where do we look for guidance, what defines an informed and active citizen?

This book argues that a well-resourced and fully independent public service television system free of political coercion offers our most reliable means of rebuilding public trust and accountability.

From time to time we glimpse the possibility of renewal, all too frequently evolving out of tragedy; we have to get better at grasping and building upon the lessons of Hillsborough, Bloody Sunday, the deaths of Milly Dowler, Dr David Kelly and the murder of Jo Cox MP.

I started out by suggesting that public service broadcasting was a noble idea. The issue is whether *we* ourselves can find sufficient nobility to nurture and protect it.

In his introduction to the White Paper on charter renewal, the former secretary of state for culture, media and sport, John Whittingdale, said of the BBC:

It is a revered national institution, and a familiar treasured companion. It is a cultural, economic and diplomatic force that touches the lives of almost all of those who live in the UK and hundreds of millions beyond these shores.[2]

Of what else in British life could a similar claim be made?

This book attempts to analyse the current strengths as well as the threats to our PSB ecology, and to offer an evidence-based argument for the conditions under which it can not just survive, but thrive.

Notes

1 Mark Thompson, *Enough Said: What's Gone Wrong with the Language of Politics?* (London: Bodley Head, 2016).

2 Quoted in Department for Culture, Media & Sport, *A BBC for the Future: a Broadcaster of Distinction*, White Paper, May 2016, p. 5.

Introduction: The Long Revolution

Des Freedman

TV Won't Go Away

Television is leading a charmed existence. After all, it is no longer supposed to exist. With the rise of the internet and the widespread availability of digital platforms, what is the point in the 21st century of a 20th-century technology that broadcasts from a central point out to millions of viewers who are increasingly preoccupied with making, circulating and consuming non-broadcast content on their smartphones and iPads? How can television, with its baggage of 'mass audiences' and one-way transmissions, compete with a digital universe that embodies the more fragmented and decentred nature of the way we live today? The American writer George Gilder noticed this development back in 1994, just after the emergence of the web. He predicted that 'TV will die because it affronts human nature: the drive to self-improvement and autonomy that lifted the race from the muck and offers the only promise for triumph in our current adversities.'[1]

But TV hasn't died. In fact it has stubbornly refused to disappear in the face of the white heat of the digital revolution. Contrary to what people like Gilder predicted, the internet hasn't killed television but actually extended its appeal – liberating it from the confines of the living room where it sat unchallenged for half a century and propelling it, via new screens, into our bedrooms, kitchens, offices, buses, trains and streets. Television has both grown and shrunk: its giant screens now adorn the walls of our shared spaces but it is simultaneously mobile and portable. Meanwhile, we are facing an epidemic of TV content described by one US executive as 'Peak TV'. Back in 2015, when FX boss John Landgraf first warned about the viability of the explosion in scripted programmes on American television, there were 'only' 400 such shows; now there are likely to be over 500.[2]

Even more puzzling than the resilience of the television experience is the fact that in the UK, the heartland of creative innovation and deregulated markets, the vast majority of the content consumed is provided by a group of people who are described as 'public service broadcasters' (PSBs)[3] and whose remit is not simply to secure profits alone but instead to pursue a range of political, social and cultural objectives aimed at maximising the public interest. This too has been dismissed as a project without a future. 'Public service broadcasting will soon be dead,' argued the former ITV chief executive Richard Eyre in 1999. 'It will soon be dead because it relies on an active broadcaster and a passive viewer.'[4] Yet millions of 'passive viewers' continue to consume, on average, just under four hours a day of material that combines, in Eyre's language, 'the wholesome, healthy and carefully crafted' with the 'easily digestible, pre-packaged, and the undemanding'.

One of the reasons for these apocalyptic visions of TV's imminent demise is the confusion between television as a specific technology and its status as a cultural form. The media commentator Michael Wolff highlights the frequent conflation of TV 'as a business model,' which he argues is incredibly healthy, and TV as a 'distribution channel' whose future is far less certain. He concludes that there is little reason to believe that 'people will stop watching TV, even if they stop watching *the* TV.'[5] So while we may not all watch *Games of Thrones* at the same time and on the set in the living room, millions of us will nevertheless still watch it – perhaps days later, perhaps as part of an all-night binge and perhaps on our tablets on the train home from work. Our routines and access points may not be the same but there is little evidence that we have lost our appetite for television-like content.

On the other hand, there is ample evidence that television is changing – and changing fast. Back in 1982, the UK had three channels, a powerful duopoly and audiences for individual programmes that were regularly in the tens of millions; now we have a multichannel landscape, fragmented audiences, more complex consumption patterns, new sources of production and a constant innovation in distribution platforms. In particular, there is the prospect of a mass exodus of young people from linear television to online video consumption that is not controlled by traditional channels and voices. Viewing of live TV by 16–24 year olds dropped by 14% between 2014 and 2016 and now makes up only a third of their total consumption of video content.[6] These are the digital natives who are less committed to watching television, given that they are now likely to consume content across a range of platforms and devices, and we cannot be sure whether they will ever return to a quiet night in front of the TV. On the other hand, even the most 'disruptive' voices are launching television *channels*, with Vice Media, a relatively new entrant to newsgathering that has millions of subscribers to its videos, launching 'Viceland' on the Sky platform in 2016. The fact remains that even the young remain voracious consumers of television content.

Television is, therefore, characterised by its durability as well as an underlying volatility and uncertainty. Just as the landscape is undergoing enormous change, it is also characterised by important continuities. The public service broadcasters (PSBs) – BBC, Channel 3, Channel 4 and Channel 5 – continue to command our attention; their share of viewing (if you include their portfolio channels) has fallen but only from 76% in 2006 to 70% in 2016. The PSBs also dominate investment in original programming and the vast majority of our viewing – some 80% according to Ofcom – still takes place live.[7] It is important to acknowledge these continuities if we are to appreciate the significance of the change that *is* taking place and then to consider how best to inspire and sustain high quality television in the UK.

Sometimes, this means going beyond the headlines. For example, a 2016 report examining the crisis affecting TV news notes the 'significant declines in traditional television in technologically developed markets' and argues that television is now facing the same collapse as the print press with audiences in the UK declining by some 3–4% per year since 2012.[8] That is true but highly selective. Viewing via the TV set has indeed fallen by 26 minutes a day in the last five years but this has simply brought it back to virtually the same level that it was in 2006: 3 hours and 32 minutes every day.[9] Meanwhile, Enders Analysis predict that the broadcast sector is likely 'to account for the greatest share of viewing for many years to come' with a scenario that sees over four hours a day of viewing in 2025 of which three-quarters continues to take place via a television set.[10]

So while it is easy (and necessary) to be absorbed by the challenge of the new, it would be foolish to ignore the grip of the old. For example, there is, understandably, a huge amount of interest in (and concern in the ranks of traditional broadcasters about) the vast subscriber base of video bloggers on YouTube given that they constitute the digital generation that is not guaranteed to return en masse to linear TV. However, there is a big difference between the *potential* audience of these vloggers and the numbers who *actually* watch an individual video. So while the British rapper KSI has, at the time of writing, some 16 million subscribers, just over 2 million watch the average programme; while the fashion vlogger Zoella managed to garner 4.1 million hours of viewing in the first 3 months of 2016 – itself an incredibly impressive feat – this hardly compares to the 76 million hours that UK audiences spent in front of ITV's *Downton Abbey*.[11] 'Buzzy, short form content fill gaps that have always existed' conclude Enders; 'yet, despite the hype, it will remain supplementary to long-form programming.'[12]

Similarly, while digital intermediaries like Facebook and Twitter have changed the dynamics of the news landscape, terrestrial news outlets remain extremely influential in shaping agendas and legitimising specific perspectives on major issues of public concern including Brexit, immigration, race, austerity and terrorism. The bulletins provided by the public service broadcasters remain by far the most popular sources of news for adults in the UK with BBC One bulletins alone dominating more than two-thirds of TV news viewing.[13] The BBC's overall reach is vastly greater than that of any other source; even for those people who rely on the internet for news, some 56% of them use BBC sites, far more than the 27% who use Facebook which, of course, continues itself to rely heavily on other legacy news outlets to generate the bulk of original news content.[14]

The Puttnam Inquiry into the Future of Public Service Television (and its Antecedents)

This relationship between continuity and change, and indeed the role and relevance of public service itself, were issues that were key to the launch of the Inquiry into the Future of Public Service Television and the report that followed and that is the focus of this book.[15] The Inquiry was very much a product of its time: it was launched in November 2015 when the BBC's decennial Charter Review was in full swing and when government ministers were speaking about the possibility of diminishing the scope of the Corporation and of privatising the publicly-owned Channel 4. With the rise of new sources of content from pay-TV broadcasters and on-demand services like Amazon and Netflix – some of whose output, it could be argued, has clear public service qualities – and changing modes of consumption to which I have already referred, the UK's television landscape was going through a particularly volatile period.

The Inquiry was initiated by researchers at Goldsmiths, University of London and chaired by the Labour peer and film producer David Puttnam. It invited submissions from broadcasters, academics, civil society groups and regulators and organised events – from a conversation between Lord Puttnam and the BBC director general at BAFTA in London to a debate about representation involving filmmaker Ken Loach and TV producer Phil Redmond at a community festival in Liverpool to a forum on the future of television in Wales held at Cardiff University – in all the nations of the UK. It sought to reflect on the place and performance of public service in an age of platforms and populism; it did *not* attempt to second-guess which platforms will dominate in the future nor to speculate, for example, on precisely when we will switch off terrestrial television and move to a wholly online system. Instead it was more preoccupied with the *purposes* of television in an era that is characterised not simply by technological transformations but also by changing cultural and political attitudes: high levels of disengagement from traditional political parties, the collapse of the centre ground, falling levels of trust in major public institutions and a willingness to identify with social groups beyond the level of the nation state. It aimed not to reproduce old debates about the shape of public service broadcasting but to inform what a future public service media system might look like.

Above all, the Inquiry and report considered the extent to which the UK's most popular television channels were successful in addressing the concerns, representing the interests and telling the stories of all the citizens of the UK. It sought to highlight the conditions that best facilitate the production and circulation of high quality, creative and relevant public service content in these complex circumstances rather than to dwell only on the specific apparatuses through which this content is likely to be consumed.

In doing this, the Inquiry drew inspiration from a previous investigation into the future of broadcasting that also set out to examine the purposes of television at a time of major social change. In 1962, the report of the Pilkington Committee recommended the adoption of colour television licences and the creation of a further television channel to be run by the BBC. That report, however, was far more than a mere list of policy prescriptions and technological missives, but a searing indictment of the direction of travel of British television under the influence of a growing commercial mindset and an increasing number of programmes imported from the USA. It advocated measures designed to revitalise the idea of public service broadcasting and to foster a more creative and robust public culture.

The Pilkington report was perhaps best known for its hard-hitting critique of the 'emotional tawdriness and mental timidity' of a new 'candy-floss world'[16] that was epitomised by commercial television. Television's power to influence and persuade, it argued, was being abused in the search for cheap thrills and high ratings, a situation from which the BBC too was not immune. This 'lack of variety and originality, an adherence to what was "safe"' was directly related to TV's 'unwillingness to try challenging, demanding and, still less, uncomfortable subject matter.'[17] Critics, however, attacked the report as elitist and moralising when, in fact, it made a very strong case for an expansion, and not a narrowing, of content. In words that resonate today given contemporary debates about whether public service broadcasters should restrict themselves to areas left vacant by their competitors, Pilkington argued that:

No one can say he is giving the public what it wants, unless the public knows the whole range of possibilities which television can offer and, from this range, chooses what it wants to see. For a choice is only free if the field of choice is not unnecessarily restricted. The subject matter of television is to be found in the whole scope and variety of human awareness and experience.[18]

Public service television, if it is to show the full diversity of its audience base, needs to make available the broadest range of content while, at the same time, it cannot afford to turn away from the responsibility to engage minority audiences.[19]

Pilkington contributed to a hugely important debate about the contribution that television could make to public life and private interests. It attempted to create an infrastructure that would allow both the ITV network and the BBC to act as a public service engaged in a 'constant and living relationship with the moral condition of society'.[20] Today, the status and definition of public service is far more fluid and we have lost the 'moral' certainties that underpinned the Pilkington Committee's investigation. For some, the whole notion of public service television suggests the paternalistic imposition of 'desirable' (for which read 'establishment') values at a time when citizens are increasingly unwilling to be the passive recipients of established belief systems. For others, public service suggests a regulated form of speech that inhibits accumulation and restricts growth. For example, in a famous speech at the 1989 Edinburgh International Television Festival, a year in which walls were coming down across the world, Rupert Murdoch tore into what he described as the 'British broadcasting elite' and demolished the 'special privileges and favours' that were associated with the 'public interest'. 'My own view', insisted the founder of the UK's new satellite service, 'is that anybody who, within the law of the land, provides a service which the public wants at a price it can afford is providing a public service. So if in the years ahead we can make a success of Sky Television, that will be as much a public service as ITV'.[21]

This neoliberal conception of broadcasting's role runs counter to the normative position that public service should not simply be measured by ratings nor should it exist simply to correct any tendency for markets to under-serve minority audiences. Public service television – and public service media as it will emerge – are not merely the medicine that it is sometimes necessary to take to counter the lack of nutrition of a purely commercial system. In many ways, public service television is – at least, it is *supposed* to be – about a specific conception of culture that is irreducible to economic measures of 'profit and loss'; it refers to the 'establishment of a communicative relationship' rather than to 'the delivery of a set of distinct commodities to consumers'.[22] Its main goal is not to sell audiences to advertisers or subscription broadcasters or to conduct private transactions but to facilitate public knowledge and connections. According to Liz Forgan, a former director of programmes at Channel 4: 'Television channels are not pork barrel futures or redundant government buildings. They are creators, patrons and purveyors of a highly popular (in both senses) variety of entertainment, information and culture to millions'.[23]

But if public service television is to be a 'public good' that has multiple objectives,[24] then it must, for example, provide content that is popular and challenging; it must be universally available; it must enhance trust in and diversity of news and opinion; it must increase the plurality of voices in the UK media landscape; and it must provide a means through which UK citizens can enter into dialogue. The problem is that, for increasing numbers of people, public service television is falling short of these lofty ideals; whether it is in relation to news coverage that reproduces neoliberal agendas on a wide range of political issues, or the inadequate representation of minority groups in their everyday lives.[25] Indeed, this is not a new phenomenon: various studies have drawn attention to the long and intimate association between the 'high priests' of PSB – notably inside the BBC – and elite interests.[26]

Given that public service broadcasting is, at its root, simply an intervention into the media market, it will therefore need constant renewal if it is boldly to serve publics and not be captured by establishment interests. So when – under political and commercial pressure – broadcasters fail to protect their own independence, to challenge falsehoods, to provide a meaningful range of opinions and to relate to the full diversity of their audiences, then it isn't enough to fall back on the *status quo*. While right-wing populist voices are very effective at attacking the failures of what they describe as the 'liberal media', it is interesting that those voices on the liberal left – including many who feature in this book – who seek to champion public service are so often wedded to an uncritical defence of a media system even when there is evident need for reinvigoration. There is, after all, a big difference between the *concept* of public service and its actual institutional forms. As Raymond Williams once pointed out: 'it is very important that the idea of public service should not be used as a cover for paternal or even authoritarian system ...The only way of achieving this is to create new kinds of institution'.[27]

For too long, most broadcasters have gravitated towards a perceived 'centre ground' and, when this 'centre ground' was coming unstuck, instead of promoting a multitude of voices and taking risks, they have too often clung to the familiar and acceptable. Yet precisely because of the political polarisation and fragmentation we see today, there is an urgent role for public service television to act both as a counterweight to a commercial system that is more likely to chase ratings and pursue formulae and as a champion of all those whose needs have not been adequately met by mainstream media organisations. The challenge, therefore, is to design both a regulatory and professional culture to support – indeed to build – a creative, spirited and independent public service media. The truth is that the *status quo* isn't really an option: technology won't allow it, markets won't stand still and, perhaps most significantly, there is a growing appetite on the part of the public for change and a desire to refresh even some of the most 'cherished' institutions in the UK. What should replace the *status quo* is precisely the subject of this collection.

Structure of the Book

This book aims to contribute to the discussion about what kind of public service media we want and to provide some blueprints for future policy action. The editors have commissioned leading practitioners and academics to write brand new chapters that both provide context for the debate and reflect on some of the issues raised in the Puttnam Report itself. Mark Thompson, the former chief executive of Channel 4 and former director general of the BBC, argues that we are 'living through an exhilarating and bloody revolution' in which public service media will be increasingly important as a corrective to the enormous uncertainties facing commercial media. The award-winning independent producer Jon Thoday provides a robust defence of public service broadcasting as a 'public good' and calls for imaginative measures to increase its visibility in the TV landscape while Tess Alps, founder of industry body Thinkbox, reminds us why watching television remains such a powerful and necessary cultural activity and highlights the role of advertising in sustaining at least a part of the public service environment.

Leading academics from the US, UK and Europe then map out some of the central dynamics of public service television as a specific intervention into the broadcast sphere. Amanda Lotz notes that the Puttnam Report remains stuck in a 'broadcast paradigm' and urges us to consider how internet protocols can inspire a whole new conception of public service media; the eminent media economist, Paddy Barwise, on the other hand, insists that public service broadcasting as it exists today continues to provide both better value for money than its commercial rivals and important citizenship benefits. Jennifer Holt, Matthew Powers, and Trine Syvertsen and Gunn Enli provide a crucial comparative dimension by focusing on a range of different television systems. Holt examines how changing technologies, metrics and policies are affecting what is left of 'public service' in the US while Powers reports back on a survey of 12 countries that shows that, despite all the challenges, public media 'remain vital components of contemporary democratic media systems'. Syvertsen and Enli address the main dimensions of what they see as a 'crisis discourse' in relation to European public service broadcasting and highlight some of the strategies and responses that have been pursued across the continent.

The final three freshly commissioned chapters of Part Two confront some specific issues that are at the heart of the Puttnam Report. Sarita Malik addresses the vexed (and very topical) debate about diversity and concludes that PSB, while having a responsibility to address multicultural audiences and to provide multicultural content, also has to recognise its own complicity in denying marginalised groups full and representative access to the airwaves. The historian David Hendy considers some of the key challenges – both external and organisational – that face the BBC in the run-up to its centenary celebrations in 2022 while James Bennett makes a persuasive case for a policy focus on public service algorithms: for new 'logics of recommendations' to leverage public service principles of exploration and serendipity into the digital age.

These new chapters are followed by extracts from the Puttnam Report that are then supplemented by evidence that was provided to and generated by the Inquiry in the shape of formal submissions and transcripts of events. We have ordered these thematically into separate parts that consider first the underlying principles that animate public service television and, next, the role of public service in a digital and on-demand environment. The book then focuses on debates on representation, both in relation to the UK as a multicultural society and also as an entity composed of different nations, regions and social groups. The penultimate part looks at the specific challenges facing genres that are traditionally associated with public service broadcasting – such as arts, news, and children's programming – before concluding with the report's recommendations and an Afterword written by two members of the Inquiry team.

Every edited collection should come with a health warning. First, while the book is designed to inform general debates on the future of public service television, it is also a product of a rather specific environment: the UK's very mixed broadcast ecology. As such, some of the discussion might seem to be rather 'local' (and even dated) even though we believe that these debates have far more widespread ramifications. Second, we are not trying to legitimise a particular narrative that will magically unify the very different contributions that populate the book but to reflect the hugely different assessments of both the problems and solutions facing PSB. After all, one thing became very clear during the course of the Inquiry: that there is very little agreement on key issues including the speed and impact of change, the performance of our major broadcasters, the measures needed to protect public service institutions, future funding solutions and, indeed, the very definition of public service content in an on-demand age. Perhaps this is a valuable lesson: that dissensus, not consensus, is the 'new normal' and that we had better get used to *difference* and not attempt to impose a consensus on unwilling and rebellious audiences. Indeed, many of the report's final recommendations (for examples its proposals on devolution and for a new digital innovation fund) are designed *specifically* to allow some of these differences – demographic, cultural and political – to be expressed in a revamped public service media environment.

In that spirit, this book – like the Puttnam Report itself – will not satisfy everyone. It will probably be accused of both underplaying the pace of change and exaggerating the need for change, of being too soft or too harsh on the BBC and public service broadcasting as a whole, of being too timid or too unreasonable in some of its prescriptions. We welcome this difference of opinion as, after all, the Pilkington report was also heavily criticised in parliament and in the main newspapers of the time. Richard Hoggart, one of the report's main authors, recalls that one ITV executive 'gave a party in his garden at which copies of the report were put to the flames ... Other [newspaper]s threw every dirty word in their box of cliché abuse at us: "nannying ... elitist ... patronising ... grundyish ... do-gooding"'.[28] The language is likely to have changed in the last 50 years but the passionate debate still continues: how best to imagine, improve and democratise what remains one of our central preoccupations: television.

Notes

1 George Gilder, *Life After Television: The Coming Transformation of Media and American Life* (London: W.W. Norton, 1994), 16.

2 Maureen Ryan, 'TV Peaks Again in 2016: Could It Hit 500 Shows in 2017?' *Variety*, 21 December 2016, http://variety.com/2016/tv/news/peak-tv-2016-scripted-tv-programs-1201944237/.

3 We use a number of different and overlapping terms in this book. Our main area of concern is television and, in particular, public service television (PST), a system of television broadcasting that continues to be subject to specific forms of public regulation in return for particular benefits. The organisations that have traditionally delivered PST in the UK are public service broadcasters (PSBs) but this is likely

to change as new sources of public service content (PSC) start to emerge from outside the traditional PST sector. Instead of looking forward simply to a future of public service broadcasting (PSB), it is vital to consider how best to secure an ecology in which public service media (PSM) – organisations that produce both linear video and non-linear, interactive digital content – will play a central role.

4 Richard Eyre, MacTaggart Memorial Lecture, *The Guardian*, 28 August 1999, www.theguardian.com/media/1999/aug/28/bbc.uknews.

5 Michael Wolff, *Television is the New Television: The Unexpected Triumph of Old Media in the Digital Age* (New York: Penguin. 2015), 28.

6 Ofcom, *Communications Market Report*, August 2016, 57, www.ofcom.org.uk/__data/assets/pdf_file/0024/26826/cmr_uk_2016.pdf.

7 Ofcom, *PSB Annual Research Report 2017* (London: Ofcom 2017), 16, 5.

8 Rasmus Kleis Nielsen and Richard Sambrook, *What is Happening to Television News?* (Reuters Institute for the Study of Journalism, 2016), 3.

9 Ofcom, *PSB Annual Research Report 2017*, 15.

10 Enders Analysis, *Watching TV and Video in 2025*, November 2015, 1.

11 Enders Analysis, *Does Short Form Video Affect Long Term Content*, 12 May 2016, 6.

12 Ibid., 1.

13 Ofcom, *News Consumption in the UK: 2016* (London: Ofcom, 2017), Figure 2.1.

14 Ibid., Figure 5.4.

15 The Inquiry was chaired by the film producer and former TV executive David Puttnam and based in the Department of Media and Communications at Goldsmiths, University of London. The report was published as David Puttnam, *A Future for Public Service Television: Content and Platforms in a Digital World* (London: Goldsmiths, University of London, 2016).

16 Sir Harry Pilkington, *Report of the Committee on Broadcasting 1960* (London: HMSO, 1962), 34.

17 Ibid., 16.

18 Ibid., 17.

19 See the chapters in this book by Julian Petley and Michael Bailey for further discussion of Pilkington's proposals concerning how television ought to relate to minority groups and interests.

20 Pilkington, *Report*, 31.

21 Rupert Murdoch, 'Freedom in Broadcasting', Speech to the Edinburgh International Television Festival, 25 August 1989, www.thetvfestival.com/website/wp-content/uploads/2015/03/GEITF_MacTaggart_1989_Rupert_Murdoch.pdf.

22 Nicholas Garnham, 'The Broadcasting Market and the Future of the BBC', *Political Quarterly*, 65, 1 (1994), 18.

23 Liz Forgan, 'Could Channel 4's Distinctive Voice and Adventurous Shows Continue if it is Sold?', *The Guardian*, 8 May 2016, www.theguardian.com/media/2016/may/08/channel-4-distinctive-voice-lost-privatisation.

24 See Appendix 3 of Puttnam, *A Future for Public Service Television*, for Onora O'Neill's thoughts on public service broadcasting as a public good with the capacity to provide 'a shared sense of the public space and of what it is to communicate with others who are not already like minded; access to a wide and varied pool of information and to the critical standards that enable intelligent engagement with other views; an understanding of the diversity of views held by fellow citizens and by others; a shared enjoyment of cultural and sporting occasions that would otherwise be preserve of the few or the privileged; an understanding of the diversity of views others hold', 174.

25 See for example Mike Berry, 'Is the BBC Biased?' *The Conversation*, 23 August 2013, https://theconversation.com/hard-evidence-how-biased-is-the-bbc-17028; Stephen Cushion and Justin

Lewis, 'Broadcasters were Biased during the EU Referendum Campaign – But Not in the Way You Think', *New Statesman*, 7 October 2016, www.newstatesman.com/politics/staggers/2016/10/broadcasters-were-biased-during-eu-referendum-campaign-not-way-you-think; Louise Ridley, 'Corbyn Study Claims TV and Online News "Persistently" Biased Against Labour Leader', *Huffington Post*, 30 July 2016, www.huffingtonpost.co.uk/entry/jeremy-corbyn-media-bias-bbc_uk_579a3cd7e4b06d7c426edff0; Jasper Jackson, 'Broadcasters Failing to Make Minorities Feel Represented on TV, Says Ofcom', *The Guardian*, 2 July 2015, www.theguardian.com/media/2015/jul/02/broadcasters-minorities-tv-ofcom.

26 See, for example, Stuart Hood, *On Television* (London: Verso, 1997) and Tom Mills, *The BBC: Myth of a Public Service* (London: Verso, 2016).

27 Raymond Williams, *Communications* (Harmondsworth: Penguin, 1968), 121.

28 Richard Hoggart, *A Measured Life: The Times and Places of an Orphaned Intellectual* (London: Transaction, 1994), 60.

Part Two

Contexts and Reflections

1

Reflection on *A Future for Public Service Television*

Mark Thompson

Public service broadcasting will become more, not less, important to British audiences over the next decade. Political support for public service broadcasting is weaker today than at any time in its history. This growing mismatch between need and support is the central problem in current broadcasting policy.

For some decades, ministers and their advisers have developed policy in the belief that public service broadcasting (PSB) would become steadily less justified as technology and deregulation opened up the range of commercial content available to British audiences.

If PSB had any continued relevance, it would be in market failure 'gaps', ensuring the continued provision of genres which were not attractive to audiences (and thus not commercially viable), but which were deemed to be of continuing civic or cultural value. Arts programmes and religious output are sometimes cited as examples of these market failure genres.

A Future for Public Service Television is right to argue that this is an impoverished account of the role and value of PSB.[1] But I want to go further: the underlying assumption is false. Digital technology has certainly increased the range of content available to consumers, but it is also undermining the economics of many commercial content providers. Market failure in the provision of high quality British content is more likely to increase than decrease over the next ten years.

Consider the central PSB genre of news. Most people would accept that broad public access to high quality news about the UK and the world is vital if we want informed citizens and a healthy democracy.

But no British newspaper has yet found a viable business model for a post-print world, and none has demonstrated sufficient willingness-to-pay on the part of its readers, or leverage and pricing-power in the digital ad market, to be likely to do so. The major global digital platforms – all of which are American, and none of which place a priority on British civic and social needs – will win most of the advertising revenue which once paid for quality newspaper journalism, as that revenue switches from print to digital.

At first glance, news on advertising- and subscription-funded TV and radio seems more secure. This is a timing issue, however. We should expect these markets to also be fundamentally disrupted by digital over the next decade; for margins to be squeezed; and the investment and airtime devoted to news progressively to shrink.

Perhaps new digital news providers will take up the slack? Alas, no such provider has yet demonstrated an ability to create a sustainable, profitable business from high quality news, given the considerable cost of a professional newsroom.

A handful of players will survive and thrive. I believe *The New York Times* is one of them, but it has advantages – a vast home market and significant global subscription potential – enjoyed by almost no British commercial news publisher. A few British national titles and magazines may also make it, and we shouldn't write off the most innovative of the new digital news companies. But – particularly when it comes to international coverage and investigative journalism – the range of British high quality sources of news available to households is likely to narrow. News from the BBC and Channel 4 will become more, not less, important.

The same story will play out with many other genres. Tighter business models, a focus on the most commercially attractive audiences and most popular genres, less financial room for cross-subsidy and creative experimentation. More choice in absolute terms, but less choice when it comes to high quality homegrown content.

British broadcasting policy has been largely blind to all of this, and is still broadly following the naïve free market play-book set out by Alan Peacock and others in the 1980s.[2] Indeed, commercial lobbying has led, as *A Future for Public Service Television* notes, to an obsession with 'market impact', as if the only problem commercial players faced was the PSBs, and in particular the BBC. This is manifestly not the case: look at the similar downward trajectories of commercial players in countries with weaker or no PSB provision. The story of most newspapers in the US is exactly the same as that of their British peers, despite the absence of the BBC.

I welcome the robust way in which the Puttnam Report challenges the lazy doctrine of market impact, and its suggestions for new rationales for targeted PSB intervention and partnership. The 'public service algorithm',[3] a serendipity engine to encourage viewers to encounter content they *didn't* know they would enjoy and value, points to an interesting dilemma – how do you encourage diversity of consumption in a personalised digital world? – without being remotely practical.

The UK has enviable natural advantages when it comes to creative talent, but it is desperately weak in the means of distributing the fruits of that talent to the world – and enjoying its economic benefits. That was always true in the feature film industry. Given that all of the major digital platforms are American, there is a real danger that the same will also become increasingly true of our television and video output as well.

The risk is not that British content will disappear. Our writers, actors directors and craft talent are simply too good for that. It is that they will work for companies owned by US or other foreign interests, on projects chosen by non-British commissioners and aimed at global audiences, and that the lion's share of the economic benefits of their work will accrue to major international players.

The result would be a thriving British production sector, but a TV/video industry that was a net importer rather than a net exporter of finished content; programmes on British TV screens which increasingly served the needs of international partners and generic global audiences; and a loss of national economic opportunity in what could be a significant growth sector for the economy. Historically, a small feature film industry that was essentially tethered to Hollywood was balanced by a powerful tradition of TV production whose primary mission was to produce outstanding and culturally relevant output for British audiences. Without intervention, our TV production industry will increasingly resemble our film industry.

Successive British governments have rightly focused on the need for a national strategy to build a world-class digital technology industry. We need a national industrial strategy for TV production, one which is far less bashful about the central role which the PSBs can and should play, both on focusing on the needs of our own audiences, and in bringing British talent to the world.

The PSBs have a particular role to play in ensuring that this industry thrives not just in London, but in the UK's other great cities, and that it fully reflects the diversity of talent and experiences in this country. Often – as in the BBC's investments in BBC North, BBC Scotland and the BBC Wales Drama Village – the presence of a PSB operation will encourage purely commercial production in non-metropolitan centres as well, creating more jobs and more opportunities for talent and local cultural expression.

Finally, the PSBs – and again especially the BBC – have the opportunity to deliver innovation at scale and with world-class creative and engineering talent. At the time of the launch of the BBC iPlayer, James Murdoch, who was then (as now) Chairman of

BskyB, argued vociferously against the launch and the BBC's right to take part in innovation in media technology. In the event, the iPlayer played the key role in introducing the British public to a new way of consuming TV and radio – and in process helped the entire industry, including Sky.

It is not that policy-makers do not hear these arguments, but rather that they attach too little weight to them compared to other considerations, and in particular to their outdated and misguided belief that, in some unspecified way, a perfect digital marketplace will solve all the problems which public service broadcasting was invented to address.

Not just in the UK but across Europe and in other western countries, the same story is playing out: fierce and significantly successful lobbying against the PSBs by struggling newspaper groups and other commercial interests; politicians who care more about their standing with key editors and proprietors than they do about national culture or the needs of the public; a slow but inexorable reduction of the PSBs' room for manouevre, through budget cuts, regulatory limitation and ever more complex and onerous governance.

In Britain in 2016, far more civil service brain-power and column inches of coverage was devoted to a debate about the best methodology for appointing members of the new combined BBC board than to the astonishing disruptive impact of the major global search and social media platforms. We are living through an exhilarating and bloody revolution; our policy-makers have their heads down trimming their fingernails.

Fourteen years ago, when I was chief executive of Channel 4, I referred in a speech to Matthew Arnold's *Culture and Anarchy* to try to capture the cultural ambition of public service broadcasting at its best:

The great men of culture are those who have had a passion for diffusing, for making prevail, for carrying out from one end of society to the other, the best knowledge, the best ideas of their time; who have laboured to divest knowledge of all that was harsh, uncouth, difficult, abstract, professional, exclusive; to humanise it, to make it efficient outside the clique of the cultivated and the learned, yet still remaining the best knowledge and thought of the time, and a true source, therefore of sweetness and light.[4]

We might add 'the great men *and women* of culture' but, other than that, this passage – written many decades before the invention of broadcasting and mass media – faithfully sets out the political, cultural and social ideals of PSB.

All is not lost. The BBC's new charter is a reasonable one, though the BBC's funding is unnecessarily and damagingly tight. Channel 4 remains a public service broadcaster, though its long-term future is still unclear. But broadcasting policy is still significantly in the grip of the people whom Arnold called the *Philistines*, people for whom the public are consumers and customers first and citizens second, who regard culture as 'elitist', and the social and educational claims of public service broadcasting as so much special pleading.

We need to fight for public service media, through patient, evidence-based argument and with thoughtful contributions like the Puttnam Report. When it comes to broadcasting policy, we need a change in the wind – but such a change is not out of the question. The public love what the British public service broadcasters do, and the public is in a rebellious mood. Woe betide the technocrat who puts market theory ahead of digital reality. Woe betide the politician who puts their relationship with the media magnate above their duty to the public. Sweetness and light may make better electoral sense than they know.

Notes

1 David Puttnam, *A Future for Public Service Television: Content and Platforms in a Digital World* (London: Goldsmiths, University of London, 2016), 33.

2 See Alan Peacock (chair), *Report of the Committee on Financing the BBC*, Cmnd 9824 (London: HMSO, 1986).

3 Puttnam, *A Future for Public Service Television*, 36.

4 Matthew Arnold, *Culture and Anarchy* (Oxford: Oxford University Press, 2006 [1869]), 53.

2

Public Service Television and the Crisis of Content

Jon Thoday

Public service broadcasting (PSB) has a long and noble tradition in the UK. We invented it and the BBC has been its finest exponent, admired and respected throughout the world. This is the case whether in the breadth and integrity of its news reporting or, at the other end of the spectrum, the brilliance of its entertainment offering. This highly successful model is now under existential threat as never before. The threats come externally from market forces, from commercial adversaries chipping away to advance their own agendas and from the political class with their concerns about bias. The BBC also faces the challenges common to all mature organisations: that of calcification and inertia consequent upon size and success. These attacks have not only had a real effect at the BBC, but also the wider industry with UK PSB content spend reduced by almost £1billion in the space of a decade.[1] Both management and government need to recognise that content must be re-prioritised. If this can't be done the industry must act to found a new organisation whose sole priority is the support of content.

It is widely agreed that British public service broadcasting (a term which principally encompasses the BBC and Channel 4) is and has been both culturally and commercially good for UK plc. Despite this consensus there is regular debate and criticism, not only of the BBC, but also recently a serious suggestion that Channel 4 is a national asset that should be sold off to the highest bidder. The main focus of attack is generally value for money or of regulatory issues borne out of perceived concerns of political bias or, in Channel 4's case, a simple desire by the government to realise value. Issues around regulation, which is a costly necessity of PSB, seem to attract more debate than the more important and fundamental question of the actual purpose of PSB.

PSB is a vital resource for nurturing talent. If you look at where creative talent both on and off screen cut their teeth, it is almost always at the BBC or Channel 4. These organisations have been almost the sole investors in future talent. They have given a start to, and allowed for creative experiment and risk-taking for, countless artists, writers and behind-the-screen talent. The rest of the commercial industry has time and time again benefitted from this investment. There has been cross-fertilisation within the industry as repeatedly key artists, producers and executives have made the move from PSB to the commercial sector. There is no sign of this kind of investment being made by commercial organisations. The most commercial TV environment in the world, the USA, regularly looks to the UK for new ideas and talent. The old US resistance to UK product on the basis of impenetrability of accent and desire to 'Americanise' UK formats has gone. Whereas ten years ago independent producers like All3 Media, Freemantle, Endemol/Shine, Tinopolis and Avalon were at best bit part players in Hollywood, today they are prominent participants. With the right support this growth is just beginning. The change in terms of trade which happened less than 15 years ago allowed independent production companies to retain the rights to their shows and was transformational in the UK TV industry, incentivising production companies to market their content leading to massive growth in international sales. PACT/Oliver & Ohlbaum analysis reveals growth from £200 million per annum in 2005 to over £1 billion per annum in 2015.[2] UKTV has gone from being admired creatively to being a sought-after asset.

British talent and British TV is also a form of soft power. Shows about the UK (witness the likes of *Downton Abbey* and *The Crown*) are great adverts for the country. One of the go-to people in the USA for a worldview on politics is a British comedian, John Oliver. The success of the industry, as well as increased exports, is also demonstrated by the amount of inward investment from foreign media groups. The Europeans (RTL Group, StudioCanal and ProSeiben), the Chinese (Hejing Culture) and the Americans (Fox, Discovery, Liberty Global, NBC Universal, Warner Bros., Viacom and Sony) are all now heavily invested in the market. This inward investment is almost all because of a hunger for the new ideas and the commercial growth which flowed from the change in terms of trade.

In light of the overwhelming evidence that PSB is a significant net contributor to a UK TV plc, it is puzzling that the attacks seem more frequent and the BBC more defensive. This is symptomatic of the fact that the original purpose of PSB has become blurred over the years and the public has become complacent, believing that PSB is a social good that government will not interfere with. There is no recognition that the continued good health of this resource is not a foregone conclusion, or that what is a success story might turn into a massive own goal if support continues to be withdrawn. Inward investment is not a sign that PSB funding can be reduced or has effectively done its job. This commercial interest is a by-product of previous investment and to maintain the success creative renewal has to continue otherwise we will see decline rather than growth.

Clearly the BBC is the main torch-bearer for PSB, but it faces significant challenges which contribute to difficulties in delivering on its core purpose. This is one faced by all mature institutions, namely that with success and power the organisational imperative becomes one of survival of the organism itself as opposed to the execution of its purpose. When founded, a public institution is largely peopled with altruistic employees who are very clear what they are there for, whether it be a health service, a national theatre or a broadcaster, and their energies go into providing something which didn't previously exist. Over time, for a number of reasons, more and more resources and intellectual time get put into survival strategies and less and less into promoting its original function. In the case of a non-commercial organisation this is compounded by the ambitions of its employees. A publicly funded organisation does not have immediately obvious markers of success. In the business world it is profit or market share, both of which are quite clear indicators. In the publicly funded world, a manager's success is judged by two things: the number of people he or she manages and the level on the management tier they achieve. This gives perverse incentives for a manager to employ ever more people to do ever-smaller jobs. It also contributes to an inability for anyone to be held to account for any one decision. Both these phenomena are very evident at the BBC.

The BBC management faces three external threats. First, under pressure from successive governments, the BBC has been forced to take on functions (and their consequent costs) which arguably do not form part of its core purpose. The costs of this mission creep are such that the money must be taken from somewhere else within the BBC. The result is that any area which is not ring-fenced by government diktat is vulnerable to having money taken from it. Second, the BBC is constantly under attack from its commercial rivals who see its successes as encroaching on their profits. Last, the BBC faces regular attacks from politicians feeling misrepresented by news reportage. These feelings are then translated into allegations of bias.

There is little realisation amongst the public that successive governments have placed increasingly onerous requirements on the BBC which are affecting its ability to carry out its core purpose: 'serving all audiences through the provision of impartial, high-quality and distinctive output and services which inform, educate and entertain.'[3] An example of this can be found during licence fee negotiations in 2015 whereby an agreement was extracted from the BBC to fund the television licences of pensioners. A cynic might say the government originally gave pensioners this perk as a sweetener for electoral advantage and then years later foisted the £650 million per annum cost of this onto the BBC.[4] Past incursions on the licence fee have included the decision in 2010 to withdraw funding of £253 million for the benefit of the World Service (this was subsequently ameliorated somewhat by the then Chancellor George Osborne), funding the Welsh-language broadcaster S4C which costs £74.5million, supporting the government rollout of fast broadband throughout the nation (£80 million), and the provision of local television services (£40 million). There is also the long-term funding of nine orchestras and choirs, which could be said to sit more appropriately within the aegis of the Arts Council. The Nations and Regions policy, whereby the BBC is required to produce a certain amount of content outside of London, has caused massive increases in production costs which are not seen on screen. Whilst these things are all on the face of it born of good intentions they have been done at the expense of funding content.

The government requirement that the BBC be at the forefront of technology is particularly outdated. This has in the past been a benefit but more recently has led to misspending of substantial sums of money. Exhibit A here is the cost of the disastrously misconceived spend on digitisation which current BBC Chairman, Tony Hall, cancelled with a loss of £100 million with practically nothing to show for it.[5] We live in a world where technology is so advanced and so readily available that there really is no need for the BBC to speculate on future developments when the commercial world is doing exactly that very successfully. We can all broadcast from our phones and many people have become stars online without the need for technological support from public funding. At the dawn of broadcasting it was easy to see why this was not only in the remit but an absolute necessity. Today, it is access to funding of quality content that is the main issue and if new talent is to be nurtured and new shows created to replace the ageing shows then much more focus needs to be given to it.

The second attack comes at the behest of commercial rivals of the BBC and Channel 4. There is a constant drumbeat from both the Murdoch press (in print and broadcast) and the *Daily Mail* seeking to denigrate public service broadcasting. Consider, for example, Paul Dacre's attacks on Michael Grade while chairman of Channel 4 as 'pornographer in chief' or James Murdoch's attacks on the BBC during his MacTaggart lecture at the Edinburgh TV festival, in which he notably compared government intervention in broadcasting with failed attempts to manipulate the international banana market in the 1950s, described the BBC's size and ambitions as 'chilling', and indicated that its news operation was 'throttling' the market.[6]

The third area of attack concerns bias. Successive governments, whether Tory or Labour, have complained of BBC bias. It has been argued that the fact that fellow travellers from both sides of the political divide feel misrepresented by the BBC is evidence in itself that the BBC has been largely successful in walking the independent line. However, new regulatory arrangements are set to increase the potential for day-to-day interference in the BBC through government-nominated board members and can be argued to be an incursion upon the BBC's independence.[7] Before this development previous governments had used the threat of more control over the BBC as a bargaining chip in licence fee negotiations.

It can be seen that the BBC is being squeezed by myriad forces. These forces have tied the hands of BBC management requiring them to spend decreasing time worrying about content and more time worrying about cost control and threats to its role. The solution to budgetary problems has not been to reduce management costs but rather to reduce the spend on programming. For example, the BBC invested some £500 million less in original content in 2013 than it did in 2004,[8] and it is predictable that if you spend less money on content, quality suffers. It's a simple calculation: if you make the same number of programmes for less money, the quality will suffer and you will have fewer successful shows. Furthermore, for every new show that succeeds there are many that fail. If cuts lead to fewer shows being made then it is going to take longer to find the needle in the haystack – new successful shows. Creative experimentation leads to innovation but fear of failure makes this experiment too risky. Creative risk should remain at the very heart of PSB – however, current circumstances make it less and less acceptable for executives to take such necessary risks.

The reduction in spend has also led to the BBC making cuts to platforms. Most notable in recent history was its decision to move BBC3 from its terrestrial platform to an online-only platform. This was significant because BBC3 was unique as a place for diversity. It gave an outlet for innovative content often by new creative talent and producers. The reason given for closing was simple: money.

The closure of BBC3 was a symptom of a wider problem, a general move away from focusing on innovation and the younger audience and a refocusing on 'tent pole' dramas with high-end established talent. This failure to engage a younger audience and develop new talent will have adverse effects for which the rest of the industry, and the UK viewer, will pay in the future. Now what is left in the free, non-commercial

UK broadcast television landscape is two ageing channels with similarly ageing audiences.[9] This is not because the young do not want to watch television, or because they are only prepared to watch it on their tablets or phones. It is because the BBC's two main channels are largely filled with old programmes for old people. They have very long-running, successful shows that appeal (unsurprisingly) to ageing audiences. The past success of these channels militates against creative renewal as they are full of shows which effectively cannot be cancelled. In some ways they have become hostage to their own success, meaning that younger viewers' interests are not being addressed.

Today, faced with the reduction in the content spend discussed above, it is unlikely that any channel controller would clear primetime schedules by, say, cancelling an episode of a soap. There would be a viewer outcry. Even if they did, then most likely the replacement show would fail because, as noted above, most new shows do. Potential failure is not a problem any executive wants. It is much easier to pick the next signature, high-budget, international co-production drama than it is to find new programmes to entertain the mass audience in primetime. One can always point at a few programmes which skew young but there has been a withdrawal from youth-centred platforms and there is no sign of any attempt to skew these remaining channels younger overall.

Weekday primetime pre-watershed, which is the time where traditionally the very large audiences can be found, is full of ancient soaps. A study of BBC1 schedules before nine o'clock during the week reveal either on the one hand a soap or, on the other, a show that is low priority because ITV is broadcasting a soap at the same time so the competition is too great. There is virtually no thought or investment in programming for the primetime mass audience in that time slot during the week and there hasn't been for decades. Controllers of BBC1 are left with relatively few slots to play with and almost none of those having a decent budget allocation are pre-watershed. This is a dereliction of both the youth audience and of creative talent at the beginning of their careers.

These cuts do not only affect the BBC. They have knock-on effects on the other broadcasters because it allows them to feel able to make their own cuts as this maximises value for their shareholders. What is good for shareholders is not necessarily good for the viewing public. Whatever the new BBC Charter[10] and BBC management say about the BBC working in partnership with the rest of the industry, it is in competition with other broadcasters and if it fails to do the things that only it can do, then commercial broadcasters no longer feel that they have to step up to the plate either.

Senior executives are able to point to many successes as a way to convince the government and the viewer that all is well, but they are in denial; not only are things not well but they are in significant decline and worryingly endangered. The lack of investment in new talent and the lack of new investment in prime time programming is at the heart of potential problems for the future and mirror the problems of the ageing institution as a whole. BBC1 is secure in its soaps and makes waves with some prestige dramas. Having signature pieces like David Attenborough programming or *The Night Manager* are get-out-of-jail-free cards, whilst the vast number of soap hours maintain the big audience. Successive industry heads have blamed the decline of broadcast television viewing on changes in viewer habits as opposed to simply a reaction to the shows that they choose to broadcast. TV watching is not in decline in general. More people than ever are watching screens and the rise of the likes of Netflix shows what happens if you launch a new service predominantly focusing on content.

Decline is not inevitable. It could be reversed with enough will both politically and from the industry. Government needs to require the BBC not just to entertain the public but also to do what commercial organisations find hard, namely to innovate, to support new talent and to take far more risks. There needs to be a requirement that investment on content must be increased, not reduced. This should be fundamental to the BBC charter. Currently BBC management faces a dilemma. It is easy to cut content budgets, hard to cut overheads, and impossible to cut ring-fenced obligatory spends. Therefore the lead must come from government which, if it was less concerned about perceived political bias and more concerned about what best served the audience, would ring-fence the content spend of the BBC.

Realistically, in the light of the difficulties facing the current guardians of public service broadcasting set out above there needs to be another way forward. If the UK is to maintain its position as a leader in the creative world both culturally and creatively, with its ensuing consequent financial advantages, then investment must be found. There must then be a way of focusing that investment on the green shoots of creativity that lead to huge popular hits.

So while doing everything possible to persuade the government to require the BBC to maintain, and even grow, its content spend and doing everything possible to persuade the government not to privatise Channel 4, there needs to be a third way to rebuild and support PSB. One possibility would be to have a publicly funded third body whose sole purpose is to fund content.[11] Companies, producers and broadcasters could apply to fund projects which are either unfunded or in need of top-up funding. This could be for start-up companies or for project development, as well as for production funding. Revenue generated from these projects would be part-owned by the new organisation as a return on investment which could then in turn be reinvested.

This could be a new body with funding of, say, £500 million per annum (being the gap left by the BBC) or it could be an existing body like Creative England which, despite being very poorly funded, has already done amazing work in funding start-ups. If its remit and funding were expanded it could be empowered to concentrate solely on UK-produced content be it radio, TV or gaming.

To be clear, this is not a proposal to top-slice the licence fee but a suggestion to replace the lost BBC content funding with a new organisation which will stand or fall by its focused decisions. Initial funding for the new PSB or public service funder (as it might be called) could come from the commercial sector and government and, once established, be partially self-funded from revenue which would flow from the projects and companies that benefit. There is precedent for this. When Channel 4 was founded it was part-funded by ITV for many years until it became self-sufficient. This new long-term investment in content and new talent would help plug the gap left by the BBC. The commercial broadcasters would benefit because they need the content and the public would benefit it because it is what makes our television the envy of the rest of the world. This body will, by necessity and remit, be forced to focus on new and diverse talent. A fiscally minded government should be part funders because the evidence of increases in exports is overwhelming. In the light of Brexit, the future economy needs all the help it can get. Receipts of £1 billion per annum should be just the beginning if we do not turn away from the success that the industry has achieved thanks largely to PSB.

Public service broadcasting is a social good, both economically and culturally. The BBC as the primary guardian of PSB is facing too many challenges to be able to protect and advance it. In the absence of political will to force the BBC to concentrate on its core function, a new organisation has to be empowered to plug the content gap and intellectual gap left by the BBC. It is imperative to find a way to generate the talent of the future and ensure that the focus remains on the production of great new British content for the benefit of the UK as a whole.

Notes

1 Oliver & Ohlbaum, *Trends in TV Content Investment (Final Report)*, 2015, www.ofcom.org.uk/__data/assets/pdf_file/0021/62706/psb_content_investment.pdf.
2 PACT, 'Our Policy Work', last accessesd 13 March 2017, www.pact.co.uk/services/policy.html.

3 BBC, *Royal Charter for the Continuance of the British Broadcasting Corporation*, December 2016, http://downloads.bbc.co.uk/bbctrust/assets/files/pdf/about/how_we_govern/2016/charter.pdf.

4 BBC, 2015/16 *BBC Full Financial Statements* (London: BBC, 2016), http://downloads.bbc.co.uk/aboutthebbc/insidethebbc/reports/pdf/BBC-FS-2016.pdf, 35.

5 BBC, 'BBC was "Complacent" over Failed £100m IT Project', *BBC News Online*, 10 April 2014, www.bbc.co.uk/news/entertainment-arts-26963723.

6 James Murdoch, 'The Absence of Trust', speech to the Edinburgh International Television Festival, 28 August 2009, available at: http://image.guardian.co.uk/sys-files/Media/documents/2009/08/28/JamesMurdochMacTaggartLecture.pdf.

7 Mark Sweney, 'Sir David Clementi Confirmed as new BBC Chair', *The Guardian*, 10 January 2017, www.theguardian.com/media/2017/jan/10/sir-david-clementi-bbc-chair-unitary-board-trust-bank-of-england.

8 Oliver & Ohlbaum, *Trends in TV Content Investment*, 14.

9 From 2005 to 2015, BBC1's reach fell from 80.3% to 72.5% and BBC2's reach from 59.6% to 47.1%. Of the main five PSBs, BBC1 and BBC2 have the oldest viewing profiles. Ofcom, *The Communications Market 2016, 2. Television and audio-visual* (London: Ofcom, 2016), www.ofcom.org.uk/__data/assets/pdf_file/0024/26826/cmr_uk_2016.pdf, 98, 100.

10 *Royal Charter for the Continuance of the British Broadcasting Corporation*, Cm 9365, December 2016, http://downloads.bbc.co.uk/bbctrust/assets/files/pdf/about/how_we_govern/2016/charter.pdf.

11 See the proposals for a new source of public service content in chapter 7 of David Puttnam, *A Future for Public Service Television: Content and Platforms in a Digital World* (London: Goldsmiths, University of London, 2016).

3

TV Advertising for All Seasons

Tess Alps

When I arrived to set up Thinkbox[1] from scratch in 2006, the name had already been chosen. There have been many times since when I have wished for a more 'Ronseal' name – the TV Advertising Association maybe – but, back then, the broadcasters weren't sure whether 'TV' would remain the word to describe best what they do. In fact, 'television' has proved itself to be the most elastic, comprehensive, comprehensible and appropriate word that exists to describe the business we are in.

The word 'television' means something very real to viewers: the word they reach for when they are watching any professionally-made long-form entertainment in-home or on a personal device. Sometimes we professionals agonise over whether the BBC iPlayer, All 4 and Netflix are truly TV or just video, but the average viewer is in no doubt – of course they are 'telly'.

Television is also a word that many other companies have ambitions to appropriate: for example, Google TV, Apple TV and Telegraph TV. It's an aspirational word; they choose it because it conveys a certain quality.

Thinkbox's stakeholders comprise both public service broadcasting (PSB) and non-PSB channels. It would therefore be inappropriate for us to engage in public debate about the desirability of any particular outcome regarding PSB or public service content (PSC). Hence, when we were asked to contribute to David Puttnam's report on the future of public service television[2] we had to consider how best to be involved and, indeed, whether it was appropriate to be involved at all. After consultation with our board, we decided that the Inquiry's Report was an important and responsible undertaking and that it would be remiss not to help it by sharing our research. Thinkbox's specific expertise in TV advertising was an important consideration. Advertising income is, after all, a major part of the whole TV economy and a totally vital one for the commercial PSBs.

The report references some of our research, in particular the work we do to contextualise the changes that are happening to how TV is being viewed. As you will read in other parts of this book, people are watching as much TV as ever but less of it is live or captured within the industry's 'standard' definition of viewing in-home on TV sets, live plus 7 day playback and on-demand. We need to stop being surprised that 'standard' TV is declining given how much effort and investment broadcasters are making to entice people to watch differently.

At some point – hopefully sooner rather than later – BARB will be able to quantify reliably all these new avenues for TV viewing. In the meantime, our work suggests that almost all of any apparent 'loss' – yes, even for younger viewers – will become reclaimable.

In addition to the various forms of TV, there are many other new sources of video from the sublime to the ridiculous. We should not assume that one form of video is necessarily a substitute for another. For example, watching Zoella's make-up tips on YouTube is more a replacement for reading a teenage magazine than watching *Hollyoaks*; watching (or listening to) music videos on Vevo is more a replacement for radio than watching Dave; relaxing with Netflix is at least as much replacing the rental and purchase of DVDs as it is the viewing of *Endeavour* or *Game of Thrones*.

The graph below gives an overview of the time spent watching all the various forms of video for which reliable research exists. Detailed data is only available for the more recent forms of video and viewing to DVD/Blu Ray for the last four years. However, we have estimated viewing using broadband penetration and DVD/VHS sales as a proxy to determine historic viewing levels. What is clear is we are watching more and more video – this is not a zero sum game.

I would like, in this chapter, to do just three things: to give my highly personal interpretation of PSB, to dissect the relationship between advertising and public service broadcasters, and finally to use some of Thinkbox's qualitative insight to imagine the future of advertising within PSB.

Public Service Broadcasting and 'Sacrifice'

I was clearly waiting for my parents to buy their first television set for the televised Coronation before I deigned to be born in the autumn of 1953. I have never lived in a home without at least one TV set and I would definitely feel rather queasy if it ever

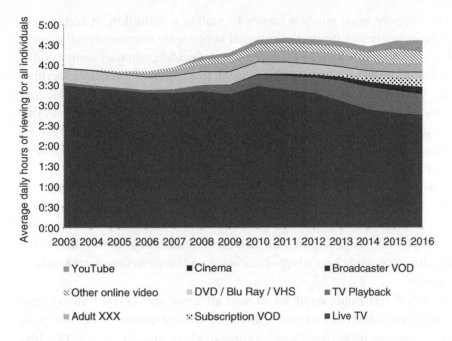

Average daily hours of viewing for all individuals

YouTube
Cinema
Broadcaster VOD
Other online video
DVD / Blu Ray / VHS
TV Playback
Adult XXX
Subscription VOD
Live TV

Figure 3.1

Total TV viewing is resilient as all video grows. *Source*: 2016, BARB / comScore / Broadcaster stream data / OFCOM Digital Day / IPA Touchpoints 2016 / Rentrak / OFCOM Communications report / Statista / IPSOS / Thinkbox estimates / data for all individuals.

happened. Growing up in a small village in the wasteland of north Nottinghamshire where it took 40 minutes to get by bus to the nearest town, which boasted one cinema, television was – in Homer Simpson's words – my 'Teacher! Mother! Secret Lover!'

Our playground games were influenced heavily by the TV we had all been watching: *The Adventures of Robin Hood* and *Doctor Who* were great favourites. I saw my first Shakespeare play, my first opera, my first political debate thanks to TV. My first crush was on Dr Kildare – and frankly I still haven't quite recovered.

The value of PSB is rightly defined by all the noble words you will read many times in this book: quality, range, diversity, accuracy, ambition, innovation, impartiality and so on. But one of those words – universality – is at least as important as the others in my opinion. The universality – the availability and accessibility – of PSB is what makes it such a unique cultural asset. It's not particularly high-minded to be able to share jokes about, say, Simon Cowell's trousers or Claudia Winkleman's eye-liner, but it provides crucial social glue.

PSB is a slippery beast when it comes to nailing a definition. It certainly is not to be found solely in a limited number of supposedly uncommercial genres, though that is part of it. My personal approach to deciding whether something is public service in character or not is bound up in the notion of 'sacrifice'. Sacrifice means more than just risk-taking. All broadcasters take risks; it's vital for any organisation in any sphere with a desire to remain successful and relevant. Sacrifice is different; it is about making a deliberate decision in the pursuit of public good which will almost certainly disadvantage the broadcaster commercially, politically or reputationally.

A very obvious example of this is Channel 4 News, an hour-long news programme scheduled at 7pm each weekday. The conscious 'sacrifice' is three-fold: quality news programmes are relatively expensive to produce; news programmes have no shelf-life and minimal residual rights value; and an entertainment programme broadcast at 7pm would almost certainly gain a higher audience and hence generate more advertising revenue.

The 'sacrifice' test holds good for almost all news and current affairs programmes and most children's output. How does quality drama fit in? Well it is certainly a hugely expensive risk. There was absolutely no guarantee that ITV's historical drama series, *Victoria*, shown in autumn 2016, would be a success. The vast cost was mitigated by a co-production deal but the international appeal of a story about a dead British queen was, on the face of it, limited. One could argue that investing in quality drama is an important element in the brand of ITV which raises its overall status. Purely from an advertising angle, one could also argue it attracts viewers from the demographic groups that command a price premium. But, without doubt, *Victoria* was a 'sacrifice' decision in advance; whether a recommission, made after the successful first series, could also be counted a sacrifice is a point for debate.

The 'sacrifice' aspect of PSB does not relate only to the nature of the content, its genre and cost. Decisions about scheduling frequently demonstrate sacrifice, Channel 5's *Milkshake* segment for young children being a prime example. And it affects marketing too. When Channel 4 uses limited marketing budgets to promote *Dispatches* instead of its latest comedy, the return on investment in terms of ad revenue will be lower, notwithstanding the high cultural return on investment.

When a commissioning, scheduling or marketing decision consciously embraces a 'sacrifice' but ends up delivering unexpected success, this is a cause for rejoicing.

Many stalwarts of PSB schedules were discovered through a 'sacrifice' decision and they pave the way for others to follow. This is one of the enormous gifts PSB gives to the entire television ecology.

There are many obvious examples, as the Puttnam Report acknowledges, of public service purposes being met by non-PSB broadcasters; Sky News, Discovery, Nickelodeon, Dave, Amazon and many others can point to various programmes that tick PSC boxes. But understanding why those companies commission such shows – when there is no regulatory obligation to do so – is a complex undertaking, involving issues such as unique reach, audience profile and the value of niche demographics for advertising, marketing, bundling and political lobbying. None of that should stop us valuing the resultant quality output, however.

The Relationship between Advertising and Public Service Content

If universality is a key tenet then any form of subscription or payment at the point of use immediately precludes pay TV providers, platforms and channels, from sitting within the PSB camp. In light of this, I conclude that PSB can be paid for in only one of three ways: a universal licence fee, direct treasury funding or advertising. I shall leave discussion of the BBC licence fee to others though it is worth saying that, were the BBC to stop receiving it and look to advertising income instead, overall TV ad revenues would be unlikely to increase sufficiently to replace it and all other broadcasters would suffer, not to mention many other media.

Total TV advertising income (including sponsorship and broadcaster video-on-demand (BVOD)) has almost kept pace with the overall growth in the advertising market, which is somewhat remarkable given the extraordinary dynamism of online formats such as search and social media. In 2016, TV ad revenue[3] totalled £5.27bn gross (which is how the ad industry talks about ad income) or £4.48bn net of agency commission (which is how Ofcom and public companies express it). There is sometimes confusion between these two ways of talking about TV ad revenue. TV's share of total advertising is estimated at 25.3% in 2016 whereas in 2000 it was 27.5%.[4] Despite TV advertising's relative health, the percentage contribution advertising makes to the TV industry's total revenue has declined, as income from subscriptions has increased. According to Ofcom, in 2000 advertising contributed 44.5% to TV industry total income; in 2015 it was just 30%.[5] In recent years, this percentage has stabilised because the percentage gathered from the licence fee has declined. Without advertising providing the third 'leg' of TV's income stool, commercial PSBs would collapse and the profitability of subscription broadcasters would be severely dented.

Those of us who work in advertising are humble enough to know that it is not universally popular but, presented with the choice of paying or seeing ads, most consumers opt for the ad route when questioned. Our challenge is to make TV advertising acceptable and ideally enjoyable. The existence of the ad-free BBC has a marked effect on British viewers' relationship with TV ads. It makes them more sceptical and more impatient. You might think that these that these sentiments are wholly negative but in fact they have acted as a brake on excessive advertising on the commercial PSBs in the UK. This is in marked contrast to the USA where the growth of cable and subscription video-on-demand services (SVOD) appears to be, at some small level, attributable to ad-avoidance.

In a world where a reported 22% – and growing – of UK online users have installed an ad-blocker,[6] the public remains generally accepting of TV spot advertising and even more positive about TV sponsorship in the UK. While the majority of ads are skipped in the 14% of standard TV that is recorded and played back, 86% of standard TV is watched live. We do not see levels of militant ad avoidance in British commercial TV. One way to assess this is to look at comparative levels of recording and playback between the BBC and commercial channels. As the chart below shows, there is no evidence of people recording more programmes on commercial channels in order to allow them to skip ads.

As you can see, the levels of recording are chiefly determined by genres, with news and sport being the least recorded and drama the most. TV ads must be watched at normal speed to be counted by the official Broadcasters' Audience Research Board (BARB); hence fast-forwarded ads are free to advertisers even though they do have a residual effect. One unfortunate consequence is that some programmes that are amongst the most expensive to make – and the most highly valued and hence recorded – lose the highest percentage of the potential value of ads within them, drama being the biggest victim here.

One of the benefits of commercial broadcasters' VOD services, versus playback, is that the ads can be made unskippable, and access denied to viewers with ad-blockers installed. This is, however, a delicate balance which I will address in my final section. But as all forms of video grow it is interesting to see that broadcasters deliver the vast majority of minutes actually watching video *advertising*, with live TV performing very strongly.

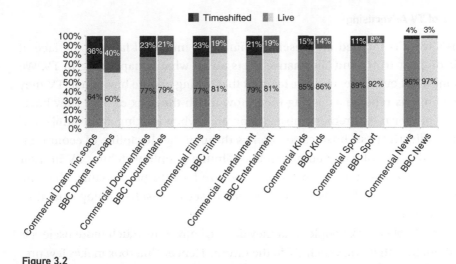

Figure 3.2
Time-shifting is driven by genre not ad avoidance. *Source*: BARB, 2016, individuals in DTR homes, commercial TV vs BBC. TV set viewing within 7 days of broadcast.

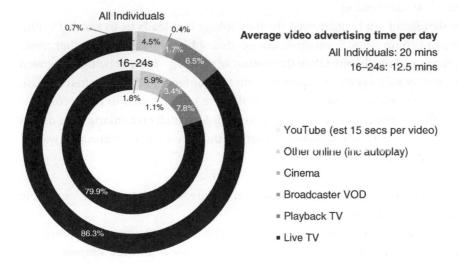

Figure 3.3
TV accounts for 94% of all video advertising time. *Source*: 2016, BARB / comScore / Broadcaster stream data / Ofcom Digital Day / IPA Touchpoints 2016 / Rentrak.

The Future of TV Advertising

Thinkbox's work is focused on presenting hard and impartial facts in the face of a decade of 'post-truth' and 'alternative' facts about what is happening to TV. We deal in empirical evidence; we have learned that asking people how much TV they watch is about as reliable as asking them how much they sex or food they have. We rely on proper measurement, observation and, when looking at the subject of advertising effectiveness, business outcomes through the discipline of econometrics. One of the joys of using ethnography – filming people watching TV in their own homes – is being able to present the footage back to participants. They are amused, bewildered and occasionally horrified to see themselves doing what they swore they never did.

If it's unreliable to ask people what they do currently, how much more useless is it to ask them how they will watch TV in the future? Hence, Thinkbox makes few predictions. Technology is one half of the equation; while some developments are predictable, occasionally a technology pops out of nowhere to change the game. But the other half of the equation is human psychology and this is an area we have found it worthwhile investigating in various ways including neuroscience, in-depth interviews and implicit attitude testing.

One significant study uncovered the psychology that leads us to watch TV, film and video in certain ways. Sometimes our needs are highly personal; at other times they are more socially driven. Often the context of viewing – such as being with loved ones on a comfy sofa – is more important than the content itself; more often though, certain programmes are so important to us that we are prepared to make compromises with the context, such as watching a live football match on a smartphone on the way home from work. The researchers identified the six core need-states that watching TV and video fulfil.

- **Unwind**: defer life's chores or de-stress from the pressures of the day, for example *Countdown* or *This Morning*.
- **Comfort**: shared family time; togetherness, rituals, familiarity and routine, for example *Coronation Street, Hollyoaks, Antiques Roadshow* and *The Simpsons*.
- **Connect**: a sense of 'plugging in' – to feel a sense of connection to society, to time or to place, for example through all forms of news, current affairs, live reality and factual entertainment such as *Big Brother, I'm a Celebrity, Gogglebox, Have I Got News For You*.

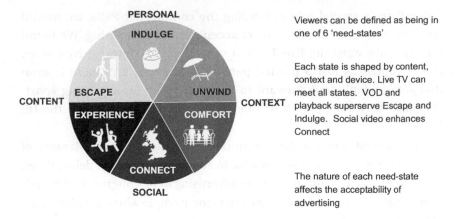

Viewers can be defined as being in one of 6 'need-states'

Each state is shaped by content, context and device. Live TV can meet all states. VOD and playback superserve Escape and Indulge. Social video enhances Connect

The nature of each need-state affects the acceptability of advertising

Figure 3.4

The six need-states of TV and video. *Source*: Screen Life: TV in Demand, 2013, Flamingo/Tapestry/Thinkbox.

- **Experience**: a need for fun and a sense of occasion to be shared e.g. major entertainment such as *The Voice, Strictly Come Dancing* and major live sport.
- **Escape**: the desire to be taken on an enjoyable journey to another time and place, for example *Game of Thrones, Victoria, Peaky Blinders*.
- **Indulge**: satisfying your (often guilty) pleasures with personal favourites, usually alone, with programmes such as *Celebrity Juice, Great British Sewing Bee* and *America's Next Top Model*. This is also the need-state that online pornography satisfies.

The study found that live TV offers a range of programming to meet all six need-states. But some of the newer forms of TV and video perform particularly well against certain of these need-states. Being able to watch TV content of your choice, at a time and place to suit, whether through VOD or playback, serves the Escape and Indulge need-states strongly, while social video, such as sharing clips of TV with friends, enhances the Connect need-state. However, four of the six need-states – Unwind, Comfort, Connect and Experience – are highly dependent on scheduled TV. This leads us to anticipate that, although the various forms of on-demand TV will grow, the need for live TV will always remain, whether broadcast or live-streamed. Technology changes, but human psychology doesn't.

Given advertising's crucial role in funding the commercial PSBs, we moved on to address how the six need-states affect acceptance of advertising. We found that where people are watching live TV in a very relaxed, context-driven state, advertising is accepted as just a normal part of commercial TV. There is even evidence that some people look forward to the advertising in, say, a big sporting event, because they anticipate it will be of high quality, much as in the US Superbowl.

When people are watching playback or on-demand, however, their tolerance of advertising is lower. They want the programme to start with minimum delay. When they are in Escape mode, they also expect the advertising to be congruent with their emotional state; holidays and perfume sit well here for example whereas toilet cleaners and car insurance do not.

Notwithstanding the need for greater sensitivity in placing advertising in on-demand programming, there is no evidence that it is rejected. Indeed, the ITV Hub and All 4 require viewers to register and to disable ad-blockers and this has generated minimal resistance. Where viewers believe advertising is a fair exchange for watching quality PSB content for free it is understood and accepted. This is very good news. It gives us confidence that advertising will be able to keep making a vital contribution to public service television as it continues its fascinating evolution.

Notes

1 Thinkbox is the central marketing body for commercial TV in the UK, similar to the Radiocentre or the Internet Advertising Bureau. Thinkbox represents close to 100% of all TV advertising revenue in the UK through our main shareholders (Channel 4, ITV, Sky Media, Turner and UKTV) and all their broadcaster partners, encompassing the entire range of formats and platforms they operate on from linear broadcast to on-demand web and mobile services. We also have many international associate members, such as Globo in Brazil and RTL in Europe, with whom we share best practice and coordinate research. Our core remit is to encourage the demand for TV advertising by keeping advertisers and agencies informed of the hard facts about TV (and other media), proving TV's advertising's supreme effectiveness and helping them make their own TV advertising even more effective. Our daily tools are all the respected and impartial industry research sources from BARB and Comscore to Ofcom and IPA Touchpoints. However, where we find an information gap, we supplement these with our own substantial studies, generally of a more qualitative and ethnographic nature, commissioned from respected third party companies. These help us not just measure *what* people are doing but also understand *why*. www.thinkbox.tv.

2 David Puttnam, *A Future for Public Service Television: Content and Platforms in a Digital World* (London: Goldsmiths, University of London, 2016).

3 Thinkbox shareholder data. www.thinkbox.tv/News-and-opinion/Newsroom/A-year-in-TV.
4 Advertising Association/WARC and Thinkbox estimates, 2016. http://expenditurereport.warc.com/
5 Ofcom Communications Market Reports, 2004 and 2016. http://stakeholders.ofcom.org.uk/binaries/research/cmr/tele.pdf, 22; www.ofcom.org.uk/__data/assets/pdf_file/0024/26826/cmr_uk_2016.pdf, 65.
6 YouGov for Internet Advertising Bureau UK, March 2016 https://iabuk.net/about/press/archive/iab-uk-reveals-latest-ad-blocking-behaviour.

4

Inventing Public Service Media

Amanda D. Lotz

The Future of Public Service Television Inquiry recognises the scale of its task. This is not another periodic moment of assessment and recommendation. Rather, what is required is nothing short of the invention of public service *media*. Although broadcasting remains, will persist and perhaps continue to play a role in the public service media project, it is clear that continuing to think only in terms of public service broadcasting is to ignore a situation of great opportunity.

Though a century of public service broadcasting experience exists, there have never been public service media – in the United Kingdom or elsewhere. Public service broadcasting had key affordances and limitations, not only because of its public service directive, but also because of the affordances and limitations of broadcasting.

Radio and television industries – public service and commercial alike – have been defined for their entire existence by their broadcast distribution technology. The technological abilities of broadcasting – sending a single message across a vast geography – revolutionised human communication previously confined to person-to-person communication or to those connected by wire.

The abilities of broadcasting in this context were so extraordinary that little thought has been given to broadcasting's limitations – in particular, its profound scarcity. Technologically, broadcasting allows the transmission of just one message, from one to many. Assumed as 'normal' and consequently rarely considered, broadcasting technology strangled radio and television as media, creating an exceptional bottleneck with its capacity constraint of transmitting a single signal. This led to the importance of scheduling and afforded profound power to the intermediaries that selected the schedule. Without technology capable of countering this limitation, such constraint was not worth deep interrogation. Over the decades, British public service television responded by launching additional channels precisely because this technological bottleneck limited its ability to service its broad citizenry.

But internet distribution of video has revealed that nearly all the features presumed constitutive of television are not, in fact, characteristic of television, but of broadcast technology. Rather, the necessity of intermediaries to order a schedule of programming and the experience of programs being 'on' at particular times are 'protocols' developed for and specific to broadcast technology.[1] The technological development of a new mechanism of audio and video distribution requires reconsideration of the practices and metrics of evaluation of broadcast public service television to account for the opportunities of other forms of distribution.

The arrival of a new distribution technology – internet distribution – with different affordances and limitations, has left those in both public service and commercial television feeling unmoored and uncertain of the present and future. The implications and transformation introduced by internet distribution have been slowly revealed. Terminology used in initial phases of its arrival, such as 'new media' and widely shared expectations of the 'death' of television, created presumptions of a medium in transition. True internet distribution of television is barely a decade old, but it is now clear that the profound change has less to do with the medium than with its distribution technology.[2]

Although the scale and nature of the implications of a new mechanism for distribution of video are just now being fully understood, the early years of internet-distributed television illustrate well the differences between commercial and public service television. While US content owners (studios) and networks/channels did all they could to hinder the development of internet-distributed video by tightly controlling intellectual property and allowing minimal authorised circulation of professional quality video in the first decade of the twenty-first century, the BBC launched the iPlayer. A truly revolutionary offering, the iPlayer was the first indication of public service *media* and clearly illustrates the difference between commercial and public service practices. US commercial broadcasters lacked reason to make desired content easier to access and sought to maintain the scarcity to access characteristic of broadcasting that supported windowed licensing revenue. By embracing a technology that made its content more accessible, the BBC acted in accord with its mandate of public service.

Commercial and public service television operators around the world are now struggling to understand a competitive environment that includes internet-distributed television. They face the challenge of adapting practices developed for the abilities and limitations of broadcasting to a distribution technology with different capabilities and hence different strategic potential. Many of the old practices have become ineffective. For example, the business of television series creation has relied on strategies of price discrimination in which rights to exhibit a series are sold again and again through a variety of distribution 'windows'. Scarce access to content supported delayed windows, but television viewers now demand access when they desire and turn to unauthorised sources if distributors attempt to make them wait. Moreover, measurements of performance, such as the number of viewers in a live airing, are of diminishing utility. US commercial broadcasters now compete with a growing number of internet-distributed services supported entirely by subscriber fees. These services curate libraries of programmes cultivated through previously unimagined data about viewer behaviour. The challenges for public service television differ but are every bit as extraordinary.

Although the final report of the inquiry on A Future for Public Service Television attends to this new environment that includes internet-distributed television, the report in many ways maintains a 'broadcast paradigm' in its focus and perspective. The report rightly identifies the necessary shift to thinking in terms of public service media, yet many of its ultimate recommendations belie the logics of public service *broadcasting*. For example, the continued attention to the electronic programme guide (EPG) fails to acknowledge the irrelevance of EPGs in an era of algorithmically-based recommendation engines and personalised user interfaces. A paradigm of traditional 'newscasts' pervades that assumes the perpetuation of this broadcast protocol largely abandoned by younger viewers. The report asserts programming produced outside of public service broadcasting to have a 'public service character', though it is not derived from a motivation of service, but embraces a strategy of distinction for commercial ends.

Rather than a regular reappraisal of public service broadcasting, the context of the development of a new mechanism of video distribution requires a more exhaustive task of identifying the ways in which the affordances of internet-distributed video require the abandonment of the broadcast paradigm and creation of a paradigm of public service that embraces the opportunities and characteristics of internet distribution.

Challenges and Opportunities of Public Service Media

Arguably, the greatest challenge for public service broadcasting has been its mandate to serve the full citizenry of a diverse nation while simultaneously creating common culture. As Owen and Wildman identified, 'the production of mass media messages involves a trade-off between the savings from shared consumption of a common commodity and the loss of consumer satisfaction that occurs when messages are not tailored to individual or local tastes'.[3] Creating content for a 'nation' offers economically valuable scope, but rarely yields programming compelling to a citizenry with varying identities, daily realities and interests.

In addition to the paradoxical directives of serving diverse populations and building common culture, these competing tasks were made even more difficult by the scarcity characteristic of broadcasting. Public service broadcasters have thus been challenged in the commissioning of programmes by the incompatible pressures to create content meaningful to and preferred by diverse populations, at the same time as fitting it into the narrow confines of schedule availability. The time-specificity of broadcasting necessitated that PSBs not only provide programming aimed to service a particular group or perspective, but to gather that intended population at a specific time.

The affordances of internet distribution offer opportunities to create and circulate media without the challenges imposed by a schedule and its associated scarcity. A public service media system utilising the affordances of internet distribution bears several commonalities to the existing broadcast system but crucially differs by cultivating a *library* rather than *schedule* of content. Internet-distributed services, what I have elsewhere identified as 'portals', must be theorised and understood as possessing both similarities to and differences from previous mechanisms of video distribution that significantly affect their capabilities.[4] As can be seen in commercial portals such as Netflix, the strategies of portal curation differ from broadcast scheduling and provide significant opportunity to both commercial and public service media.

Exploring the case and strategies of Netflix is a valuable exercise for imagining public service media. Netflix's subscriber-funded business model leads it to different strategies from advertiser-supported television – whether broadcast, cable, or internet distributed – and make it more comparable to a public service mandate. Subscriber-funded media must provide content of such value that subscribers are willing to pay for it, rather than the aim of advertiser-funding to attract as many eyeballs as possible to increase the audience that can be sold to advertisers. The commercial success of subscriber-funded services is not derived from attracting the most viewers at every moment of the day, but from creating programming of value to the viewer. The need to create content of value parallels the mission of public service media. Technologically, internet distribution allows the creation of a library – or rather several sub-libraries of content in the case of Netflix – which provides viewers with a broader range of viewing options than the schedule of a handful of channels could supply.

Different strategies are possible with internet-distributed libraries of content. For a service such as Netflix, a recommendation algorithm built from vast data about usage and preferences enables the service to know considerably more about what programmes are valued and by whom than traditional measures. This allows the service to offer a better value proposition to viewers and more strategic programme development. Such capabilities also have public service applications. For example, Netflix's recommendation algorithm aims to provide personalised recommendations – which more effectively services a public service mandate aim of helping diverse populations find content of interest. As the report notes, algorithms might also be designed to offer programming most watched so as to simultaneously provide viewers with access to programming that facilitates the public service mandate of creating common culture.[5]

Of course the affordances of internet distribution come with limitations. The asynchronicity characteristic of offering viewers a library rather than a schedule works contrary to public service goals of common culture. Outside of public service systems, the exceptional convenience facilitated by on-demand, internet distribution has proven very compelling. Freeing viewers from time-specific viewing may decrease the possibility of simultaneous cultural conversation, but it is arguably impossible to enforce time-specific viewing in a technological context in which it is no longer required. Trying to enforce such viewing behaviour risks being ignored.

The affordances of internet distribution also offer tools to balance asynchronous engagement. A key advantage of creating a public service portal such as the iPlayer is archival capability. The value of content accessible in a library is derived over a much longer period of time than has been characteristic of broadcasting. There has been considerable consternation among US linear television services about Netflix's refusal to share data about how many view its series. But the number of viewers, as I have already argued, is not a measure of success for a subscriber-funded service in the same way it is the currency for advertiser-supported services. Assessing the number of viewers in the first week, month or even year that content is available is only a limited indication of the value of that content. Content owned by the portal is contracted in near perpetuity and thus provides value to the library long after its first arrival. This can be beneficial to public service media because it better enables viewers to derive value from content.

Although internet-distributed public service media do not face the constraints of broadcasting's distribution capacity, they are not without constraint. Rather, programming budget becomes the primary limitation of the system. Put another way, the broadcast schedule provided the most stringent limitation on public service broadcasters' aspirations to serve their constituencies due to the technological constraints of broadcasting. The cost of programme development becomes that limitation in an era of on-demand, internet-distributed television.

The affordances of internet distribution allow for and encourage the evolution from public service broadcasting to public service media. It must be understood that much of the paradigm of thinking that has developed for public service broadcasting – about where value derives, about how viewers are served – must be significantly adapted in a transition to an environment of public service media. Many of the strategies that have developed from norms of the linear environment are ineffective in the cultivation of libraries of content, while libraries likewise allow for strategies previously unavailable.

Facing the Contemporary Technological Landscape

Shifting from deeply ingrained practices, approaches and expectations derived from a century of broadcasting is a significant task. Among the necessary paradigm shifts are understanding the changed competitive environment in which PSBs operate, reconsidering the norms that derive from a schedule of programming, and reflecting on whether the remit of public service media includes convenience or simply access.

PSBs compete in an environment in which they are not alone in seeking distinction. The development of distinctive and valued content is core to the business strategy of subscriber-funded services such as Netflix and HBO, and even some advertiser-supported services now use distinction to stand out in what has become a video programming marketplace that has shifted from abundance to outright surplus. In such an environment, brand identity and awareness is crucial. The BBC possesses a brand with considerable domestic and international awareness as a provider of distinctive programming. Maintaining that brand is crucial, and as internet distribution allows the creation of truly international distribution services – seen clearly in Netflix's recent global moves – it provides opportunities as well.

Existing international awareness of the BBC is valuable and enables British public service media to develop a multifaceted strategy that allows it to leverage itself internationally for commercial gain to generate revenue from outside of Britain that can augment further innovation for the domestic audience. Such a strategy only adds to the complexity of the demands on programme development, as it is crucial to maintain a focus on serving the core domestic market while also developing enough content capable of travelling abroad to maintain an international audience that delivers revenue that supplements that generated by the funding mechanisms of the future (including those suggested in the Inquiry report).

A second paradigm shift must come from rethinking programme practices tied to scheduling programming. A good example of this is the persistence of 'newscasts' as a category of focus. Video news need not be constrained to particular times of day, nor to an aggregation of stories into a 30 or 60 minute block. Public service media needs to create rigorous journalism that is made accessible to viewers in the ways they want to access it. Continuing to focus on time-specific newscasts risks losing audience to the several developing services that may offer an inferior product but embrace changing media use behaviours. In an era of internet and social media, journalists remain crucial for insight into how and why the events of the day matter even if countless sources now immediately proclaim what has happened.

There are many other ways that the paradigm of scheduling and time specificity continues to dominate the imagination of what is possible for public service media. Indeed, this time of multiple, coexisting distribution technologies is most complicated. It remains unclear whether internet distribution will completely overtake broadcasting or two separate distribution norms will persist for some time, but it is clear that practices designed for distribution outside of a schedule should be the focus moving forward.

Finally, PSBs have been driven by a mandate to provide accessible services. Given broadcasting's limitations, these services were not the most convenient: they required that viewers accept time specificity to access programmes. Greater consideration of whether internet-distributed public service media is a matter of access or convenience should be pursued. The opportunities of internet distribution are significant – but so long as broadcasting persists, it is reasonable to consider whether a stronger public service media system can be achieved by requiring additional payment for the convenience it provides, as convenience is not clearly part of the public service remit. Research on why viewers subscribe to Netflix found that 82% do so because of the 'convenience of on-demand streaming programming', 67% because it is 'cost-effective', and 54% because of its 'broad streaming content library'. In other words, content is only the third most compelling reason for subscription.[6] Such information is telling of viewers' deeply held frustration with the limitations of broadcast and multichannel technology and an important consideration for those assessing programming and distribution strategies in this new environment.

Public service broadcasters are not alone in this experience of transformative change. The US commercial television industries have been slowly pivoting toward strategies that account for multiple distribution technologies. A key adjustment has been diversifying revenue models. Even services still characterised as advertiser-supported (broadcast networks and basic cable) are now significantly reliant on subscriber funding paid by multichannel providers and by licensing revenue derived from owning their own intellectual property. Increasingly, they distribute this self-owned intellectual property through multiple self-owned outlets, changing a business that historically utilised licensing rather than extensive vertical integration of content and distribution.

Such strategic revenue diversification might also help fund public service media to an extent that viewers identify greater value from the service without substantially increasing the costs of providing the service. Internet-distributed media pose several challenges to media operations developed based on previous technologies, but offer considerable opportunities as well. Such opportunities are most clear when imagining these media anew – as a call for public service media suggests – instead of limiting possibility to the practices and norms constituted by broadcasting.

Notes

1 Lisa Gitelman, *Always Already New: Media, History and the Data of Culture* (Cambridge, MA: MIT Press, 2008).

2 Amanda D. Lotz, 'The Paradigmatic Evolution of U.S. Television and the Emergence of Internet-Distributed Television', *Icono 14 Journal of Communication and Emergent Technologies* 14, 2 (2016): 122–42, www.icono14.net/ojs/index.php/icono14/article/view/993/566.

3 Bruce Owen and Steven Wildman, *Video Economics* (Cambridge, MA: Harvard University Press, 1992), 151.

4 Amanda D. Lotz, *Portals: A Treatise on Internet-Distributed Television* (Ann Arbor, MI: Maize Books), 2017.

5 David Puttnam, *A Future for Public Service Television: Content and Platforms in a Digital World* (London: Goldsmiths, University of London, 2016), 36.

6 Statista, 'Leading Reasons Why Netflix Subscribers in the U.S. Subscribed to Netflix as of January 2015'; data reported in eMarketer from a Cowan & Company study, methodology not specified; www.statista.com/statistics/459906/reasons-subscribe-netflix-usa/.

5

Does Public Service Television Really Give Consumers Less Good Value for Money than the Rest of the Market?

Patrick Barwise

Why do we still have public service television (PST) when commercial broadcasters and online TV companies now offer consumers so much choice? The obvious answer is that people are citizens as well as consumers: for policy reasons, we want to ensure the availability of public service programmes that offer social, cultural and political benefits and economic externalities but are not commercially viable. Despite disagreements about scope, scale, governance and funding, the idea that there should be *some* PST for 'citizenship' reasons is not seriously disputed in most countries, the USA being perhaps the main exception.

Among some commentators, however, this 'market failure' argument – the market's under-provision of some kinds of programme – is now the only continuing justification for PST. In the words of British economist Helen Weeds, '[t]he rationale for public intervention in broadcasting must now rest on citizen concerns.'[1] Many others would broadly agree.[2]

The Conservative government's position also seems to reflect this view. A 2016 White Paper argues that '[t]he BBC has faced questions in recent years, including about ... its distinctiveness, the market impact of its more mainstream services ... and its efficiency and value for money... [It should] focus its creative energy on high quality distinctive content that differentiates it from the rest of the market.'[3] The White Paper frames broadcasting policy as a kind of balancing act between citizen and consumer interests: for citizenship reasons, we need some PST to address gaps in provision; but for consumer reasons, we should minimise its cost and market impact.

One practical problem with this view is that it assumes a clear-cut distinction between popular/commercial and minority/non-commercial programmes. The reality is much fuzzier and less predictable: what could be more 'minority interest' than a baking competition? Yet *The Great British Bake Off* turned out to be a huge hit.

At a deeper level, the key assumption underpinning the 'market failure' argument is that, whatever its citizenship value, PST offers less good *consumer value for money (VFM)* than the rest of the market. This chapter explores this assumption in the UK context. The main analysis completely ignores the citizenship benefits of PST, treating it as if it were just a consumer product like baked beans. At the end of the chapter, I briefly return to the citizenship issues and discuss the policy implications.

The UK Market Context

I here define a public service broadcaster (PSB) as a broadcaster governed, managed and regulated to achieve a different or broader set of public interest goals than maximising shareholder value. This is not a black-and-white concept. For instance, all UK broadcasters, apart from online-only TV services such as Netflix, operate under – and almost always comply with – the Ofcom Broadcasting Code, designed to protect children, avoid undue harm and offence, ensure accurate, impartial news, and so on.[4] As well as the BBC – publicly owned and largely funded by compulsory licence fees – the UK's diverse and highly competitive TV system includes two other sets of broadcasters:[5]

- Commercial PSBs: the publicly owned Channel 4 (C4) and privately owned ITV and Channel 5 (C5), all mainly funded by advertising[6]
- Non-PSBs: a combination of platforms, channels and online-only services, mainly funded by subscriptions and advertising.

The BBC and C4 are 'pure' PSBs with detailed public service remits; ITV and C5 are also defined as PSBs because they have agreed to deliver some public service objectives (in addition to those in the Ofcom Broadcasting Code) in exchange for privileges such as access to spectrum at a lower price than they would have to pay in a competitive auction.

Method

This chapter is about the relative consumer value for money of PST (BBC TV and the commercial PSBs) and the non-PSBs. Consumer VFM is a familiar concept in marketing and consumer policy, usually measured by simply asking consumers, who typically have no difficulty interpreting the question and relating their responses to their own buying behaviour, perhaps with some post-rationalisation. In thinking about the different brands, they know that they are broadly comparing like with like or, at most, trading off price and quality within a category, for example when comparing a premium brand with an economy brand.

Evaluating the consumer VFM of television is less straightforward because it is rarely bought one programme, or even one channel, at a time; much of its funding comes from advertising; and different broadcasters have quite different revenue models:

- BBC TV is mainly funded by a compulsory licence fee which also funds BBC Radio, BBC Online, the BBC World Service, and much of the cost of broadband rollout, the Welsh public service channel S4C and local TV. Also, all households with one or more members aged over 75 get a free TV licence (regardless of household size and income).
- The commercial PSBs are mainly funded by advertising.
- Pay TV is mainly funded by monthly subscriptions, supplemented by advertising, for a package of channels, increasingly bundled with apps, DVRs, catch-up services, telephony, broadband, etc.
- Online-only TV services are funded by a mixture of subscriptions, advertising and one-off payments (pay-per-view, rentals, download-to-own).

We can, however, infer a lot from a combination of consumer behaviour ('revealed preference'), consumer costs and selected attitudinal data. To illustrate, consider the overall consumer VFM of UK television.

The Overall Consumer VFM of UK Television

Consumers' revealed preference suggests that most see television as excellent VFM – so much so that almost every household chooses to have a TV set (95.5% of all households in late 2016[7]) and/or access to an online TV service (including an unknown proportion of the other 4.5%). At a minimum, the direct cost per household is the 40p/day BBC licence fee[8] and the cost of a TV set and electricity.

Watching TV is extremely cheap by any standards. Robert Picard and I estimated that, in 2012, the direct consumer cost per viewer-hour (CPVH) – subscriptions, the TV proportion of the BBC licence fee, and one-off payments – was

9.2p. Including the indirect cost of TV advertising (discussed later), it was still only 13.5p. On a comparable basis, the cost per consumer-hour was roughly 50p for fixed and mobile telephony, tabloid newspapers, paperback books and 'free' advertising-funded online services (including the cost of broadband). For magazines, quality newspapers and DVDs it was significantly higher than that; for most out-of-home leisure activities (restaurants, pubs, cinemas, etc) much more again.[9]

Of course, these figures do not mean that these other activities represent poor VFM – the experiences are not closely comparable with everyday TV viewing. But the low cost of television clearly helps explain its huge and continuing popularity. Only radio listening worked out even cheaper, at only 1.9p per listener-hour.

Testing the 'Market Failure' Assumption

Using the same broad approach, we can start to test the 'market failure' assumption that UK PST – certainly the BBC and perhaps the commercial PSBs – offers less good consumer VFM than the rest of the market. We can observe consumers' revealed preference at two stages:

- Adoption: households deciding whether to pay for access to any TV; and, if so, whether also to subscribe to (basic or premium) pay TV
- Usage: individuals then deciding which programmes to watch.

These are very different. The first is an occasional household choice involving money. The second is a constant series of individual choices (albeit often negotiated with other household members) and rarely involves money. Both throw light on the 'market failure' assumption about the relative consumer value for money of PST and the non-PSBs.

On the first point, adoption, as already noted, almost all households choose to have access to TV content, including the small but growing minority who watch only online. The proportion of households with access to TV content who, over any extended period, watch no PST is unknown but certainly very small, skewed towards light-viewing, young, upscale online-only households without children.

Among the over 95% of UK households with TV sets able to receive broadcast TV, in late 2016:[10]

- 45.1% had no pay TV (including a majority of low-income households, who rely disproportionately on PST)

- 33.6% had basic pay TV costing £20–30/month on top of the £12/month BBC licence fee
- 21.3% had premium pay TV, i.e. sport and/or movie packages costing £20–40/month on top of the cost of basic pay TV and the BBC.

Because consumers legally have to have a TV licence in order to have pay TV, and a basic pay TV package if they want premium pay TV, these figures do not show their willingness to pay for each option separately. But they do show that, although over 95% of households regard TV as good VFM, a large minority of these (45% of TV households) do not regard even basic pay TV as cost-justified as an addition to PST; and only about 21% think premium pay TV offers good enough VFM to justify a subscription.

Turning to the second type of revealed preference, usage, viewers switch between their favourite channels at no additional cost and seamlessly – although nudged by their EPG[11] – with little or no distinction between PSB and non-PSB channels.

In 2016, BBC TV had a total viewing share of 32% among all UK individuals aged 4+ while the commercial PSBs (including their portfolio channels) had a combined share of 38% and the non-PSBs the remaining 30%. PST was therefore still extremely popular, accounting for 70% of viewing.[12] To assess its relative VFM, however, we also need to take account of, first, costs, i.e. the direct and indirect consumer cost per viewer-hour (CPVH) and, second, perceived quality.

The Cost per Viewer-Hour (CPVH)

We can estimate the CPVH of a specific (type of) broadcaster by dividing its direct and indirect cost to consumers by its viewing hours. We have good data on viewing hours but estimating the consumer cost involves two assumptions:

1. The consumer cost of BBC TV is the proportion of licence fee revenue allocated to it, including proportionate overheads
2. The indirect consumer cost (or opportunity cost) of TV advertising is equal to commercial broadcasters' net advertising revenue (NAR) – that is, the revenue they receive from media agencies.[13]

Using these assumptions, Robert Picard and I estimated the following CPVH figures in 2012: BBC TV 9.2p, commercial PSBs 8.0p, non-PSBs (excluding online-only services) 24.9p.[14] We can now update these estimates to 2016 – see Table 5.1. The first two columns show weekly revenue (£m/week) in 2012 and 2016. Column 3 then shows the ratio between these: 0.96 for BBC TV (a 4% reduction in nominal revenue resulting from the 2010 funding settlement), 1.10 for the commercial PSBs (a 10% increase, mainly reflecting the recovery in NAR) and 1.17 for the non-PSBs (mainly from higher revenue per subscriber).[15] Columns 4–6 show the equivalent figures for viewing. Total viewing (scheduled programmes watched on TV sets, live or up to seven days after broadcast) decreased by 12%, from 28.1 to 24.7 hours/week.[16] Most of this reflected reduced PST viewing, leading to ratios of 0.86, 0.85 and 0.96, respectively, for BBC TV, the commercial PSBs and the non-PSBs.

Table 5.1

Nominal Revenue and Viewing Hours per Week (Rounded): 2012 vs 2016.[21]

	Revenue (£m/week)			Viewing (Hours/week, all inds 4+)		
	2012	2016	2016/2012	2012	2016	2016/2012
BBC TV	52	48	**0.96**	9.3	8.0	**0.86**
Commercial PSBs	56	62	**1.10**	11.1	9.4	**0.85**
Non-PSBs	133	156	**1.17**	7.7	7.4	**0.96**
Total	**240**	**265**	**1.10**	**28.1**	**24.7**	**0.88**

Assuming changes in total consumer costs are proportional to those for broadcaster revenue, we can use these ratios to update the 2012 CPVH figures to 2015:

BBC TV $9.2p \times (0.96/0.86) = 10.2p$
Commercial PSBs $8.0p \times (1.10/0.85) = 10.4p$
Non-PSBs $24.9p \times (1.17/0.96) = 30.3p$

These estimates are approximate (+/– 10%) because they are based on rounded revenue figures, but the qualitative picture is clear: as in 2012, the 2016 CPVH for the non-PSBs was almost three times as high as for the PSBs. The reasons are:

1. The high cost of sport (and, to a lesser extent, movie) rights for premium pay TV channels. The £1,712m annual cost of live TV rights for Premier League football (ie excluding all other football rights and production costs) is now marginally more than the *total* programme budget of BBC TV (£1,702m in 2015/16).[17]

2. The much higher non-programming costs of pay TV versus PST (marketing, distribution, consumer equipment, installation, customer service).
3. The non-PSBs' significantly lower availability. As already discussed, TV household penetration is only about 55% for pay TV and, within that, 21% for premium pay TV, versus 100% for PST.[18]
4. Finally, the market is highly competitive, forcing the PSBs to be much more efficient than they are sometimes portrayed.

The difference in total CPVH is probably somewhat less than this analysis suggests because it excludes the unknown opportunity cost of the PSBs' DTT (Digital Terrestrial Television) spectrum. However, even if we incorporated this, the general pattern – with the non-PSBs' CPVH being much higher than for the PSBs – would be unaffected, because of the above points.

Basic Multichannel, Premium Multichannel and Online-Only TV

Ideally, we would split the non-PSBs in Table 5.1 into two groups: (a) basic satellite, cable and DTT platforms and free-to-air non-PSB channels (accounting for over 95% of non-PSB viewing); and (b) the *additional* cost and viewing of premium sport and movie channels, since these are not sold separately from basic pay TV.

Because of the high cost of the premium channels' content – point 1 above – and their relatively low availability and viewing levels, their CPVH is much higher than that of basic multichannel pay TV. But basic pay TV still has much higher non-programming costs and lower availability than PST – points 2 and 3 – so its CPVH is almost certainly significantly higher than PST's.

Unfortunately, Robert Picard and I were unable to find any published data to enable us to separate the CPVH of the basic and premium non-PSBs and I am still unable to do so.

Similarly, I have been unable to find reliable published data on the consumer cost and viewing of the online-only TV services. But, to illustrate, a household paying Netflix's or Amazon's entry-level £5.99/month, with two adults each watching the service, on average, five hours/week, would equate to a direct CPVH of 13.8p.[19] The total CPVH would be significantly higher if we include the indirect consumer cost of advertising and perhaps additional equipment and broadband costs as these video-on-demand services are extremely bandwidth-hungry.

Download-to-own box sets are slightly more expensive. For example the complete *Mad Men* costs £39.99 from Sky for 68 hours of content, which works out at

58.8p/hour. The CPVH depends on how many people watch it how many times but if, say, two people watched every episode once, on average, the CPVH would be 29.4p.[20] For pay-per-view sport and movies, the CPVH is likely to be significantly higher again.

In summary, although we lack the data to make precise estimates, the evidence is that, far from PST being more expensive than the rest of TV, it is almost certainly significantly cheaper per viewer-hour than basic non-PSB multichannel TV, entry-level online-only TV or online box sets, and very much cheaper than premium services such as sport and movie pay TV channels and online pay-per-view.

Perceived Quality: Audience Appreciation

The cost per viewer-hour is only part of VFM, however. The other is perceived quality, typically measured as audience appreciation. Until the mid-1980s, with only four public service UK channels, audience appreciation was routinely measured using self-completion diaries that asked respondents to say how 'interesting and/or enjoyable' they found each programme they watched. The results were reported as a 0-to-100 Appreciation Index (AI) for each programme. Most AIs were between 60 and 80. Among the general run of entertainment programmes, there was no evidence of 'niche' programmes attracting small but especially loyal and appreciative audiences. Instead, there was a 'double jeopardy' pattern under which, for a given channel and time of day, lower-rating programmes tended to have lower repeat-viewing rates and audience appreciation than more popular ones. A secondary pattern was that, other things being equal, more demanding programmes tended to have smaller audiences but higher AIs because only viewers who liked them a lot were willing to invest the extra effort needed to watch them.[22]

As far as I know, there has been no published research comparing the average audience appreciation of PSB and non-PSB programmes. But, based on the earlier studies, I would be surprised if AIs were significantly higher, on average, for programmes on the basic multichannels than on the PSB channels. There are two reasons why they might be *slightly* higher, however. First, their viewers are people who have invested in pay TV, presumably because, other things being equal, they like television more, on average, than do those who do not subscribe to pay TV. Second, as with demanding programmes on the PSBs, viewers will typically make the extra effort of switching to a small multichannel only if they expect to like the programme more than those showing on the main channels. Typically, this will happen when the multichannel is showing a predictably enjoyable favourite programme.

The appreciation of premium sport and movie channels and online pay-per-view (among those who choose to pay for them) may well be much higher than for either the PSBs or the basic non-PSB channels in order to justify their much higher CPVH. For entry-level online-only TV, I have no solid basis on which to hypothesise.

The Relative Consumer VFM of PST

Based on the above analysis of adoption, usage, CPVH and audience appreciation, both BBC TV and the commercial PSBs most likely offer most consumers *better* VFM than the basic non-PSB channels (including platform costs) that account for the great majority of non-PSB viewing – the exact opposite of the assumption underlying the 'market failure' view. This provisional conclusion is based on the likelihood that, relative to the PSBs, the basic non-PSB channels' significantly higher CPVH (even after allowing for the opportunity cost of the PSBs' access to spectrum) is not compensated for by commensurately higher audience appreciation. This tentative conclusion is researchable. The relative consumer VFM of premium sport and movie channels, online pay-per-view and entry-level online-only TV is unclear: their much higher CPVH may or may not be fully compensated for by higher audience appreciation among those who subscribe to them. However, this too is researchable.

What about the significant minority of consumers who have always said, in response to surveys, that the compulsory BBC licence fee represents poor VFM?[23] A 2015 study[24] focused on this minority: the sample included 24 households saying they would prefer to pay nothing and receive no BBC services (representing 12% of UK households) and a further 24 households saying they would prefer to pay less for a reduced BBC service (representing 16% of UK households). The total sample of 48 therefore represented the 28% of households who, at least to some extent, saw the licence fee as poor VFM. These households then lived with no BBC services for nine days, after which they were re-interviewed and given £3.60 (nine times the 40p/day cost of the licence fee). Over two-thirds (33 out of the 48) changed their minds, deciding that the licence fee did, after all, represent good VFM. In contrast, only one out of a control sample of 22 households who, in the initial interview, had said the BBC represented good VFM went the other way, saying in the second interview that they now felt it represented poor VFM.

This study was based on limited samples but, if it generalises, the proportion saying the licence fee is good VFM increases from 72% in the initial survey to 88%[25] once respondents experience life without the BBC for just over a week. There has been no equivalent study for other broadcasters but the results are certainly consistent with the above evidence that PST represents good VFM for UK consumers.

'Citizenship' Benefits

Of course, PST also offers citizenship benefits beyond those provided by the rest of the market. This chapter is about consumer VFM so I will discuss these only briefly. Ofcom's *PSB Annual Research Report 2017* found that, in 2016, most of the UK public valued the ten defined PSB purposes and five PSB characteristics highly – and increasingly over the five years 2011–2016. The highest importance ratings were for providing 'high-quality UK-made programmes for children' (89% among those with children) and 'trustworthy news' (89% among all UK adults). The lowest, at 68%, was for distinctiveness ('The style of programme is different to what I'd expect to see on other channels'). A majority also said the PSBs were delivering on all these purposes and characteristics, ranging from 87% for high-quality UK children's programmes down to 61% for 'distinctiveness'.[26]

In summary, the UK public values the 'citizenship' purposes and characteristics of PST; does so increasingly; and believes that the PSBs are doing a good job delivering them. The characteristic on which they are least convinced on both importance and delivery is 'distinctiveness'. As I have written elsewhere in a more general context, consumers see no value in a product being distinctive as an end in itself. Instead, they value products that are distinctive because they are 'simply better'.[27] In line with this, a recent study for the BBC, unpublished at the time of writing, confirms that audiences mainly interpret 'distinctive' as 'distinctively good'. Unfortunately, the government's White Paper appears to use 'distinctive' in the quite different sense of 'distinctively different'.

Policy Implications

The above analysis shows that the key assumption underpinning the 'market failure' view – that PST offers consumers less good value for money than the rest of the market, so that its only continuing rationale rests on citizen concerns – appears to be simply wrong, at least in the UK. PST does, of course, give citizens public service benefits over and above those provided by the non-PSBs and online-only TV players, and these 'citizenship' benefits are highly valued by the public. But the numbers suggest that PST *also* offers consumers better value for money because the non-PSBs' significantly higher cost per viewer-hour seems unlikely to be compensated for by commensurately higher audience appreciation.[28]

The main policy implication is simple: there is no necessary trade-off between citizen and consumer benefits: pound for pound, PST appears to deliver both sets of benefits better than the rest of the market. This does not mean we should return to a world with only PST: the addition of non-PSB platforms and channels (now including the online-only services) to what has so far been a strong, well-funded PST system has hugely increased viewer choice and competition. It does, however, mean that the relentless current reduction in the BBC's real income, in particular, is now unambiguously against the public interest from both a citizenship *and* a consumer perspective.

In 2014, Robert Picard and I showed what would happen if this reduction were continued until the BBC were reduced to nothing or a minor sideshow like PBS in America – the logical conclusion of the 'market failure' view. Even, optimistically, assuming commercial broadcasters significantly increased their investment in content, including first-run UK content, in response to the BBC's removal from the market, we showed that the net impact would still be to reduce the range, quality and VFM of television for most households, as well as the income of UK producers and, of course, the citizenship benefits of PST.[29] Since then, the cuts in BBC income have, if anything, accelerated.

The assumption that PST offers consumers less good VFM than the rest of the market is the cornerstone of the 'market failure' view: without it, the argument for further reducing the role of PST simply collapses. From a rational policy perspective, the onus should therefore be on those advocating this view to provide evidence that – contrary to the analysis here – this key assumption is correct, i.e. that PST does offer less good VFM than the rest of the market.

Notes

1 Helen Weeds, 'Digitisation, Programme Quality and Public Service Broadcasting', in *Is There Still a Place for Public Service Television?: Effects of the Changing Economics of Broadcasting*, ed. Robert G Picard and Paolo Siciliani (Oxford University, Reuters Institute for the Study of Journalism, 2013), 19. http://reutersinstitute.politics.ox.ac.uk/publication/there-still-place-public-service-television.

2 Alan Peacock, *Public Service Broadcasting without the BBC?* (London: Institute of Economic Affairs, 2004); Mark Armstrong, 'Public Service Broadcasting', *Fiscal Studies*, 26, 3 (Sept 2005) 281–99; Mark Oliver, *Changing the Channel: A Case for Radical Reform of Public Service Broadcasting in the UK* (London: Policy Exchange, 2010). Historically, a proviso was that, with advertising-funded television, the programme mix that maximises profit (net advertising revenue minus cost) may not be optimum for consumers:

R. H. Coase, 'The Economics of Broadcasting and Government Policy', *American Economic Review*, 56 (1 March 1966): 440–47. Today, with lower entry barriers and many ways for consumers to pay directly, this 'two-sided market' problem is less important.

3 White Paper, *A BBC for the Future: A Broadcaster of Distinction* (London: Department for Culture, Media & Sport, 2016), www.gov.uk/government/uploads/system/uploads/attachment_data/file/524864/DCMS_A_BBC_for_the_future_rev1.pdf, 10.

4 See www.ofcom.org.uk/tv-radio-and-on-demand/broadcast-codes/broadcast-code. A rare exception was Fox News, simulcast from the USA, which was found in breach of the impartiality codes three times in two years: www.thedrum.com/news/2016/08/22/fox-news-building-record-itself-it-receives-third-breach-ofcom. (In contrast, Sky News has an excellent impartiality record). 21st Century Fox has now withdrawn Fox News from the UK, perhaps as part of its bid for the rest of Sky.

5 This excludes the Welsh language S4C, funded by the BBC and advertising.

6 Throughout this chapter, ITV (which, for simplicity, here includes all holders of Channel 3 licences), C4 and C5 are defined as commercial public service *broadcasters* (PSBs). This includes their portfolio channels, although, for most regulatory purposes (eg minutes per hour of commercial airtime) only their main channels are designated public service *channels*.

7 Ofcom, *Communications Market Report 2017* (London: Ofcom, 2017) https://www.ofcom.org.uk/_data/assets/pdf_file/0017/105074/cmr-2017-uk.pdf, 77.

8 Unless the household (i) includes at least one person aged 75+ or (ii) watches only online, free-riding BBC-TV (the 'iPlayer' loophole).

9 Patrick Barwise and Robert G. Picard, *What If There Were No BBC Television?: The Net Impact on UK Viewers* (Oxford University: Reuters Institute for the Study of Journalism, 2014), http://reutersinstitute.politics.ox.ac.uk/publication/what-if-there-were-no-bbc-television, 34–35.

10 Sources: BARB Establishment Survey 4Q16 and company websites, accessed 10 March 2017.

11 FEH Media Insight, *EPG Prominence and Channel Performance: It Still Matters*, March 2017. www.feh-mi.com/blog/epg-prominence-and-channel-performance-it-still-matters/.

12 Ofcom, *PSB Annual Research Report 2017: TV Viewing Annex*, July 2017, https://www.ofcom.org.uk/__data/assets/pdf_file/0015/103920/annex-b-tv-viewing.pdf

13 For an explanation of why this is an appropriate measure, see Barwise and Picard, *What If There Were No BBC Television?* 33–34. Note that advertisers' total expenditure on TV advertising is about 15–20% higher than NAR as it also includes commercial production costs and agency fees.

14 Barwise and Picard, *What If There Were No BBC Television?* 36–37.

15 To adjust for 8% cumulative CPI inflation 2012–2016, these ratios should be divided by 1.08. For instance, the inflation-adjusted ratio for BBC TV was 0.96/1.08 = 0.89, an 11% reduction in real revenue.

16 Note that this excludes the viewing of programmes time-shifted by more than seven days and/or viewed on other devices. Both of these grew significantly over this period, accounting for most of the reduction in 'TV viewing' in Table 5.1.

17 Toby Syfret and Gill Hinds, *BBC Green Paper: Red Alert on Funding* (London: Enders Analysis, 2015) (report 2015-064). *BBC Annual Report* 2015/16.

18 The availability of non-PSB *channels* varies from a small minority (individual premium sport and movie channels) to almost 100% (non-PSB Freeview channels).

19 £5.99/month is £1.38/week. With two adults/household each watching for 5 hours/week, the direct CPVH is £1.38/(2x 5.0 hours) = 13.8p/hour.

20 58.8p/2.

21 Nominal revenue: Ofcom, *Communications Market Report 2017*, Figs 2.11, 2.14. Viewing (scheduled programmes on TV sets, live or up to seven days after broadcast): *PSB Annual Research Report 2017: TV Viewing Annex*, p26. Assumes 'other' revenue split 38/62 between commercial PSBs and non-PSBs, as in Barwise and Picard, *What If There Were No BBC Television?* 23.

22 Patrick Barwise and Andrew Ehrenberg, *Television and its Audience* (London: Sage, 1988). Chapter 5 summarises the relevant research on audience appreciation. For 'double jeopardy', see William N McPhee, *Formal Theories of Mass Behavior* (New York: Free Press, 1963) and Andrew Ehrenberg, Gerald Goodhardt and Patrick Barwise, 'Double Jeopardy Revisited', *Journal of Marketing*, 54, 3 (July 1990): 82–91.

23 Because the commercial PSBs are mostly funded by advertising, their consumer VFM has never been questioned in the way that the BBC's often is.

24 MTM, *Life Without the BBC: Household Study* (London: MTM, 2015), https://downloads.bbc.co.uk/aboutthebbc/reports/pdf/lifewithoutthebbc.pdf.

25 72% x (21/22) plus 28% x (33/48) = 88%.

26 Ofcom, *PSB Annual Report 2017: Audience Opinions Annex*, (July 2017), 4–5 https://www.ofcom.org.uk/__data/assets/pdf_file/0023/103919/annex-a-audience-opinions.pdf.

27 Patrick Barwise and Seán Meehan, *Simply Better: Winning and Keeping Customers by Delivering What Matters Most* (Boston, MA: Harvard Business School Press, 2004).

28 A possible exception is premium pay TV for the 21% of households who subscribe to it (but not the 79% who do not). Premium sport and movie channels are priced much higher than even basic pay TV but may well generate exceptional audience appreciation among their subscribers.

29 Barwise and Picard, *What If There Were No BBC Television?*

6

The Future of Television in the US

Jennifer Holt

Among the comprehensive recommendations in *A Future for Public Service Television,* the 2016 study of television culture, economics and politics in the UK, is the overarching vision statement that public television in an emerging digital media landscape should be governed not by market forces or funded in relation to their financial impact, but instead by '[p]rinciples of independence, universality, citizenship, quality and diversity ...'[1] These values for media culture have long been hallmarks of a public service mission but are much more elusive in the US system, existing only in part and often merely in name. While diversity has been a long arc principle underlying broadcast regulation and licensing requirements in the US since the Radio Act of 1927, it is a value that is largely unquantifiable and unenforceable as written; thus, it functions more conceptually as an ideal or a goal to strive for than as a practical or operational standard. The concept of universality/universal access is a tenet traditionally applied to the dissemination of public utilities, but the internet and its various delivery pipelines that also distribute digital television have yet to be classified as such. In fact, the pressures bearing on television's future in the US context are a far cry from those related to concerns of programming quality, affordability and accessibility, or the impact of television on a citizenry or culture. They are instead circulating around issues of monetisation, the evolving viewing practices of the digital audience, and the politicised regulation of distribution infrastructures. As such, the major debates about the future of television in the US context have focused primarily on digital technologies and their impact on business models, policies for broadband pipelines and Internet Service Providers (ISPs), and the need for new metrics that correspond to contemporary modalities of viewing and engagement.

Accordingly, this chapter will examine those issues that are most significant to the developing digital landscape of television in the US, particularly as they relate to the transition from an industry deeply rooted in the legacy structures of analog delivery born in the radio era to one navigating the new rules and protocols of digital distribution and 'connected viewing'. In an environment of ubiquitous mobile screens, third party providers, the fragmentation of the mass audience, increased demands on creative workers as shows proliferate across multiple media platforms and the enhanced surveillance of the digital audience, the television industry is experiencing a host of pressures previously unimagined, even in the multi-channel universe of cable in the 1980s and 1990s. As a result, debates about the future of television in the US are proliferating, as are the complications involved as content providers are forced to look in multiple directions at once to deliver their programming; they must cultivate the traditional televisual space as well as on-demand viewing practices, tend to their linear schedules, home audiences and armies of affiliates while concurrently developing online distribution strategies and rolling out their content on myriad new set top boxes, platforms, services and screens. In the end, as *Los Angeles Times* entertainment reporter Joe Flint has noted, this is an industry that will 'innovate at gunpoint'[2] and it is clear that whatever innovation takes place is going to be focused on marketplace priorities rather than cultural or civic ones.

Digital Distribution

One of the largest areas of consequence for the future of television in the US lies in the shifts to digital distribution and the imposition of new business models. Chief among the industry's problems – particularly the cable industry – is the threat posed by 'cord cutting' which is accelerating rapidly of late: as of July 2016, one quarter of US homes no longer subscribe to a pay-TV service.[3] This is due to the expanding range of digital platforms for accessing television programs (e.g. Hulu, iTunes, Amazon Prime Video, Netflix, HBO Go) and also to the skyrocketing price of a cable subscription, which is currently averaging over $100 a month in the US and perpetually climbing. Furthermore, the ability for viewers to leave the cable box behind and go 'over the top' has become much easier in the past decade. Broadband has achieved 80% penetration in US households, and the popularity of smart TVs and new delivery technologies for streaming television – including Roku, Apple TV, Amazon Fire, Google Chromecast and gaming consoles such as the Xbox and Sony Playstation – is on the rise. Presently, almost a quarter of US homes own a digital streaming device, 42% own gaming consoles, and 29% own Smart TVs.[4] Additionally, half of US homes now use subscription based video-on-demand (SVOD) services such as Netflix, equaling DVR penetration for the first time.[5]

As a reaction to the threats from new services and platforms, the cable industry began to offer different versions of 'TV Everywhere', which gave audiences the ability to stream or download content via multiple platforms and mobile devices for authenticated subscribers. Introduced by Time Warner Cable in 2009 and quickly imitated by other major cable providers and networks, this has become the primary form of paywall for most television content including the broadcast networks, which – in stark contrast to the UK and other territories with public service traditions – require a cable subscription for online access. The lack of a uniform log-in system between providers and channels, the restriction of broadcast content online (including news and sports and special events like the Olympics) to only those who pay for cable, and the frustration of geo-blocking content that was supposed to be available 'everywhere' but is often only able to be viewed on devices inside the subscriber's own home has led to consumer disenchantment with the promises of 'TV Everywhere'. It has also given fuel to activists challenging the strategy as 'TV Nowhere' that has further entrenched oligopolistic control in the cable industry while limiting competition and crushing online content diversity.[6] In addition to authenticated access, the cable and satellite industries have been forced to further reevaluate ways to entice subscribers, and they have begun to create lower-priced options of 'skinny bundles' of programming that are being rolled out in a limited way thus far.[7]

These changes in distribution have also focused attention on the evolving spectrum of what are now considered content providers and television 'channels', and the shifting relationships between studios, networks, and online platforms. For example, the National Football League – now television's most valuable product,[8] which has been shown on broadcast networks since the 1940s, on ESPN since 1987 and on TNT since 1990 – sold Twitter the rights to livestream ten games in 2015, and then sold Amazon Prime Video the rights to stream ten games in 2016. The Amazon deal was worth $50 million – five times the price of the Twitter deal sealed just one year earlier.[9] The broadcast networks CBS and NBC will also be broadcasting those games that Amazon is streaming, and Amazon will have some rights to advertising slots. Sharing rights costs with digital platforms does take some of the pressure off the networks, but it also shrinks everyone's slice of the revenue pie. Disney's exclusive deal with Netflix for first streaming rights to all of their content, including their production subsidiaries Marvel, Lucasfilm and Pixar, is another case in point. This deal represents a change in long-established 'windowing' practices – release strategies prioritising particular markets and platforms for specific lengths of time – and marks the first time that a major studio chose a streaming platform before a cable service as the designated 'Pay TV' window in post-theatrical distribution.[10] As relationships between streaming venues and heavyweight content providers evolve, unique release strategies, and revenue-/cost-sharing arrangements are likely to continue in this era of transition and transformation affecting television.

'Liveness' is one of the last remaining draws of linear television. It mostly endures in the form of breaking news, sports and major events like Presidential inaugurations, awards shows or highly anticipated series finales. This has kept many connected to their cable cords, but with ESPN now part of YouTube TV (the new streaming service offering live television from broadcast and major cable networks for $35/month) and most news and sports channels offering apps for mobile devices, there is new life for live television beyond traditional viewing practices. Further, as platforms like Facebook Live and Twitter's collective raw feeds increasingly stream live news events (often to the chagrin of journalists and critics[11]), 'liveness' is being commuted into the digital space via social media platforms as well.

What all of this means for viewers is a host of ever-expanding sites to access television programming outside of being tethered to a cable subscription. And there are more shows being produced than ever before in history – in 2016, there were 362 scripted shows on television, 455 if you include online services like Amazon, Hulu and Netflix.[12] Netflix has announced plans to spend $6 billion on original content in 2017 alone. In an era when audiences want to select content using an *a la carte* model – acting as their own programmers, unbundling and disaggregating network schedules to create their own evening's entertainment – what role will networks play in the future of television, particularly that of the broadcast networks? Will the network-affiliate structure soon be merely debris of the analog age? How will those relationships survive negotiations over the creation of the online brand experience and digital rights/streaming deals?[13] These questions lead to yet another pressure facing the future of television: the challenge of getting noticed and connecting with audiences in such a content-rich environment. Creating – and monetising – a hit in this supremely cluttered media landscape where a #1 show commands an audience of just over two million people gets harder with every passing year.

Metrics

For a show to survive, getting noticed by audiences is not enough; it also requires that those viewers are counted and valued by the television industry. This issue of metrics is another fundamental element of the television business model under siege, as a splintering audience, shifting viewing habits and changing measurement standards have had serious implications for the advertising market and, in turn, the business model for television. In 2016, the US television industry took in roughly $72 billion from advertising – a little more than the total for online advertising. However, the market for ad spending on linear television is currently stagnant while internet advertising has been steadily increasing following the popularity of online video, social media and mobile viewing. It is predicted that the internet will edge ahead of television to be the dominant advertising medium in 2017.[14]

As audiences and advertisers begin favouring digital spaces, metrics are failing to keep up with viewing practices and ratings (along with revenues) continue to fall for the networks. Nielsen includes C3 and C7 ratings (for programmes viewed three and seven days after their initial live airing) to account for the role of DVR and catch-up viewing, but the growing number of audience members watching programmes outside of those windows are not recognised by the ratings industry as significant. In other words, audiences viewing on computer and mobile screens are simply not valued equally to those watching on linear television. At the same time, new algorithms and analytics are emerging that demonstrate the value of 'engagement' to advertisers and challenge traditional metrics that support the premiums paid for rates on linear TV. Understandably, content providers, traditional networks and even digital platforms are losing patience.

Consequently, it has been suggested that US television is entering a 'post-Nielsen era' as the main ratings provider has failed to meet industry demands, and networks and their parent companies have begun designing their own methods to measure the new millennium audience and second-screen usage. Many of the big media conglomerates such as Viacom, Time Warner and Comcast NBC have started to develop new proprietary techniques and tools for measurement that incorporate set-top boxes, browsing behaviour, and online shopping data.[15] Additionally, Nielsen has faced considerable competition from ComScore in the digital space and both companies are battling it out to set the new standard in both linear TV and digital video measurement. The 2016 merger of ComScore and Rentrak, two of the biggest data analytics and audience measurement companies in the digital space, will further consolidate (and complicate) measurement for behaviour across computer screens, mobile phones, tablets and television sets. Just like the viewing landscape, the metrics industry is becoming quite fractured, and the unified ratings system that has endured for over 90 years is rapidly approaching its expiration date.

Ultimately, the US television industry, particularly insofar as it focuses on the multiscreen, app-driven, digital market, is becoming progressively folded further into the core of what Mark Andrejevic and Hye-Jin Lee have called the 'commercial surround' in which one's activities online are 'recorded, stored, and mined for marketing purposes'.[16] In this spirit, and following the lucrative strategies of streaming services like Netflix and Amazon Prime Video, the television industry has started to see some salvation for its struggling business model in the form of 'big data' related to their digital audience. Scholars have argued that the television industry has 'become enamored' with big data, and that this data has now become 'an integral part of the televisual culture; an essential tool for survival in the increasingly fragmented, crowded and competitive marketplace of digital TV'.[17] Indeed, the personal information and viewing habits of the online viewer that can easily be exploited and sold to hungry advertisers in search of their specifically targeted 'ideal' audience have proven to be extremely valuable commodities. Accordingly, the art of surveillance is being integrated into the realm of digital delivery and advertising, and simply repackaged as 'personalisation' for viewers. This snowballing invasion of digital privacy has multiple entry-points, and streaming platforms are but one.

Policy

The policy landscape is presenting its own considerable pressures for the television industry and its audience, many of which are centred around the issue of access – access to content, to internet services, to viewers, and ultimately to their personal information. What is at stake is everything from preserving diversity and localism in television to securing digital rights and maintaining a free and open internet that delivers streaming media, information and communication to all citizens equally. The importance of digital media policy extends beyond media culture into the fabric of democracy and the policy issues for television actually illuminate the much larger picture of digital rights presently at risk. The privacy concerns related to advertising and social media platforms mentioned above, for example, are even more troubling when one stops to consider how they also extend to the pipelines for internet access. In the US, Congress recently passed a bill allowing all ISPs to track, share and sell data on their subscribers' personal data and online behaviour – including people's viewing habits, internet browsing activities, and app use – *without permission.* This dismantles previous privacy protections in the digital space and further deregulates telecommunications policy that affects streaming media. Indeed, thanks to policymakers, engaging in online viewing has viewers sinking even deeper into 'the commercial surround', and our television experience has become even more 'big data' for sale to advertisers by companies providing internet access. Our online television viewing has become fully imbricated in surveillance culture.

Net neutrality is another major policy issue bearing down on the future of television, particularly in the US where there is not a strong tradition of public service affordances or values to sustain alternative visions of the medium beyond the most brutally commercial. As policies for the treatment of broadband are debated, the unrestricted provision of digital media services also hang in the balance. Will broadband pipelines be treated as a 'telecommunications service' and therefore be protected as a 'common carrier' which must prioritise public accessibility and equal treatment for all data? Or will they be categorised as an 'information service' which would strip those protections and allow for discriminatory practices by internet service providers that want to slow down ('throttle') certain content and charge for tiered levels of service?[18] While the Federal Communications Commission's 2015 Open Internet Rules[19] determined that all ISPs – including cable companies and mobile phone providers – were 'telecommunications services' ensuring that all data must be sent at the same speed regardless of its origin, these public interest safeguards are far from secure. Several ISPs have already demonstrated their willingness to flout the spirit and the letter of the net neutrality rules as they currently exist. For example, Verizon, AT&T and Comcast have exempted their own video services from mobile data caps (i.e. assigned them a 'zero rating') on their distribution pipelines while charging their competitors for data usage.[20] Such anti-competitive behaviour in a different political climate would be cause for an FCC investigation at the very least, but the Trump administration has instead signaled that Obama-era regulations protecting the 'Open Internet' and equal access to all data will not stand. The result for television in the digital space will be a marked decrease in competition and diversity as conglomerate-owned distributors will be allowed to privilege their own content services and stifle others. Sadly, the public and the 'public interest' are the ultimate losers in this particular policy fight over what values will prevail in digital media pipelines.

Policies related to protecting digital rights, intellectual property, and privacy are also at the forefront of the US television industry's concerns. Piracy and rights infringements are multiplying as are the problems they pose for the industry, all the while becoming more complex to define, track and deter as television streams globally.[21] The vulnerabilities for media companies are indeed quite extensive, and have been dramatically illustrated in events such as the 2014 hack on Sony Pictures, and the 2017 ransomware attack on Netflix that targeted then-unreleased episodes from Season 5 of *Orange is the New Black*. These issues of digital privacy also present growing problems for viewers, even extending to the hardware designed to bring television into their homes: Samsung TVs, for example, have recently been revealed to be vulnerable to CIA hacking and capable of fooling the owner into thinking the device is off, when it is actually on and recording conversations in the room, acting as a remote surveillance device.[22]

The trend of consolidation and media concentration also continues in the US context and is no longer just limited to horizontal mergers in the content industry. Since the Comcast–NBC merger in 2011,[23] there have been takeovers emanating from the distribution sector as they either merge with one another and take control of even more media infrastructure, or buy production companies to add to their expanding empires. AT&T is leading the pack on both fronts. Their deal for $49 billion to buy satellite provider DirecTV in 2015 and their pending $85 billion takeover of Time Warner will put them in new territory for media conglomerates. The attitude in the regulatory sector that has allowed for these developments is reminiscent of 'the whorehouse era' in the FCC of the 1950s, when 'federal regulators and industry leaders develop[ed] a relationship too friendly to be of honorable service to the public.'[24] This does not bode well for citizens and consumers of digital media, nor for the future of television, should the industry hope to hold onto any trace of the public service values that represent the medium's greatest potential.

Notes

1 David Puttnam, *A Future for Public Service Television: Content and Platforms in a Digital World* (London: Goldsmiths, University of London, 2016), http://futureoftv.org.uk/wp-content/uploads/2016/06/FOTV-Report-Online-SP.pdf, 155.

2 Interview with Joe Flint, conducted 27 June 2012, Los Angeles, CA. Available at www.carseywolf.ucsb.edu/mip/article/interview-joe-flint.

3 Oriana Schwindt, 'Cord Cutting Accelerates...' *Variety*, 15 July 2016, http://variety.com/2016/biz/news/cord-cutting-accelerates-americans-cable-pay-report-1201814276/.

4 'TV Connected Devices Pave the Way....' Nielsen Insights, 8 March 2017, www.nielsen.com/us/en/insights/news/2017/tv-connected-devices-pave-the-way-for-new-ways-to-watch-content.html.

5 'Milestone Marker: SVOD and DVR Penetration....' Nielsen Insights, 27 June 2016, www.nielsen.com/us/en/insights/news/2016/milestone-marker-svod-and-dvr-penetration-on-par-with-one-another.html.

6 Marvin Ammori, 'TV Competition Nowhere', *Free Press*, 20 January 2010, www.freepress.net/sites/default/files/fp-legacy/TV-Nowhere.pdf.

7 For an interesting look at the anti-competitive nature of skinny bundles, see Cynthia Littleton, 'Why Most Early Channel Packages Aren't So "Skinny" After All', *Variety*, 21 March 2017, http://variety.com/2017/tv/news/skinny-bundles-carriage-deals-early-channel-packages-1202012826/.

8 The league generated $7 billion in TV rights during 2016, $1.9 billion from *Monday Night Football* on ESPN alone, and has most of its contracts locked in for the next five years. See Kurt Wagner, 'How the NFL Juggles the Future of Streaming ...' *Recode*, 1 May 2017, www.recode.net/2017/5/1/15386694/nfl-live-stream-amazon-prime-thursday-night-football-ratings.

9 Peter Kafka, 'Amazon Will Stream Thursday Night NFL Games This Year', *Recode*, 4 April 2017, www.recode.net/2017/4/4/15184100/nfl-amazon-football-games-thursday-streaming-watch-live-prime-twitter.

10 See Alisha Grauso, 'Netflix to Begin Exclusive Streaming of Disney ...' *Forbes*, 24 May 2016, www.forbes.com/sites/alishagrauso/2016/05/24/netflix-to-begin-exclusive-streaming-of-disney-marvel-star-wars-and-pixar-in-september/#1d577146135d and 'Netflix Shares Soar After Announcing Disney Deal', *Deadline Hollywood*, 4 December 2012, http://deadline.com/2012/12/disney-netflix-deal-movies-exclusive-382271/.

11 See Charlie Warzel, 'Facebook, Twitter, and Breaking News' Special Relationship', *BuzzFeed News*, 11 July 2016, www.buzzfeed.com/charliewarzel/facebook-twitter-and-breaking-news-special-relationship?utm_term=.ocx9mX6Jn#.lfAp14q9l.

12 Rick Porter, 'The Peak of Peak TV Keeps Climbing', *TV By the Numbers*, 21 December 2016, http://tvbythenumbers.zap2it.com/more-tv-news/the-peak-of-peak-tv-keeps-climbing-455-scripted-shows-in-2016/.

13 See Betsy Skolnik interview in Michael Curtin, Jennifer Holt and Kevin Sanson, eds. *Distribution Revolution* (Oakland, CA: University of California Press, 2014), 111–20.

14 Jemma Brackebush, 'How TV Ad Spending Stacks Up Against Digital Ad Spending ...' *DigiDay*, 20 July 2016, http://digiday.com/marketing/tv-ad-spending-stacks-digital-ad-spending-4-charts/. Also see Mike Snider, 'Online Ad Spending to Top TV Ads in 2017', *USA Today*, 8 June 2016, www.usatoday.com/story/tech/news/2016/06/08/online-ad-spending-top-tv-ads-2017/85594160/.

15 Brian Steinberg, 'TV Industry Struggles to Agree on Ratings Innovation', *Variety*, 11 April 2017, http://variety.com/2017/tv/features/nielsen-total-content-ratings-1202027752/.

16 Mark Andrejevic and Hye-Jin Lee, 'Second-Screen Theory', in Jennifer Holt and Kevin Sanson, eds. *Connected Viewing* (New York: Routledge, 2014), 53.

17 J. P. Kelly, 'Television by the Numbers: The Challenges of Audience Measurement in the Age of Big Data', *Convergence*, 30 March 2017: 3–4. Also see Allie Kosterich and Philip Napoli, 'Reconfiguring the Audience Commodity', *Television and New Media* 17, 3 (2016): 254–71.

18 For in-depth analysis of the topic, see the special section on Net Neutrality in the *International Journal of Communication* 10 (2016) http://ijoc.org/index.php/ijoc/issue/view/12.

19 2015 Open Internet Order, FCC 15–24 (2015) https://apps.fcc.gov/edocs_public/attachmatch/FCC-15-24A1.pdf.

20 See Klint Finley, 'The FCC OKs Streaming for Free ....' *Wired*, 3 February 2017, www.wired.com/2017/02/fcc-oks-streaming-free-net-neutrality-will-pay/.

21 For excellent discussions on the complexities of digital piracy, see Ramon Lobato and Julian Thomas, *The Informal Media Economy* (Malden, MA: Polity Press, 2015); Ramon Lobato and James Meese, eds. *Geoblocking and Global Video Culture* (Amsterdam: Institute of Network Cultures, 2016); and Patrick Vonderau, 'Beyond Piracy: Understanding Digital Markets', in Holt and Sanson, eds. *Connected Viewing*: 99–123.

22 Janko Roettgers, 'New WikiLeaks Document Dump Suggests the Use of Smart TVs for Surveillance,' *Variety*, 7 March 2017, http://variety.com/2017/digital/news/wikileaks-smart-tv-surveillance-1202003656/.

23 For an outstanding analysis of the Comcast-NBC merger's stakes for media, telecommunications, and the American public, see Susan Crawford, *Captive Audience* (New Haven, CT: Yale University Press, 2013).

24 Patrick Parsons, *Blue Skies: A History of Cable Television* (Philadelphia, PA: Temple University Press, 2008), 144.

7

Pressures on Public Service Media: Insights from a Comparative Analysis of 12 Democracies

Matthew Powers

This chapter identifies three key pressures experienced by contemporary public service media. The first – funding – pertains to debates about whether public service media should continue receiving public funds, and if so, how much and through what means they should receive it. The second – oversight – details legal and administrative measures that threaten the independence of public service media or make it difficult for them to fulfill their civic obligations. The third – audiences – highlights competitive pressures on public service media to cater to audiences, especially socio-demographically elite ones, rather than serve the needs of a broad, diverse population.

In each domain, I suggest that the public service media best equipped to deal with these pressures– and therefore to fulfil their civic obligations – are those that deepen, rather than depart from, long-standing public service principles. Public service media that receive generous funding through a universally paid fee tend to provide robust news coverage, as well as innovative programming in arts and culture. Those with strong legal charters and arms-length oversight enjoy also relatively strong protections from undue government influence. And public service media that thoughtfully integrate public input are relatively well-positioned to serve as a broad forum for diverse voices and viewpoints. By contrast, public service media that rely on government appropriations struggle to assert their independence from partisan influence and engage in long-term planning. Those with weak charters are vulnerable to changes in government policy. And public service media that chase after niche audiences tend to skew their programming to reflect the demands of culturally and economically advantaged populations.

The data for this chapter come from research conducted with Rodney Benson and Tim Neff on public service media in 12 democracies.[1] While hardly representative of public service media in their entirety, the countries studied vary in their funding, oversight and audience size in ways that make it possible to identify distinct responses to the key pressures identified above. For each country, government source documents were used identify the pressures experienced by public service media, both online and off. This data was then complemented by comprehensive literature reviews of public service media in each country, as well as e-mail correspondence with scholars and government regulators whose expertise on public media helped to confirm analyses of the pressures public media face in each country.

In what follows, I briefly outline the key pressures experienced by public service media in the three domains highlighted above: funding, oversight, audiences. For each, I describe the key pressures, and draw on the experiences of different public service media to identify potential solutions. I conclude with a brief discussion of how this research can inform debates about the role of public service media in shaping contemporary public communication systems.

Funding

One set of pressures concerns funding. In general, funding for public media has not kept pace with increasing costs, thus leading some to explore advertising, philanthropy and others forms of revenue to support their operations. While such funds boost revenue, they also tend to dilute public service missions. In France, there are only minor differences between the evening news of the private TF1 and public (but advertising reliant) France 2 evening news.[2] In the United States, paltry government funding leads public media to seek corporate and philanthropic support, thus creating pressure to align content with donor demands. In 2012, for example, a multi-part series on the American economy sponsored by Dow Chemical tracked closely with the company's major business interests.[3] By contrast, public service television in Denmark, Finland and the United Kingdom – all of which rely primarily on public funding – gives more attention to public affairs news than commercial competitors.[4] Several content analyses also suggest these more publicly-funded channels tend to be offer more in-depth, diverse and critical reporting than their commercial counterparts.[5]

While most public media see declining or stagnant revenues, the most admired and most popular (in terms of audience share) public media outlets remain some of the best funded. At the United Kingdom's BBC, Sweden's SVT, Findland's YLE,

Demark's DR, Germany's ARD/ZDF and Norway's NRK, per capita public spending ranges from $100 to $177 US dollars annually. By contrast, the worst funded – Canada ($31), New Zealand ($25) and the United States ($3) – also tend to have difficulty attracting broad, diverse audiences and providing independent, civically oriented programming.[6] Greater funding, which often comes in multiyear increments, boosts the capacity for long-term planning and the delivery of online services. One scholar argues that the BBC's 'huge resources' have enabled it to 'offer wide-ranging outputs, from educational and cultural programmes for the web, to new, public affairs and interactive forums'.[7] Along with Finland's YLE, the BBC leads on many platforms as the most popular provider of news and information.[8]

Historically, the licence fee has been the primary mechanism for funding public media in much of Western Europe. Despite longstanding issues with viewers avoiding payment, such fees have a 'social dimension' in that 'by contributing to their national public broadcaster, citizens felt it was more accountable to them than to the politicians'.[9] These fees are also set aside solely for public media, and thus do not compete for direct government appropriations with other programs. In Canada, by contrast, the annual appropriations process keeps the Canadian Broadcasting Corporation (CBC) 'on a short leash', which makes 'long-term planning difficult'.[10] In 2000, the Netherlands replaced the licence fee with government appropriations, and one result has been a gradual decline of funding in recent years.

Many countries are now adjusting licence fees, which previously were determined simply by the presence of a television in the home, to include digital devices that provide multiple access points for content. Denmark has changed its definition to include any device that can display television content.[11] Finland has replaced the licence fee with a general media tax that citizens pay regardless of the device used.[12] Germany has moved to a flat-rate, per-household licence fee. This 'one residence, one fee' system is promoted as enabling access to content across multiple devices without paying multiple fees.[13] These and related efforts deepen, rather than depart from, the 'public good' concept of public service media and can thus be seen as 'an extension of the traditional licence fee'.[14]

Oversight

Oversight, both administrative and legal, is a perpetual issue for public service media which seek to maintain their independence from partisan interference while fulfilling their civic missions. The public service media most subject to undue influence tend to be those with weak or vague charters, poor safeguards for mitigating political interference, and overly constraining regulations that limit the ability of public media to develop online offerings. Here, too, public service media best positioned to deal with these pressures seem to be those that deepen extant public service principles by implementing arms-length oversight that simultaneously ensures independence while maintaining accountability.

Strong legal charters provide the basis for non-interference from government, while mandating that public service media provide high-quality, diverse programming. They do this by explicitly restricting government interference, and in some cases – Germany, for example – setting forth technical criteria by which funding decisions are made. To ensure public media fulfill their civic obligations, some charters also set forth mandates around educational content, cultural programming and the inclusion of diverse voices. In Norway, for example, the NRK must submit an annual report to the Norwegian media authority detailing how it fulfilled these mandates. By contrast, weak charters contain overly generic language that leave public media vulnerable to changes in government policy. In New Zealand, the centre-right party replaced the existing charter – which included language to ensure public media provide content neglected by commercial providers – with vague language about the need for a range of quality content. The resulting 'strategic ambiguity' means that the public service media provider can be more easily directed by 'the government's transitory policy priorities'.[15]

While all public service media have oversight agencies, those best able to function independently of political influence have staggered term limits and dispersed appointment power. In Sweden, the foundation's board consists of 13 members whose terms are staggered: half the board leaves every four years.[16] The German KEF is comprised of 16 members, each Land (German state) appoints one expert from a given field.[17] Protections against partisan political meddling are less robust in France, Japan and the United States. In 2008, the French administration of Nicolas Sarkozy asserted presidential authority to directly appoint the public broadcaster's director[18] which, although reversed in 2013, suggests the fragility of public media's independence. Similarly, in Japan politicians have been accused of attempting to influence NHK news.[19] Finally, in the United States the president has sole appointment authority for the CPB's board of directors. Historically, these appointments are based on political patronage rather than expertise.[20]

Questions abound as to how the remit of public service media ought to evolve in a digital era. In even the best-funded systems, like Germany, public media have sometimes been slow to embrace a pro-digital organisational culture.[21] Where public media have not expanded online, it has largely been due to legal constraints stemming from commercial media opposition claiming unfair-state sponsored competition against market actors. In Denmark, associations representing commercial media have gained political support for banning or restricting DR's online services. In the most recent policy agreement, which runs from 2015–2018, the public broadcaster's online services remain intact, though a panel of experts is being asked to clarify the future role of public service media.[22]

More broadly, throughout Europe oversight agencies have introduced 'public value tests' as a way to evaluate the impact of digital services prior to their implementation. Since 2008, the BBC has regularly submitted to such tests, and it has received negative decisions as a result of them, most notably a 2008 proposal to provide additional local video news, sports and weather services in 60 areas of the United Kingdom on local BBC news sites. The BBC Trust rejected the proposal on the grounds that the service would not be the best use of licence fee funds and might negatively impact commercial media at the local level. The Puttnam Report suggests revisiting this proposal as a way to 'help to address the immense local democratic deficit in English regions'.[23] This suggestion accords with the idea that extending – rather than abolishing – pre-existing public service values not adequately captured in extant value tests.

Audiences

Public service media around the world face strong competition from commercial channels for audiences. Those that maintain and build audiences seem to be those that integrate public input so as to ensure that public media remain a forum for diverse and broad voices and viewpoints. In the United Kingdom, audience councils publish an annual report assessing how well the BBC meets licence payers' needs. In Denmark, an eight-person regulatory authority by law must include one person nominated by the Cooperative Forum for Danish Listeners and Viewers Association. In some cases, opinion surveys also help provide public service media with an additional buffer against government intervention, as public broadcasters typically fare better in opinion polls than does the incumbent government.[24]

Which audiences count most is sometimes an issue for public service media, and this problem is growing in the digital environment. In the United States, audience members who donate effectively have more input over programming than those that do not. Only a small portion of the citizenry contributes, and they tend to be more educated and wealthier than the general population, thus creating pressures to orient programming toward more affluent groups.[25] The Puttnam Report expresses a similar concern that the BBC will sacrifice 'underserved audiences' (both regional and socio-economic) while 'superserving the literate, articulate and wealthy.'[26] Here as elsewhere, the pressure to depart from long-standing public media principles is strong, yet doing so risks undermining the historical mission of public service providers to serve as a forum for broad, diverse audiences.

Conclusion

Public service media face challenges on a number of fronts. These challenges include long-existing concerns about how to protect public media from partisan meddling, as well as more recent problems posed by increasing commercial pressures and the difficulty of balancing the need to appeal to a broad audience while upholding public service values. This review suggests that the public service media best equipped to deal with these pressures are those that deepen, rather than depart from, long-standing efforts to provide high-quality programming, and in-depth news and information that serves to promote an inclusive version of democratic citizenship across diverse populations.

This research highlights the relative strengths of public service media in the United Kingdom, Germany and the Nordic countries, all of which have media policies aimed at ensuring public media excellence. By contrast, public media have been weakened in recent years in the Netherlands and New Zealand due to the erosion of procedures for ensuring arm's-length autonomy from direct government control. Canada's public service media, already comparatively weak, remains vulnerable to political pressures because of its reliance on government appropriations. Likewise, the American public media system continues to struggle with both partisan and philanthropic pressures due to its weak institutional autonomy, government underfunding, and reliance on donors to make up the shortfall.

It is important to stress, however, that even in countries where public service media operate under less-than-ideal conditions, they retain a certain distance from commercial pressures and are thus able to provide content not found on commercial networks. Conversely, even the best funded public media do not go as far as small-scale alternative media in challenging entrenched power relations. Public service media may thus provide important content but they are unlikely to replace long-term influences that stem from the shared class interests and social networks of many politicians, regulators and public media professionals.[27]

Opponents often argue that public service offerings are obsolete in the age of cable television and the internet. They also oppose public funding on the grounds that it amounts to unfair state-sponsored competition against market actors. Yet in many countries public service media are the only media providing locally produced, innovative and experimental content online.[28] And while there are pressures to compete with commercial outlets and monetize online offerings, it is also important to emphasise that they continue to find ways to fulfil their core missions. This includes helping audiences navigate digital networks in which content is abundant but civic affairs content can get lost in the noise. In short, even as public media face challenges, they remain vital components of contemporary democratic media systems.

Notes

1 These countries are Canada, Denmark, Finland, France, Germany, Japan, the Netherlands, New Zealand, Norway, Sweden, United Kingdom and the United States. See Rodney Benson and Matthew Powers, *Public Media and Political Independence: Lessons for the Future of Journalism from around the World* (Washington, DC: Free Press, 2011), and Rodney Benson, Matthew Powers and Tim Neff, 'Public Media Autonomy and Accountability: Best and Worst Policy Practices in 12 Leading Democracies', *International Journal of Communication* 11 (2017): 1–22.

2 Rodney Benson, *Shaping Immigration News: A French-American Comparison* (Cambridge, UK: Cambridge University Press, 2013). In order to redress this blurring of lines, extra public service demands have been placed on other public channels that receive less or no advertising, such as France 3 and ARTE (a jointly funded French-German channel).

3 David Sirota. 'The Wolf of Sesame Street: Revealing The Secret Corruption Inside PBS's News Division', *Pando*, 12 February 2014, https://pando.com/2014/02/12/the-wolf-of-sesame-street-revealing-the-secret-corruption-inside-pbss-news-division/.

4 James Curran, Shanto Iyengar, Anker B. Lund and Inka Salovaara-Moring, 'Media System, Public Knowledge, and Democracy: A Comparative Study', *European Journal of Communication* 24, 1 (March 2009): 5–26. Denmark does mix public funds with advertising revenues.

5 See, e.g. Toril Aalberg and James Curran, *How Media Inform Democracy* (New York: Routledge, 2011); Stephen Cushion, *The Democratic Value of News* (London: Palgrave, 2012); and Stephen Cushion, Justin Lewis and Gordon Neil Ramsay, 'The Impact of Interventionist Regulation in Reshaping News Agendas: A Comparative Analysis of Public and Commercially Funded Television Journalism', *Journalism* 13, 7 (October 2012): 831–49.

6 For a list of all figures on public service media, as well as audience share, see Benson, Powers and Neff, 'Public Media Autonomy and Accountability', 5.

7 Benedetta Brevini, *Public Service Broadcasting Online: A Comparative European Policy Study of PSB 2.0* (Houndmills: Palgrave Macmillan, 2013), 147.

8 Ibid.

9 Stylianos Papathanassopoulos, 'Financing Public Service Broadcasters in a New Era', in *Media Between Culture and Commerce*, ed. Els de Bens (Bristol: Intellect Books, 2007), 156.

10 David Skinner, 'Television in Canada: Continuity or Change?' in *Television and Public Policy*, ed. David Ward (New York: Taylor & Francis, 2008), 16.

11 Trine Syvertsen, Gunn Enli, Ole J. Mijøs and Hallvard Moe, *The Media Welfare State: Nordic Media in the Digital Era* (Ann Arbor, MI: University of Michigan Press, 2014).

12 Ibid., 78–79.

13 Benson, Powers and Neff, 'Public Media Autonomy and Accountability', 5.

14 Syvertsen *et al.*, *The Media Welfare State*, 78–79.

15 Peter Thompson, 'Last Chance to See? Public Broadcasting Policy and the Public Sphere in New Zealand', in *Scooped: The Politics and Power of Journalism in Aotearoa New Zealand*, ed. Martin Hirst, Sean Phelan and Verica Rupar (Auckland: Auckland University of Technology Press, 2012), 109.

16 Benson, Powers and Neff, 'Public Media Autonomy and Accountability', 10.

17 Ibid.

18 Benson and Powers, *Public Media and Political Independence*, 31.

19 Ellis S. Krauss. 'Introduction'. *NHK vs. Nihon Seiji*. Japanese language version of *Broadcasting politics in Japan* (G. Jumpei, Trans.). Tokyo: Tōyō Keizai Shimpōsha, 2006.

20 Jerold M. Starr, *Air Wars: The Fight to Reclaim Public Broadcasting* (Philadelphia, PA: Temple University Press, 2001), 26.

21 Annika Sehl, Alessio Cornia and Rasmus Kleis Nielsen, *Public Service News and Digital Media* (Oxford: Reuters Institute for the Study of Journalism, 2016).

22 Henrik Søndergaard. 'Denmark Passes New Media Policy Agreement'. LSE Media Policy Project Project, 29 September 2014, http://blogs.lse.ac.uk/mediapolicyproject/2014/09/29/denmark-passes-new-media-policy-agreement/.

23 David Puttnam, *A Future for Public Service Television: Content and Platforms in a Digital World* (London: Goldsmiths, University of London, 2016), 127.

24 Jeremy Tunstall. 'The BBC and UK Public Service Broadcasting', in *Reinventing Public Service Communication: European Broadcasters and Beyond*, ed. Petros Iosifidis (Houndmills Palgrave Macmillan, 2010), 149.

25 Robert K. Avery, 'The Public Broadcasting Act of 1967: Looking Ahead by Looking Back', *Critical Studies in Media Communication* 24, 4 (October 2007), 361.

26 Puttnam, *A Future for Public Service Television*, 59.

27 Benson, Powers and Neff, 'Public Media Autonomy and Accountability', 15.

28 Syvertsen *et al.*, *The Media Welfare State*.

8

Public Service in Europe: Five Key Points

Trine Syvertsen and Gunn Enli

Introduction: The State of Public Service Broadcasting in Europe

Europe is the heartland of public service broadcasting (PSB), not least because of the leading role of the BBC, but there is of course more to European public broadcasting than the BBC. In this chapter, we will contextualise the 2016 report *A Future for Public Service Television: Content and Platforms in a Digital World* through a review of relevant research on public broadcasting in Europe as well as our own studies of PSB in the Nordic region.

We have structured our observations around five key points that condense research as well as challenges, debates and prospects: 1) the crisis discourse; 2) innovation and public service media (PSM); 3) distinctiveness; 4) editorial independence; 5) national differences and politics. In line with the report that is the centre of this volume, studies from other European countries observe that an overall narrative of decline dominates the debate on public service. However, they also note that public broadcasters, in particular publicly funded institutions with long traditions, are managing better than the overall narrative describes.[1] As we enter a more politicised and polarised social climate, there are new opportunities as well as risks for European public service broadcasters.

Challenges and Crisis

The starting point for many contemporary studies, as it has been since the mid-1980s, is the ongoing challenges to PSB; literature inevitably refers to various forms of 'crisis'. In a sense, the entire research tradition on public service broadcasting is defined by

the mapping of challenges and problems and, to some degree, on normative suggestions for remedies and improvements. The challenges described have shifted over the
last three decades, but continue to be discussed under broadly similar headings: technological, political, economic and changes in the social and cultural climate.[2]

In the 1980s and 1990s, the emphasis was on commercialisation, cable and satellite competition and a political and cultural shift to the right. In the early 2000s, it
was the introduction of new digital platforms, participatory formats and the concern
that public broadcasters could be left out by new portals and gatekeepers. Nowadays,
the challenges of audience fragmentation stem from new competitors reluctant to be
regulated as media businesses, ongoing challenges related to TV's position as content
provider, problems for established business models, and problems related to the very
definition of television. The traditional understanding of television – as an advertise-
ment or licence fee funded system distributing mixed schedule programming simultaneously to a mass audience watching in their homes – is in flux.

Specific challenges come from so-called digital intermediaries: a group of services that have in common their function as algorithm-based gatekeepers. The most
disruptive digital intermediaries for linear television are content aggregators such as
Netflix, HBO, Amazon and YouTube, in addition to AppleTV, which together represent a significant gatekeeper to, and third party provider of, TV content. The impact
of digital intermediaries vary, but they have in common that they liberate consumers from schedules, produce and distribute content based on more specific user data
than traditional TV companies, and encourage so-called 'cord-cutting' where viewers
can watch individual programmes online and avoid TV advertising and paying the
licence fee.[3]

The economic and technological challenges to public service television are similar across borders but the impact is mediated by cultural and political forces and, in
particular, the degree of trust and support for public institutions. The studies from different European countries are perhaps most varied on this point. While some describe
a polarised political climate, others describe political cultures where there are still
some form of consensual politics surrounding PSB.

A point where studies across Europe largely overlap is on their emphasis on the
increasing complexity of the political context of public broadcasting. While there has
always been a multitude of social, political and cultural interests in broadcasting,
the term 'multi-stakeholderism' describes the erosion of sector boundaries that has
brought new stakeholders into the debate, such as technology platforms, online services and distributors. A multi-stakeholder environment implies that not just public

broadcasters and the government are partners in negotiations over PSB, but that a variety of private operators can influence the debate.[4] Criticism of PSB now comes from all sides and pertains to all aspects and concerns that are external to PSB operations and increasingly affects both how it is understood and regulated.

From Public Service Broadcasting to Public Service Media

Paralleling the attention to crisis and challenges, research on public service emphasises the potential for renewal and change for the original public broadcasters. Many contributions passionately urge broadcasters to change their ways, and much attention has been given to the role of PSBs in digital environments.[5] Central to this research is a conceptual transition from 'public service broadcasting' to 'public service media' to reflect the fact that the institutions do more than just radio and television.[6]

The reconceptualisation demonstrates that public service broadcasters share characteristics with other types of media, online experiments and cultural institutions. While such a change in focus is understandable and necessary, there is also a risk that research on public service broadcasting may neglect the large-scale and mainstream broadcasting activities that continue to distinguish these types of institutions. As the research interest shifts from PSB to PSM, the most marginal of experiments may get more attention than steady and stable programme formats that may run for years and attract millions of viewers. It is still vital to investigate the mechanisms that PSB institutions use to build relationships with mass audiences and a broad public sphere.

Cross-platform formats, which combine tradition and innovation, are among the most expansive strategies of public broadcasters. Case studies from across Europe demonstrate that the cross-platform operations of PSB institutions have reached a certain maturity. First, the classic PSB genre news production has changed significantly to include online platforms.[7] A second example are innovative services for young audiences, such as the Norwegian online drama series *Skam [Shame]*, which updates randomly according to the narrative rather than following a predetermined schedule, and includes text-messages between the characters. Third, public broadcasters have invented formats with mass appeal and a focus on national identity such as *Test the Nation* and *Great Britons*, offering interactive services such as online voting. Despite the idea central to PSM research that PSB could be a node, linking together amateur and professional activities, it may appear that cross-platform experiments are more successful where the traditional broadcaster is in control and has designed the whole process.[8]

Broad Remits are Challenged Everywhere

The move to online platforms has intensified ongoing debate about the broad remit enjoyed by public broadcasters. Although there are variations, the criticism of the PSB remit rests on three interlinked premises across borders: that a broad remit is no longer needed as provisions have proliferated in the multi-platform universe, that public broadcasters use their broad remit as an excuse for transmitting crowd-pleasing programming indistinguishable from commercial broadcasters, and that the expansion of services on new platforms threaten the revenue of private media businesses.[9]

Commenting on this debate, public broadcasting scholars have nuanced the argument that PSB is becoming indistinguishable and argue that it constitutes a central quality-enhancing mechanism in a cross-media ecosystem. First, scheduling studies show that genre pluralism is more extensive on public service channels, particularly in prime time, and that their distinctiveness from commercial channels is increasing.[10] Furthermore, studies have demonstrated that even if public and commercial channels buy similar formats, the PSB productions remain distinct.[11] Third, studies of PSB activities, such as online news and weather services, dispute that these constitute an economic threat to competitors' online activities.[12]

Instead of exerting a negative influence on the rest of the market, scholars argue that PSB has a positive influence, sometimes described as an *ecosystem effect*, meaning that by influencing their commercial competitors, PSBs can, in practice, 'act as regulators of the television industry as a whole.'[13] Indeed, public broadcasters are constantly expected to take on new tasks to improve the overall sector, whether it is to act as a 'digital locomotive' or to support creative industries and private competitors.[14]

The Twin Challenges of Commercialisation and Politicisation

The operations of the public broadcasters reflect their relationship with other domains of power, particularly the market and state. Historically, there has been great emphasis on the autonomy of broadcasters and on securing governance and funding models that may guarantee a high degree of political and financial independence. The intention has been to impose mechanisms guaranteeing that broadcasters are more accountable to the general public than to commercial interests or the government. The licence fee has been an important source of funding for many public broadcasters, and people's willingness to pay it has been seen as an indication of the general legitimacy of public service broadcasting.

However with funding models in flux, a changing political climate, and TV audiences migrating online, the licence fee is increasingly being debated and challenged across Europe. In the Nordic region, several countries have abolished the licence fee system (Iceland in 2007, Finland in 2013), and replaced it with a direct unconditional income tax, while others have so far continued the system (Norway, Sweden), or extended it to a media licence, including also the internet and mobile phones (Denmark). In spite of changing funding mechanisms, the Nordic countries' public service fees and taxes remain some of the highest in the world.[15] The Nordic public broadcasters, and in particular the NRK, have been provided with a 'generous leeway for launching commercial initiatives'[16] and a high level of autonomy based on the 'arm's length principle.'

Across Europe, the level of funding, how that funding can be spent, and what else it is obliged to cover vary significantly. A less stable financial situation leaves broadcasters more open to both commercial and political pressures. Studies reveal correlating patterns between the level of funding and the level of politicisation of public broadcasting funding; the countries with high and stable public service funding, such as Sweden and Germany, are also the countries with the lowest influence of politics; in less stable funding systems, such as France, Italy and Spain, there are higher levels of political influence over PSB; in Poland, the only European Broadcasting Union member state where advertising is the main revenue source, there exist levels of politicisation and commercialisation to a degree not found elsewhere in Europe.[17] Generally, PSB institutions have less autonomy in countries where public service broadcasting was established later and in different historical conditions. For example, in Poland, public broadcasting was established only after the revolutionary changes of 1989, and in Greece, where the public broadcaster was shut down by the Government in 2013, both radio and television were introduced under political dictatorships.[18]

In spite of differences, political and financial pressures are also found in stable PSB countries, not least because multi-stakeholderism leads to vigorous debate and new challenges. In addition to the factors already discussed, the degree of compromise or conflict over PSB may also relate to differences between consensus-oriented and more polarised political cultures more generally. In smaller, more consensus-oriented cultures, it may be easier to reach a compromise to support and sustain PSB.

The Significance of National Differences

In this chapter, we have pinpointed similarities and differences, general challenges as well as diverging paths. The fifth and final observation addresses the overall perspective on public broadcasting and its relationship to political culture. Despite a common discussion and familiarity of issues across borders, studies increasingly point to significant differences between public broadcasters across different countries that are systemically related to the domain of national policies.[19] Despite all the talk about globalisation and the role of the European Union, the primary context for public service broadcasting remains at the level of the national.[20]

The emphasis on path-dependency points to factors that lie outside of public service broadcasting and, to some degree, outside the context of media policy. The fate of public broadcasters is only partially determined by how well each broadcaster succeeds in attaining its goals, and largely determined by developments at a more general political level where basic policy solutions fall in and out of favour. Few politicians are media policy specialists; rather, they are usually generalists preferring particular types of policy solutions over others. Stable political compromises in overall policy continue to be important – such as the Nordic welfare state contexts where there is broad consensus for state intervention and support for universal solutions to reduce inequalities in society.[21] As political debates over public broadcasting across Europe demonstrate, public broadcasting may be seen as both cause and indication of, as well as a solution to, problems in the social and political sphere, whether these refer to issues of diversity or pluralism or a more profound loss of trust in media and public institutions.

Notes

1 Minna Aslama Horowitz and Jessica Clark 'Multi-Stakeholderism: Value for Public Service Media', in *The Value of Public Service Media*, ed. Fiona Martin and Gregory Lowe (RIPE@2013. Gothenburg: Nordicom, 2014), 165–83; Damian Tambini, 'Problems and Solutions for Public Service Broadcasting: Reflections on a 56 Country Study', in *Public Service Media in Europe: A Comparative Approach*, ed. Karen Arriaza Ibarra, Eva Nowak and Raymond Kuhn (London: Routledge, 2015), 41–52.

2 For example, Trine Syvertsen 'Challenges to Public Television in the Era of Convergence and Commercialization', *Television New Media* 4, 2 (2003): 155–75.

3 Gunn Enli and Trine Syvertsen, 'The End of Television – Again.How TV is Still Influenced by Cultural Factors in the Age of Digital Intermediaries', *Media and Communication* 4, 3 (2016): 142–53.

4 Horowitz and Clark, 'Multi-Stakeholderism'; Hilde van den Bulck and Karen Donders, 'Of Discourses, Stakeholders and Advocacy Coalitions in Media Policy: Tracing Negotiations towards the New Management Contract of Flemish Public Broadcasters VRT', *European Journal of Communication*, 29, 1 (2014): 83–99; Karen Donders *Public Service Media and Policy in Europe* (London: Palgrave 2012).

5 For example, Graham Murdock, 'Building the Digital Commons. Public Broadcasting in the Age of the Internet', in *Cultural Dilemmas in Public Service Broadcasting*, ed. Per Jauert and Gregory Ferrell Lowe (Gothenburg: Nordicom, 2005), 213–30.

6 See Horowitz and Clark, 'Multi-Stakeholderism'; Murdock, 'Building the Digital'; Donders *Public Service*; Gregory F. Lowe and Jeanette Steemers, eds., *Regaining the Initiative for Public Service Media* (RIPE@2011. Gothenburg: Nordicom, 2012); Fiona Martin and Gregory F. Lowe, eds., *The Value of Public Service Media* (RIPE@2013 Gothenburg: Nordicom, 2014); Petros Isofides, ed., *Reinventing Public Service Communication. European Broadcasters and Beyond* (Basingstoke: Palgrave, 2010); Ágnes Gulyas and Ferenc Hammer, eds., *Public Service Media in the Digital Age: International Perspectives* (Newcastle: Cambridge Scholars, 2013); Benedetta Brevini, *Public Service Broadcasting Online. A Comparative European Policy Study of PSB 2.0.* (London: Palgrave, 2013).

7 Ainara Larrondo *et al.*, 'Opportunities and Limitations of Newsroom Convergence', *Journalism Studies* 17, 3 (2016).

8 See, for example, James Bennet, 'Interfacing the Nation. Remediating Public Service Broadcasting in the Digital Television Age', *Convergence*, 14, 3 (2008), 277–94; James Bennett, 'The Challenge of Public Service Broadcasting in a Web 2.0 Era: Broadcast Production Culture's View of Interactive Audiences', in *Public Service Media in the Digital Age: International Perspectives*, ed. Ágnes Gulyas and Ferenc Hammer (Newcastle: Cambridge Scholars, 2013), 3–22; Gunn Enli, 'Redefining Public Service Broadcasting. Multi-Platform Participation', in *Convergence* 14, 1 (2008), 105–20.

9 See, for example, Tambini, 'Problems and Solutions'; Isofides, *Reinventing*; Lars Nord, 'Why is Public Service Media Content as it Is? A Comparison of Principles and Practices in six EU Countries', in *Public Service Media in Europe*, ed. Ibarra *et al.*, 170–88.

10 e.g. Karoline Ihlebæk, Espen Ytreberg and Trine Syvertsen 'Farvel til mangfoldet? Endringer i norske tv-kanalers programlegging og sendeskjemaer etter digitaliseringen', *Norsk Medietidsskrift* 18, 3 (2011): 217–40; Karoline Ihlebæk, Espen Ytreberg and Trine Syvertsen, 'Keeping Them and Moving Them: TV Scheduling in the Phase of Channel and Platform Proliferation', *Television and New Media*, 15, 5 (2014): 470–86; Hanne Bruun, 'The Prism of Change: "Continuity" in Public Service Television in the Digital Era', *Nordicom Review* 37, 2 (2016), 33–49.

11 e.g. Pia Majbritt Jensen, 'Television format adaption in a trans-national perspective – An Australian and Danish case study' (PhD diss., Aarhus University, 2007).

12 e.g. Helle Sjøvaag, Hallvard Moe and Eirik Stavelin, 'Public Service News on the Web: A Large-Scale Content Analysis of Norwegian Broadcasting Corporation's Online News', *Journalism Studies*, 13, 1 (2011): 90–106.

13 Adrian Blake, Nicholas C. Lovegrove, Alexandra Pryde and Toby Strauss, 'Keeping Baywatch at Bay', *McKinsey Media* (1999), 20.

14 Enli, 'Redefining'; Tim Raats and Caroline Pauwels, 'Best Frienemies Forever? Public and Private Broadcasting Partnerships in Flanders', in *Private Television in Western Europe: Content, Market, Policies*, eds. Karen Donders; Caroline Pauwels, and Jan Loisen (Houndmills: Palgrave 2013), chapter 14.

15 Jessica Yarin Robinson, 'Statistical Appendix' (2016), Avaible online www.hf.uio.no/imk/personer/vit/trinesy/mws_appendix_final.pdf; Trine Syvertsen, Gunn Enli, Ole J. Mijøs and Hallvard Moe *The Media Welfare State: Nordic Media in the Digital Era.* (Ann Arbor, MI: University of Michigan Press 2014), 77.

16 Hallvard Moe, 'Commercial Services, Enclosure and Legitimacy: Comparing Contexts and Strategies for PSM Funding and Development', in *From Public Service Broadcasting to Public Service Media*, eds. Jo Bardoel and Gregory Lowe (Gothenburg: Nordicom 2008), 57.

17 Nord, 'Why is Public'; Iosifidis, *Reinventing Public*; Beata Klimkiewicz, 'Between Autonomy and Dependency: Funding Mechanisms of Public Service Media in Selected European Countries', in *Public Service Media*, eds. Ibarra *et al*, 111–25.

18 Stylianos Papathanassoupoulos, 'The "State" of 'Public' Broadcasting in Greece' in Iosifidis, *Reinventing Public*, 222–32; Pawel Stepka, 'Public Service Broadcasting in Poland: Between Politics and Market', in Iosifidis, *Reinventing Public*, 233–44.
19 Ibarra *et al.*, eds., *Public Service Media in Europe*.
20 Donders, *Public Service*; Moe, 'Commercial Services'; Brevini, *Public Service*.
21 Syvertsen *et al.*, *The Media Welfare State*; Enli and Syvertsen, 'End of Television'.

9

Diversity: Reflection and Review

Sarita Malik

Introduction

This is a particularly fertile moment to discuss the relationship between diversity and Public Service Television (PST). In late 2016, the issue of boardroom diversity was raised when the UK government took an unprecedented intervention to block a BME female candidate's application for a position on Channel 4's board. The BBC's coverage of Brexit in 2016 demonstrated its struggle to diversify the debates around immigration that framed the campaign. Meanwhile, Channel 4, with its mainstreaming of populist politics through documentaries such as *Things That We Won't Say About Race That Are True* (Channel 4, 2015) (which probed the value of multiculturalism) and *The Trouble With Political Correctness* (2017)[1] (which linked a fear of offence with the conditions for extremist politics), mark a departure from its confidently multicultural origins. Each of these examples is symptomatic of a media sector that is still contending with how to manage cultural difference in the public space. My focus in this contribution is to reflect on what the *A Future For Public Service Television* report[2] says about 'diversity' and consider more broadly the utility of PST diversity policy in direct relation to its role in mediating lived multiculture.

The timeliness of the report in relation to questions of diversity is striking. Minority communities in the UK are currently experiencing new forms of hostility as an intense backlash against immigration and multiculturalism presides, alongside a resurgence of right-wing populisms and xenophobia.[3] Within an intensified climate of ethnic separatism and border anxieties, it would be easy to assume that spaces seeking to foster multiplicity would be shut down and that public service modes of governmentality around diversity, would be on fragile ground. And yet 'diversity' is gaining traction within PST like never before.

In 2016, the government announced that, for the first time, diversity would be enshrined in the BBC Royal Charter, asserting a commitment in its Diversity and Inclusion Action Plan and Strategy 2016–2020 to embed diversity more deeply into the organisation's identity and to better include underrepresented groups (such as women, disabled, LGBT and black, Asian and minority ethnics (BAME)) in the workforce and content, both on-air and on-screen. Meanwhile, Channel 4's 360° Diversity Charter, established in 2015 is, according to its Creative Diversity Manager, 'a game changer' because 'every production has to go through a diversity tick-box process for on-and off-screen'.[4] In late 2016, the British Academy of Film and Television Arts (BAFTA) announced new initiatives to boost the numbers of ethnic minority and socially disadvantaged filmmakers, which includes plans for more diverse membership and reworked eligibility criteria for some of its award categories. The Cultural Diversity Network launched the industry-wide monitoring system, Diamond (Diversity Analysts Monitoring Data) which, from 2017, measures the diversity of those on-and off-screen. Ofcom, now the BBC's first external regulator, is further developing its monitoring programme after identifying a problem both with the under-representation of women, ethnic minorities and people with disabilities and a 'dearth of data' around diversity composition.[5] For its CEO, Sharon White, the priority for the regulator is transparency.[6] These are just a few of the many diversity interventions that seek to boost more diverse participation, representation and engagement – whether through target-setting, monitoring, mentorship or training.

For Channel 4's CEO, David Abraham, quoted in a special section of the Royal Television Society's magazine, *Television*, in January 2017, diversity 'is going to be a lifetime's effort but we can begin to look back now on the first steps to progress'.[7] So what is it about this current moment, after several decades of shifting modalities of PST diversity policy[8] and parallel sector inequalities that offers grounds for such optimism? And how is it that this ostensible proclivity *towards* diversity is being assembled and normalised alongside broader exclusionary nationalisms related not just to issues of race and ethnicity but also to other aspects of social identity including class, religion, sexuality, age and nationality?

The report states that issues of diversity 'based on the recognition that the population consists of multiple and overlapping sets of minorities'[9] are central to the relevance and legitimacy of any public service media system. Problematically, PST has also proven to be implicated in the social processes of exclusion that are enacted more widely across the creative sector, revealing a deep correlation between social and cultural inequality.[10] As I reflect on the report, I also want to briefly examine how 'diversity' is operationalised within these contexts and consider its potential efficacy in producing a less unequal, more diverse public service media culture as part of a functioning public sphere.

The Report on Diversity

Diversity – beyond questions of diversity of ownership or content (for example genre) – is normally tagged on to wider debates about public service and its remit in scholarly and industry debates. Significantly, the report positions diversity as a core concern (as did the Pilkington Report in 1962 and the Annan committee in 1977) alongside other issues such as representation, accountability and independence. Chapter 8 focuses specifically on television's environment in relation to diversity, asserting that struggles over visibility and representation will continue 'as long as different social groups are not adequately addressed.'[11] At the same time, the apparent tension that PST is tasked to deal with, involving the negotiation of common, universal approaches (that underpin the ethos of public service) alongside meeting the particular needs of ethnic minority and other under-represented groups (in an already complex scenario of a multi-platform digital age) is also acknowledged.

Public Service Broadcasting's (PSB) ostensibly unifying project, based around a national public culture and identification is tasked therefore with being entirely inclusive and representative, whilst grappling with the nuances of living with difference – cultural, racial or otherwise – in a multicultural, if not multicultural*ist*, society. Developments such as devolution, inequality, immigration and the various protected characteristics addressed by the UK's 2010 Single Equality Act[12] further complicate the picture. One specific recommendation within the report is that an amendment be made in the Equality Act to include PST commissioning and editorial policy, as part of its public service equality duties. Equality under law seems a reasonable proposal because it asserts the link between PST and wider social contexts of equality. The trickier issue, and one which I will now go on to discuss, is of how such equality might be achieved within a sector that has repeatedly been identified as deeply unequal in terms of minority access, opportunity, representation and engagement.

Following Napoli's work on broadcast diversity,[13] the report contends that PST requires a multi-pronged approach to address not just content diversity (that is, the range of representations that end up on screen and how these are perceived) and source diversity (the diversity of those who produce and supply content), but also what Napoli calls 'exposure diversity'. This, Napoli explains, is, 'the degree to which audiences are actually exposing themselves to a diversity of information products and sources'.[14] Napoli's work on diversity and PSB overlaps with some of the other ways in which we can think about diversity, for example in relation to issues of ownership and control, media plurality and democratic participation. The report further explicates how diversity might be understood, for example in relation to 'voice, representation and opportunity'.[15] These are useful routes to analysis because they open up the meanings of 'diversity', and insist on recognising the overlapping orders of mediation that include representation, production and reception. An implicit suggestion is that the realisation of 'diversity' might depend on an integrated approach factoring in such interdependencies, rather than building distinct solutions that do not take into account other orders of mediation.

Taking each of Napoli's elements of broadcast diversity – content, source and exposure – the report outlines some of the outstanding problems that have led to a deep unevenness in terms of, for example, satisfaction levels based on visibility and portrayal across different social and geographical groups. It uses as an example the representation of working class lives, particularly in the reality television genre and the strong responses to Channel 4's 2014 series, *Benefits Street*. Exposure diversity is, importantly, tied to the values placed on PST, a point that has also been backed up by industry data. For example, Ofcom's PSB 2015 Audience Opinion report demonstrates that where there is an increase in satisfaction with PSB provision, it has clearly mapped onto the ways in which PSB purposes are valued, for example an increasing satisfaction in PSB in 2014 for 'showing different cultures within the UK'.[16] All of this underscores the report's emphasis on how diversity is interwoven with the relevance and legitimacy of PST.

Ring-Fenced Funding

The report notes the lack of research that has been conducted around content or on-screen diversity compared to research on audience reception, a point that has been taken up by Ofcom in its 2017 statement about the industry's lack of diversity on- and off-screen.[17] One of the key recommendations is for the BBC and other public service broadcasters to ringfence funding specifically aimed at BAME productions to both evidence commitment and more successfully build real change. A potential funding model for this might be how the BBC has funded nations and regions, which resulted from a crisis in representation and funding allocations that it first identified in 2003. Marcus Ryder, BBC Scotland's former editor of current affairs, reminds us that, under the nations and regions template, if money is not spent on particular kinds of regional productions (for example, a Scottish production), then the money gets lost, so in the case of diversity, funding could be earmarked as 'BAME' (as one example of a social group that has hitherto been marginalised within PST and as a demographic that will make up one-third of the total audience by 2050). The upshot is that 'the actual accountability is within the structure',[18] making it difficult to ignore.

The report's two main recommendations (the first to link PSB with public service equality duties and the second, to ring-fence funding) urge a more robust commitment to reducing inequalities and providing more tangible support. Ring-fenced funding is a recommendation that has a historical context, both within the structures of PST, and as a source of critique, with claims that it risks building separateness into organisational structures (or ghettos), whilst allowing more pervasive racialised inequalities and regimes of representation to remain intact.[19] For example, the 1970s and 1980s witnessed a broader conception of public service representation that was built into the very structure of early Channel 4 (in the form of the Multicultural Programmes Unit) and the BBC and LWT had their own Black and Asian 'specialist' minority programming strands. For all the potential gains of dedicated funding, such as increasing visibility, any lasting impact on wider institutional culture or on subsequently making PST less 'hideously White'[20] has been put in doubt. The question arises of whether ring-fending money really is the solution or, indeed, enough in itself. Would separate funds, even if they were conceded to by the broadcasters, further fuel a politics of resentment (both institutionally and publicly) because of the accentuated difference that such funding would help produce at a time of anti-multiculturalism in public discourse? What kinds of representations might BAME audiences want or expect from this investment and how and where would these programmes be screened, scheduled and valued?

Inequalities in the Workforce

These concerns around representation also connect to a more prevalent method within past and current diversity strategy making, which proposes boosting minority workers in line with proportion to the population or what Herman Gray has called 'representational parity'.[21] Diversifying the workforce is an area that is potentially more straightforward to regulate through policy intervention, partly because it is easier to monitor personnel (rather than, for example, the more contested terrain of content and representation). The 2015 Creative Skillset Employment Survey found that only 4% of executive positions were held by BAME staff.[22] Creative Skillset's 2012 Census had shown a decrease in BAME employment in television, with figures falling from 9% in 2009 to 7.5% in 2012.[23] Although BAME is only one dimension of a broader under-standing of cultural diversity alongside other modes of demographic diversity, these figures indicate that, even in spite of (hypothetically 'easier') employment-targeted diversity strategies, marginalised and privileged access persists. A more recent study commissioned by BAFTA and Creative Skillset with the BFI in 2016[24] also emphasised the 'class ceiling' as a further barrier to opportunity for underrepresented groups working in the film, television and the gaming industries who have to contend with a culture of 'fitting in' together with homogenised recruitment practices and mind-sets. Whilst such findings are depressing, they also function as an important 'reality check', apparently influencing the recent intensification of institutional diversity ini-tiatives. The 2012 Creative Skillset survey in particular, by evidencing the extent of the problem, has elicited a range of high profile discussions, including those led by the actor and writer, Lenny Henry. However, there has also been serious criticism of the Diamond equality monitoring initiative launched in 2016 (and a boycott from Bectu, the NUJ and the Writers' Guild). There is a concern that measuring diversity patterns and progress at genre (rather than programme) level allows broadcasters to avoid the kind of transparency and accountability required to drive improvement.

We can say quite confidently that the culture of PST is partly reflected in the composition of the workforce and that the low representation of minority demo-graphics serves as evidence of social exclusion. However, is it a logical consequence that a more diverse workforce will lead to more diverse content and exposure for audiences? An emerging body of literature within media industry studies research contests the simple idea that diversifying those who produce and supply content (source diversity) will, in turn, diversify the range of representations that end up on screen or, indeed, how these might be perceived by audiences (when determin-ing content diversity). For Anamik Saha, the cultures of production that minority workers have to negotiate, also need to be taken in to account, as well as repre-sentational politics because 'the reproduction of neo-colonial discourses around race, ethnicity and religion does not merely spring from the values of individual gatekeepers but is embedded within the production process itself through what appears as a common-sense economic/commercial rationale.[25]

If we agree that these additional facets unavoidably shape broadcasting culture, the inevitability of 'social parity' as a result of 'representational parity' is rightly queried. This is not to suggest that labour inequalities are harmless or that more diversity at the point of production can never diversify editorial power or influence, but it is also not tantamount to chalking up the numbers in the hope that all representational problems will be solved at once. Besides, assuming that certain cultural 'types' will inevitably produce certain kinds of cultural work and from only certain perspectives, reifies essentialist tropes of cultural identity.

Gray's analysis also encourages us to consider 'the assumption, micropractices, social relations, sense about the nature of social difference and the practices of inequality' that exist.[26] This matters, not just because diversity policy often fails to mediate lived multiculture in meaningful ways, but more ominously because diversity is implicated precisely as a 'technology of power, a means of managing the very difference it expresses.'[27] Thus, it is suggested that 'diversity' not only co-exists with inequality (as is repeatedly proven to be the case), but that 'diversity' helps produce a racialised social system, in order for racialised – and other social – hierarchies to be, more or less, held in place. The enactment of diversity in institutional life – of which current PST presents us with a brilliant example because it is what Gray detects as 'a key location where diversity is practiced materially and symbolically'[28] – is therefore coming under scrutiny, rendered an ideological and discursive mechanism designed to manage rather than address cultural difference.[29] These interpretations help us to make sense of the ongoing relationship between social and cultural inequality, as well as the potential disconnect between a fervent culture of institutionalised diversity-*isms* and the effects they promise to deliver.

Diversity Strategy as Ideological Counterpoint

The report welcomes the various diversity strategies within PST, but also critiques them for not meeting their desired aims.[30] It observes that, whilst the current diversity impetus relocates diversity as a universal point of reference rather than positioning itself in relation to specific minority communities, it is precisely the prevalence of marked social inequalities that have led to such strategy building. The symbiotic *raison d'être* for the latest diversity imperative is that growing social diversity and demands for fairer representation from those who are routinely marginalised are occurring alongside acute social inequalities.

David Abraham's comment that diversity for Channel 4 is 'a lifetime's effort' is an expectation that resonates with Sara Ahmed's argument that diversity discourse is an ongoing phenomenon that rarely actually implements change and thus depends on what Ahmed calls 'non-performativity'.[31] For Abraham, 'the first steps of progress' that he discerns, signify the great promise of diversity, a tangential hope, and one that sees disadvantage and discrimination as transitory rather than systemic[32] or, perhaps more cynically, requiring systematic maintenance. It is in such ways that 'diversity' within public service media functions as what Collins has identified in the context of liberal higher education, as an 'ideological counterpoint to the race-based policy and practice of affirmative action'[33] (in the US) or anti-racism (in the UK) that came before. Returning to the wider political and popular contestations around the value of multiculturalism, one concern is whether, rather than being antithetical to the wider retreat from multiculturalism that we are experiencing, current industry-led diversity mediations are actually constitutive of it and its enactment. The supposition is raised precisely because of the underlying language and logic of institutionalised diversity that suit the valorisation of post-multiculturalism (if not an outright denunciation of multiculturalism), by evading a direct engagement with the institutional discrimination that exists within the very structures and systems of PSB.

Such evasion – that 'diversity' work facilitates – works on multiple levels. For example, the specific ways in which different minority communities are marginalised is not constructively engaged with (as an example of the predominant rejection of 'identity politics'). Resource is spent on renewing diversity policy goals rather than on holding to account those earlier misdirected diversity campaigns, meaning that organisational responsibility is lost. The risk is that the idea of tolerance underpinning 'diversity' goes untested against practice and that recurring, imitative policies are expected to yield different results. Meanwhile, diversity reproduces positive articulations – *vis-à-vis*, for example, updated scripts around creativity, talent and innovation – that are actually dependent on the social inequalities that it fundamentally helps produce.

One example of the uncritical renewal of diversity is the phenomenal rise of 'creative diversity' policy making; now the dominant way in which diversity is presented in PST contexts. As I have argued elsewhere, 'creative diversity' shifts the paradigm of the multicultural problem (in PSB), enables the 'marketisation' of television and multiculture and ultimately continues to safeguard PSB interests.[34] 'Diversity' simply becomes therefore a self-narrative requiring visibility (as a public discourse of validation), and promotion (as a discourse of liberal tolerance). One further characteristic is its economic utility because 'diversity' is also financially driven, securing the universal assurance requirements of public funding and also bringing a potential boost to profitability through, for example, a diverse workforce, leadership or customer orientation. This financial characteristic is especially prevalent in the 'creative diversity' paradigm. It is in these ways that 'diversity' is underpinned by market, regulatory and social motivations. Market-oriented diversity arguments, whilst they may obscure the social regulatory basis of public service television, can provide obvious incentives for broadcasters. Commercial broadcasters such as Sky, for example, have responded to declining BAME audiences by recently pledging to take 20% of the stars and writers of its UK-originated television shows from a BAME background. For Stuart Murphy, director of Sky's portfolio of entertainment channels (to which the pledge applies), this is 'a way of kickstarting the process.'[35]

Concluding Reflections

Even within this critique, there is a strong defence of diversity in practice as the basis of PST, coupled with a call for greater equality and a valuing of diversity in the public sphere. How PST positions itself in relation to rising populisms and new social tensions, and the steps it takes to shift the axis of debate towards a more democratic, diverse and sometimes discordant space, seems especially pressing. For broadcasters, this means constructing different ways of talking about these issues in a move that potentially challenges public service media tendencies to invite consensus rather than disagreement. For media and communications studies, but also one hopes for the broadcasters themselves, it means engaging with 'the operations of power/knowledge and the role of media in the making of racial inequality (and its potential for the making of racial justice)'.[36] All of this might seem like an abstraction from the nuts and bolts of 'doing diversity', but without interrogating the meanings, functions and utility of 'diversity' itself, it becomes impossible to curate a future that not just accommodates, but also actively defends, multiculture. How we understand PST's approach to diversity has to go beyond calling for renewed sets of remits and targets, to instead understand it as a process that, through its very existence, sustains deep inequalities that avert mutual recognition in public space.

Notes

1 Both documentaries were fronted by the former Equalities and Human Rights Commission Chief, Trevor Phillips.

2 David Puttnam, *A Future for Public Service Television: Content and Platforms in a Digital World* (London: Goldsmiths, University of London, 2016).

3 See for example, Alana Lentin and Gavan Titley, *The Crises of Multiculturalism: Racism in a Neoliberal Age* (London: Zed Books, 2011).

4 This quote comes from a wider discussion about diversity and television at the 'Diversity: Job Done?' event hosted by the Royal Television Society in 2015, https://rts.org.uk/article/diversity-job-done-don%E2%80%99t-get-me-started.

5 Ofcom, Diversity and Equal Opportunities in Television – Steps Taken by Broadcasters to Promote Equal Opportunities, 14 September 2017, www.ofcom.org.uk/_data/assets/pdf_file/0019/106354/diversity-report-steps.pdf.

6 Sharon White, quoted in Steve Clarke '2016: TV's Defining year of Diversity?' *Television*, January 2017, 22.

7 David Abraham, quoted in ibid., 23.

8 Sarita Malik, '"Creative Diversity": UK Public Service Broadcasting After Multiculturalism', *Popular Communication*, 11, 3 (2013): 227–41. Also see Gavan Titley, 'After The End of Multiculturalism: Public Service Media and Integrationist Imaginaries for the Governance of Difference', *Global Media and Communication*, 10, 3 (2014): 247–60.

9 Puttnam, *A Future for Public Service Television*, 103.

10 See Kate Oakley and Dave O'Brien, 'Learning to Labour Unequally: Understanding the Relationship between Cultural Production, Cultural Consumption and Inequality', *Social Identities*, 22, 5 (2016): 471–86. Also see Dave O'Brien and Kate Oakley, *Cultural Value and Inequality: A Critical Literature Review* (London: AHRC, 2015).

11 Puttnam, *A Future for Public Service Television*, 106.

12 The Equality Act covers age, disability, gender, gender reassignment, pregnancy and maternity, race, religion or belief and sexual orientation.

13 Phil Napoli, 'Deconstructing the Diversity Principle', *Journal of Communication* 49, 4 (1999): 7–34.

14 Napoli, 'Deconstructing', 25.

15 Puttnam, *A Future for Public Service Television*, 104.

16 Purpose 4, Representing the UK, its nations, regions and communities, http://stakeholders.ofcom.org.uk/binaries/broadcast/reviews-investigations/psb review/psb2015/PSB__2015_audience_impact.pdf.

17 Ofcom, Diversity and Equal Opportunities in Television – Steps Taken by Broadcasters to Promote Equal Opportunities.

18 Marcus Ryder, comment made at the 2014 BAFTA lecture delivered by Lenny Henry at which Henry outlined his industry proposal, http://static.bafta.org/files/lenny-henry-speech-proposal-2200.pdf.

19 Simon Cottle, *Ethnic Minorities and the Media: Changing Cultural Boundaries* (Buckingham: Open University Press, 2000).

20 This was the famous assertion made by BBC director general Greg Dyke in 2001. See Amelia Hill, 'Dyke: BBC Is Hideously White', *The Guardian*, 7 January 2001, www.theguardian.com/media/2001/jan/07/uknews.theobserver1.

21 Herman Gray, 'Precarious Diversity: Representation and Demography', in Michael Curtin and Kevin Sanson, eds., *Precarious Creativity. Global Media, Local Labour* (Berkeley, CA: University of California Press, 2016), 242.

22 Creative Skillset, 2015 Employment Survey, March 2016, https://creativeskillset.org/assets/0002/0952/2015_Creative_Skillset_Employment_Survey__March_2016_Summary.pdf.

23 Creative Skillset, 2012 Employment Census of the Creative Media Industries, http://creativeskillset.org/assets/0000/5070/2012_Employment_Census_of_the_Creative_Media_Industries.pdf.

24 BAFTA/British Film Institute, *Succeeding in the Film, Television and Games Industries*, www.bafta.org/sites/default/files/uploads/baftareportsucceedinginfilmtvandgames2017.pdf.

25 Anamik Saha, 'From the Politics of Representation to the Politics of Production', in Georg Ruhrmann, Yasemin Shooman and Peter Widmann, eds., *Media and Minorities* (Göttingen: Vandenhoeck & Ruprecht, 2016), 46.

26 Gray, 'Precarious Diversity', 246.

27 Ibid., 242.

28 Ibid., 242.

29 Malik, 'Creative Diversity'.

30 Puttnam, *A Future for Public Service Television*, 157.

31 Sara Ahmed, *On Being Included: Racism and Diversity in Institutional Life* (London: Duke University Press, 2012).

32 Ruth Eikhof and Chris Warhurst makes this argument in 'The Promised Land? Why Social Inequalities Are Systemic in the Creative Industries', *Employee Relations*, 35, 5 (2013): 495–508.

33 Sharon Collins, 'From Affirmative Action to Diversity: Erasing Inequality from Organizational Responsibility', *Critical Sociology*, 37, 5 (2011): 517–20.

34 Malik, 'Creative Diversity'.

35 Stuart Murphy, quoted in Mark Sweney, 'BSkyB to take 20% of talent from black, Asian or other minority backgrounds', *The Guardian*, 18 August 2014, www.theguardian.com/media/2014/aug/18/bskyb-20-percent-talent-black-asian-ethnic-minority.

36 Gray, 'Precarious Diversity', 252.

10

The BBC: A Brief Future History, 2017–2022

David Hendy

The report into the 'Future for Public Service Television' is out. So, what next for the biggest and most influential player in the public service firmament, the BBC? The document's diagnosis – and its prescriptions – are clear and persuasive. Will it be transformative in the ways we hope?

I'm reluctant, as all historians should be, to engage in too much futurology. In any case, the Puttnam Report is admirably wide-ranging and thorough. Its authors have ensured that almost all the runes have already been read. It's a shame, perhaps, that radio wasn't in their original remit. When Radio 4's breakfast programme still does so much to shape the daily news agenda and its drama-serial brings to greater public attention the issue of gas-lighting in domestic abuse, when Radio 3 offers such significant patronage to the music industry and 1Extra recruits a younger and more diverse audience than many commercial stations, I'd say the medium easily justifies at least a walk-on part in this topical drama. But that single caveat aside, the key issues – content, universality, variety, benchmark-setting in quality, the nurturing of talent, above all perhaps the facilitating of 'public knowledge and connections' – have all been identified and analysed with precision and fairness.[1] Deeper currents of history have been addressed, too, not just more immediate matters of policy and structure. The report makes clear, for instance, that the most profound threats to public service television have not only come from the 'hyper-commercial, market dominated media environment' unleashed in the 1980s,[2] or from the prospect of a 'mass exodus' of younger viewers from old-fashioned TV boxes in the family living room to various forms of individualised online, on-demand, mobile video services such as YouTube and Netflix. They have also come in response to longer-term changes of behaviour and attitude in British society – a healthy decline in deference, for instance, and a gathering reluctance to allow only a narrow elite to pronounce on matters of newsworthiness or artistic value. Quite rightly, the report has made clear that despite everything, reports of the death of public service broadcasting have been premature: not only does it persist in having a firm emotional grip on audiences, it is still *needed* – perhaps more than ever in the age of the internet 'filter bubble' – as a bulwark to a 'plural and informed democracy'.[3] And in their closing recommendations the authors show decisively that ensuring an old institution such as the BBC remains fit for purpose will nevertheless involve more than a few tweaks: it will have to show even more determination to support productions that don't just feature Black and Minority Ethnic people, but are actually conceived and made by them; it will have to devolve even more production to Wales and Scotland; it will have to contemplate replacing the licence fee with a household levy; appointments to its unitary board will need to be independent from government; and so on.

As I say, all the detailed policy analysis and the prescriptions that follow are self-evidently sensible. The unresolved issue, though, is this: the ability of such a vision to take hold – to *actually* influence government policy or the BBC's own sense of what it needs to do to reform itself. In this respect, public service television in Britain faces two enormous challenges: one political, the other cultural; one external, the other internal.

First, then, the political challenge: the inconvenient fact that we currently reside in a kakistocracy – rule not by the best people, but by the worst. As with Brexit, as with Trump, as with all the 'fake news', as with the whole damned political ferment we find ourselves mesmerised and horrified by as we lurch dazed and confused towards the BBC's Centenary year of 2022. We are confronted by one apparently immovable obstacle: the repeated failure of sound empirical data and rational argument to gain purchase in government policy-making circles. It's hard to know whether this is a matter of ignorance or design – both, perhaps. But the pattern is clear enough. Scientists warned very clearly that a cull of badgers would do little to control the spread of bovine TB – but the government went ahead anyway. An avalanche of sociological data has been produced to prove that reintroducing selective secondary schooling will depress both educational achievement and social mobility – but the government is still allowing new Grammar schools to open. Convincing, well-researched case-studies, whether uttered calmly or screamed out loud, have been met with a studied indifference from a government less flashily confrontational than that of Thatcher but almost certainly more ideologically extreme. So, one is forced to ask, what hope is there that even the best arguments in favour of public service television will gain traction in the corridors of power? Why, indeed, should we expect the kakistocrats – most likely in power for several more years – to act in the public's interests at all?

There's worse. The survival of public service broadcasting, and of the BBC in particular, in a form we would recognise – universal, diverse, large enough to have cultural force – now has to deal with a very malevolent Catch-22 which the political class has constructed for it. In the 1980s and early 1990s, the BBC could be attacked for being weak – for its falling share of the national TV audience, for perceived lapses in editorial judgement and the like. Now, even more perversely, it's being attacked for being strong: the range and quality of its online presence is said to be blocking new players from entering the field, the size and popularity of its news service is said to be killing off local newspapers, and so on. Its public usefulness in providing a service – and in doing something well – is discounted in favour of its impact on the ability of private companies to make a profit. In other words, so the reasoning goes, the better the BBC is at doing television, the more it's a danger to competitors, real or imagined. It's the kind of reasoning that translates into a pervasive government belief that the BBC is going beyond its 'true' remit in broadcasting pop music on Radio 1 or the hugely successful prime-time show *Strictly Come Dancing* on BBC One.

It's difficult to know if this complaint is simply the language of a government acting as proxy for the commercial companies who stand to benefit from a smaller BBC, or if it comes from a genuine though misplaced belief in some mythical past when the Corporation really was dedicated solely to minority programming, with not a whisper of popular dance music or comedy on air – a belief that its broad appeal and the broad range of programming it offers is merely a result of that steady imperialistic, bloated ambition perceived by the Right as typical of 'statist' bureaucracies. Either way, one feels despondent. One can doggedly point out to our politicians that back in 1924 – that is, when the BBC was less than two years old – it was John Reith himself who declared that 'it is most important that light and "entertaining" items' be broadcast. Or one can show that a couple of years later popular or dance music occupied a whopping 35% of the BBC schedule. Or one can simply remind politicians that even in Reith's vision, 'pleasing relaxation after a hard day's work' was as vital a building block of the rounded, balanced citizen as any programme of 'edification and wider knowledge'.[4] But the only response we're likely to get is a by now deeply rehearsed demand: that the BBC becomes more 'distinctive' – a benign-sounding phrase in which terrible danger lurks. The comedian David Mitchell put it most succinctly. What the Right mean by 'distinctiveness', he pointed out, is programmes that are simply not popular. This idea hasn't been arrived at in order to improve the BBC, but 'specifically to make it do less well'. In effect, Mitchell suggested, the government attitude is that, like telling a boxer to throw a fight, the BBC will 'be expected deliberately to perform less well than it's capable of'.[5] Charter review might have been completed and a licence fee settlement reached. But a highly resistant strain of ideological malevolence and historical illiteracy shows no sign of abating among those currently in power.

And then there is that second challenge – this time, internal. It concerns the culture that seems to have taken root within the BBC – or, more specifically, inside BBC *News* – over the past few years. Perhaps we should say, not 'culture' but *ideology* – a term I assiduously used to avoid deploying when discussing the BBC because it always felt too strong a word to describe the rather ad hoc, accidental way that policy has seemed to evolve inside the Corporation for most of its life. Any length of time spent in the BBC's written archives poring over the minutes of endless editorial meetings shows that what's ended up on air for most of the past 95 years or so has usually been the result of a complex negotiation between individual personnel with hugely varied – never entirely coherent, always debated, sometimes fiercely contested – opinions and prejudices. Naturally, some voices will have been more powerful than others: the BBC is not without its hierarchies and chains of command. But the cut and thrust of disagreement has palpably been there. I think back to the late 1920s and early 1930s, for instance: the supposed high tide mark of Reithian orthodoxy and timidity. In the offices and studios we would find individuals like the documentary-pioneer Geoffrey Bridson, a Manchester poet who loved jazz and folk and the voices and opinions of 'the Common Man' and who proudly declared himself to be a 'Bolshevik'; or Lionel Fielden, aristocratic and conservative in background, though also pacifist and instinctively anti-establishment; or his immediate boss, Hilda Matheson, well-connected to the Westminster political elite but also to the socially liberal Bloomsbury set and fiercely committed to filling the airwaves with what she called 'all the most important currents of thought' while preserving a 'balanced diversity'.[6] Fast-forward to the 1970s, and we can sense in the written records a similarly agreeable discordance of character and viewpoint around the editorial tables – amplified, perhaps, by the cosmopolitan background of many of those gathered there: the Columbian-British journalist George Camacho, the Viennese émigré Stephen Hearst, the Hungarian George Fischer, the Anglo-French radio executive Gerard Mansell. Mansell was born in Paris and educated there and at the Chelsea School of Art. He would regularly arrive at Broadcasting House after a morning listening to France-Inter and with the French newspapers tucked under his arm. Back in the 1970s, the Controller of Radio 4 said that editorial meetings represented the notion of a 'Republic of Ideas' in action. It was at this time, too, that the BBC's director general Charles Curran, wrote to the Bishop of St Albans (and future Archbishop of Canterbury), Robert Runcie, to explain that although the Corporation claimed to be neutral between the different sections of British society it was also committed to what Curran called a 'Miltonic freedom of opinion and its expression.'[7]

If freedom of opinion did indeed make its way to the airwaves, it was surely only possible in the first place because the BBC had, at the heart of its programme-making machinery, a body of staff with an array of cultural hinterlands and a predisposition towards challenging received opinion. This meant that their eyes were open and their ears alert to all the ideas floating around out there in the world, including those that might be unexpected and novel and sometimes perhaps even unsettling – a body of men and women ever curious and formidably well-read. If, as Stuart Hall once argued, broadcasting has had a major role in 're-imagining the nation,'[8] it is only because those who have worked its levers have shown imagination themselves.

The approved BBC line now is that nothing has changed. In April 2017, for instance, the *Today* programme presenter and former political editor Nick Robinson, used a column in the *Radio Times* to dismiss all talk of favouritism in the Corporation's reporting of Brexit, arguing essentially that it was a case of confirmation bias – listeners and viewers finding it 'hard to accept that on the BBC they will often hear people they disagree with saying things they don't like' – that inside Broadcasting House the tradition of weighing arguments, assessing evidence, asking difficult questions in the spirit of 'due impartiality' remained as strong as ever.[9] The director general, Tony Hall, has also asserted confidently that the BBC remains 'independent ... impartial ... brave.'[10] To which, I say: only up to a point. It's hard to dispute the data showing that overall the BBC remains a highly trusted news provider as far as the public are concerned. And I think the BBC's senior managers and journalists both believe in *and* attempt to serve the principles of 'due impartiality'. But on the shop-floor there are discomfiting signs that even if impartiality remains a governing value it is being applied to a more restricted range of viewpoints than we have a right to expect. Why? Not because of government interference as such, I think. Or even, I would argue, an innate corporate bias – though, as Tom Mills has recently demonstrated very persuasively, the BBC, if only as a matter of survival, has always been uncomfortably close to the establishment.[11] Rather, it's more a question of working culture and mental habits. It's because among the Corporation's journalists, some cultural horizons have diminished and a certain 'group think' taken hold.

In 2014, the *Guardian*'s Charlotte Higgins wrote a series of lengthy, well-researched articles about the state of the Corporation as it faced the prospect of Charter Review, and all the government scrutiny that would come with it. In her analysis of the BBC's news output, she spoke to several of the Corporation's most experienced journalists – both named and unnamed. One, asked about whether BBC coverage had moved to the right, laughed out loud. 'Undoubtedly. You're not supposed to read the *Guardian* at the BBC, because it confirms everyone's prejudices. For years it has been more important at the BBC to be seen reading the *Telegraph* or the *Times*'. Higgins also spoke to Robert Peston, who was at the time the BBC's economics editor. 'What actually sends BBC news editors into a tizz', he told Higgins, 'is a splash in the *Telegraph* or the *Mail*. Over time the criticism of the *Mail* and the *Telegraph* that we are too leftwing has got to us. So BBC editors feel under more pressure to follow up stories in the *Telegraph* and *Mail* than those in the *Guardian* ... There is no institutional bias to the left – if anything, it is a bit the other way'.[12]

So: fewer papers are being read; the same papers are being read time and again. In a very real sense, the BBC newsroom's 'intake' – the range of perspectives and ideas to which its journalists are exposing themselves – is diminishing. Higgins' report implicitly laid the blame for this at the door of insistent political pressure of some vague but omnipresent kind. But another cause – for me, just as significant – is a deeper cultural shift: an overcompensating swing of the pendulum that's been gathering momentum since the 1960s, as those inside the BBC anxiously counter their inherited guilt at what they've been told for decades, namely that they are too elite and liberal and cosmopolitan. John Reith's old assertion, that 'only those who have a claim to be heard above their fellows on any particular subject' should reach the microphone, has long been unsustainable – and was seen as such by many of Reith's employees at the time.[13] But the perfectly proper desire to be more inclusive has now mutated, it would seem, into rather too partisan a commitment to the demotic, the voice of 'common sense', the 'man on the street': a distorted version of democracy that dispenses with its more deliberative dimension, leaving something more, well, *reactionary* – in the literal sense and, consequently, also in the political sense. It's not that there's anything wrong with hearing the voice of Everyman (or better still, Everywoman). It's just that the journalist's sense of what such a voice is *likely to say* – what its role in any debate is *constructed to be* – is already shaped by that journalist's habitual reading of a set of newspapers that demonstrably has a highly selective view of what public opinion is in the first place. Media theorists have mapped this process of framing and filtering and agenda-setting for years, so I claim no original insight here. Nonetheless, witnessing it unfolding before your eyes and ears – and to something you care about and generally think well of – is salutary.

Let me give two examples of how this manifests itself on air. First, *Today* – a radio programme, so not strictly speaking part of the report on the 'Future for Public Service Television', but, as I say, undoubtedly an agenda-setter for the British news media at large. Its apparent commitment to keeping a chair warm for climate-change sceptics such as the former chancellor Nigel Lawson – a man whose biography reveals no scientific training whatsoever – seems inexplicable. Inexplicable, that is, until one reads an interview, in 2014, with the programme's editor, then Jamie Angus – a former researcher and press-officer for the government's junior coalition partners, the Liberal Democrats – who suggested that Lawson earned his place because 'if you go into a pub on Oxford Street ... you will probably find a couple of people who are unconvinced by the science on climate change'. Now it's true that not everyone believes in climate change. But the consensus among the scientifically literate that this is happening and is largely human-made is extraordinary, unprecedented, almost, one is tempted to say, unarguable – much more than what Angus describes as 'a relatively settled view'. His aim, ensuring that 'alternative points of view' are not squeezed out, is admirable. But in defending the demotic he seems to have chosen a singularly unconvincing test-case.[14]

A second example: BBC TV's *Ten O'Clock News*. As it happens, I'm a bit of a fan of this programme. It usually embraces a good range of stories and makes excellent use of the Corporation's pool of foreign correspondents. It has provided some insightful reports, in particular, on the refugee crisis engulfing southern, eastern and central Europe. But even on the *Ten O'Clock News* one finds certain internalised attitudes leaking out. In early April 2017, it featured the BBC's economics editor, Kamal Ahmed, discussing the fall-out from Britain's referendum in favour of Brexit. In his report, he claimed that over 25% of the workforce in parts of Britain's hospitality industry was 'from the EU' – a figure that immediately struck me as strange because we in Britain are also *in* the EU, a simple but evidently overlooked fact that meant the correct figure should surely have been nearer 100%. I didn't think this choice of words was a deliberate attempt to skew the debate. Nor did I see it as a sign of direct government interference. Rather, and just as dangerous, it was unconscious – a cultural assumption, perhaps, from reading too many editions of the *Telegraph* and the *Mail*, and one that went uncorrected by anyone else on shift that night. The effect, in this case, was surely to reinforce subtly the notion that we are already separate from the EU, not part of it – that there is a 'them' and an 'us'. It was, in a very real sense, a report biased in favour of Brexit. I haven't attempted to measure how often such subtle inflections of meaning sneak under the radar, but it's reasonable to assume it wasn't an entirely isolated incident. And perhaps the proof of this comes in a recent article by the conservative commentator Simon Jenkins, in which he declares that the BBC's 'past bias' in favour of the European Union has at last been corrected to his satisfaction.[15]

This, I know, is all a bit anecdotal. But sometimes anecdote piled upon anecdote starts to build a picture that is highly suggestive. So, here's just one more. Among my own acquaintances – admittedly, and perhaps predictably, a fairly leftish, cosmopolitan bunch, though not in any uniform way, for we certainly argue a lot – I find the vast majority no longer tuning in to the BBC for their daily news. This cannot be a good sign.

Even more worrying, many of these critics are failing to distinguish BBC News from the BBC as a whole. They complain of bias – and perhaps rightly so. But their diagnosis only really applies to one part of the machine. They overlook, perhaps even start to forget, that there have been some brilliant and enlightening programmes on the BBC which throw new light on current problems and genuinely challenge prevailing political orthodoxies – programmes found not in the category of news or even current affairs, but rather in documentary, features, drama, even comedy. Take, for example, *Exodus*, a three-part TV series broadcast in 2016, which gave cameras to refugees fleeing to Europe. One episode of this BBC–Open University co-production dropped us in the most visceral way right into the midst of the packed human cargo of a dilapidated boat as it drifted somewhere between Greece and Turkey. The abstraction of so many news reports was immediately replaced on screen by the tangible reality of what it's like to be on a sinking boat facing the prospect of losing your child at any moment. In other episodes, we gradually got to see the workings of people smuggling, the vital importance of mobile phones in helping to hold families together, the hurt and distress of rejection on arrival in host countries. More recently, there's been *Welcome to Zaatari*, a Radio 4 drama series made in collaboration with the UNHCR, which portrayed life in a sprawling refugee camp.

Both *Exodus* and *Welcome to Zaatari*, then, were as informative as anything I've seen or heard on the news; neither of them deserve to be discounted by too sweeping an accusation of BBC bias. And they go to show that for the best insights into current affairs, it's now often best to look beyond the area of journalism.

Yet wouldn't it be better if we didn't *need* to? I'd like to think that Simon Jenkins could agree with this point. Like me, he worries about a 'narrow monoculture' having taken hold inside the Corporation. Unlike me, however, he believes it is 'left liberal'.[16]

So, I fear we won't be able to agree on the way forward – except, perhaps, to say this: that whatever happens in the way of licence fees or regulation or policy and planning or content and platforms, one of the key features of the BBC's near future will be a vigorous, perhaps vituperative, but always *necessary* debate about something deeper, namely the values and mind-sets of those thoughtful though fallible men and women who make it all possible to begin with.

Notes

1   David Puttnam, *A Future for Public Service Television: Content and Platforms in a Digital World* (London: Goldsmiths, University of London, 2016), 12.

2   Ibid., 4.

3   Ibid., 4.

4   John Reith, *Broadcast over Britain* (London: Hodder and Stoughton, 1924), 133–34.

5   David Mitchell, 'The Trouble with Getting the BBC to be Less Popular', *Observer*, 6 March 2016.

6   Michael Carney, *Stoker: The Life of Hilda Matheson OBE 1888–1940* (Llangynog: Pencaedu: 1999), 75–76.

7   Letter from Curran to Runcie, October 29, 1973, Papers of Michael Meredith Swann, Edinburgh University Library, File E105.

8   Puttnam, *A Future for Public Service Television*, 30.

9   'The referendum is over – now the BBC must fight a new Brexit bias', *Radio Times*, 4 April 2017.

10 Charlotte Higgins, 'BBC's long struggle to present the facts without fear or favour', *The Guardian*, 18 August 2014.

11 Tom Mills, *The BBC: The Myth of a Public Service* (London: Verso Books, 2016).

12 Higgins, 'BBC's long struggle'.

13 Asa Briggs, *History of British Broadcasting Volume 1* (Oxford: Oxford University Press, 1995), 232.

14 John Plunkett, 'Today Editor Jamie Angus: "You Can't Mess with Thought for the Day"', *Observer*, 13 July 2014.

15 Simon Jenkins, 'The Best Way to Tackle BBC Bias is Make it Plain for All to See', *The Guardian*, 5 April 2017.

16 Ibid.

11

Public Service Algorithms

James Bennett

As in many households, live television is becoming a rarity in our house. Most programmes are consumed on-demand at times that suit the rhythms of our life. We have become accustomed to this system from a broadcast heritage, but it is the norm for our five-year-old son. His viewing is no longer dictated by the times at which he can watch, like the very old *Watch with Mother* time slot or the CBeebies *Bedtime Stories* programme before switch-off at 7pm, but rather the algorithms that serve him up content he'll like. Increasingly this is more of the same, based on the recommendation algorithm that works within the marketing logic of 'if you liked this, you might also like …'. As William Uricchio argues, such algorithms posit the past as 'prologue, as the data generated through our earlier interactions shapes the textual world selected for us. No "surprises", or "unwanted" encounters, just uncannily familiar themes and variations.'[1] The results of this are evident in my son's Netflix profile which appears currently to be very keen on a particular form of 'girly' programming. Once logged in he is greeted with a wall of pink idents and other brightly or pastel branded shows featuring prototypically feminine characters. With a choice between different on-demand services in our house, Netflix has quickly established itself as his favourite, 'go to' app for its ability to respond to his current viewing preferences: offering him hundreds of shows and films to choose from, all of which will provide him reassuringly familiar pleasures within the pastel universe of princesses, pre-teen beauty queens and *Glitter Force*.

The choice of programming itself is not an issue: I'd be equally concerned if all he watched was 'boy-ish' programmes of robot battles, knight quests and the like. But what is striking about the way the Netflix algorithm works is this wall of pink and purple that greets him: ingraining his viewing preferences as if these were the entire televisual world on offer to him. Equally, my own Netflix profile offers up a database of black and muted colour palettes, often punctuated with visceral red titles that reveal my penchant for dark dramas and horror. My concern here, as viewer, parent and academic, is how to broaden our viewing horizons so that he and I might have the opportunity to discover other equally pleasurable viewing experiences in other genres, modes and moods. Whilst I must take responsibility for my own choices, I'm aware that this is not always done in my best interests: 'choice fatigue' and 'time famine', as John Ellis described these emergent problems over a decade ago,[2] from the thousands of options on offer can easily return me to familiar and comfortable pleasures rather than exploring something new. As a parent, however, I'm able to employ a range of tactics to broaden my son's viewing horizons, including using different Netflix accounts to browse children's content that is not so algorithmically determined, researching what friends are watching (often the same thing!), drawing on the archive of my own children's television experiences, and even very occasionally insisting he watches the 'flow' of a live broadcast channel airing children's content, such as CBeebies. After some strongly articulated resistance to the notion of watching something he has so little choice over, he succumbs and finds himself swept up in the madcap world of *Justin's House* or the educational adventures of *Go Jetters* or the simple slapstick humour of *Dip Dap*, which he finds utterly hilarious. None of these programmes are *better* per se, certainly not to his critical judgement, than what is *available* in Netflix's huge catalogue – but they are more varied than what is *offered* by Netflix's algorithm. Without this variety, we descend into our own echo chamber and develop an only partial perspective, 'in which our already existing views of the world are reinforced but rarely challenged'.[3] The outcomes of recent elections around the world might alert us to the dangers of such an approach developing, especially as future generations may take the world presented to them through an algorithmic lens for granted.

This would be not only bad for democracy, but also counter to the role that television has played in society throughout its history. A prevailing metaphor for television throughout its history has been as a 'window on the world', which enables us to explore a variety of different content, viewpoints, debates and landscapes. This was a function largely fulfilled in the broadcast era by scheduling: providing viewers with a mixed diet of programming, albeit at the scheduler's behest. Crucially, within a PSB remit, this window on the world offered viewers the chance to broaden their horizons – taking them from comedy, to news, to drama, to a music documentary to a current affairs programme and not to just another moody drama or pink cartoon. In what follows I argue that this variety of offering is a crucial part of what public service algorithms should aspire to offer us, a proposition supported by the Puttnam Inquiry report.[4] This requires thinking differently about the data that is collected and measured for public service broadcasting (PSB) within our 'algorithmic culture' and using it to set different objectives that escape some of the bounded thinking of a commercially-driven, on-demand digital television market.

The TV Market: Algorithms, Choice and Paternalism

The current era of television is one largely defined by an abundance of choice – from the services one can access, the channels that are available through to the individual programme titles one can consume. The proliferation of choice has been brought about by a range of socio-political and technological factors that are well documented elsewhere.[5] For the purposes of this essay, I want to focus on the rise of choice in the context of PSB and algorithms.

Choice is a lynchpin of the marketisation of television, which has been underpinned by the regulatory approach to the digital TV landscape in not only the UK but in most countries around the world. At the outset of the digital television era, the then 'digital tsar' of the UK's switchover programme, Barry Cox, underlined this market and consumer choice model for the future by recasting the television landscape in the image of a high street retail store: 'our homes [would] become an electronic retail outlet, the equivalent of a video version of WH Smith. ... we would have the ability to choose – and pay for – what we wanted from that wide range'.[6]

As Cox's position makes clear, the idea of choice is attached to the idea of 'empowering' the audience to choose which services and programmes suit their own needs and desires. Indeed, the idea of 'choice' has a long regulatory history having been continually invoked since the 1988 White Paper, 'Broadcasting in the 1990s: Competition, Choice and Quality'.[7] This period marked the arrival of new competitors in the form of cable and satellite and the beginning of a challenge to European models of PSB that were formerly based on the audience as citizen, with one based upon consumer choice. But with digitalisation, and the technological capacity to expand the TV marketplace that accompanied it, this emphasis on choice redoubled, also having a profound impact on how the BBC positioned its own services and policy. Thus the blueprints laid for the BBC's digital future in the mid- to late- 1990s emphasised choice as a key public value to be offered by the Corporation; documents like *Extending Choice in the Digital Age* (1996) and *The Future Funding of the BBC* (1999) promoted the BBC as an institution capable of supporting 'the public policy aims of quality, diversity, choice and accessibility'.[8] In the 2000s choice became a watchword for the BBC's own strategic vision, *Building Public Value* (2005)[9] and even marked the Corporation's move into digital broadcasting, with the launch of BBC Choice in 1998 (to become BBC3 in late 2003). It is within such a milieu that the BBC iPlayer emerged in 2006, immediately positioned by then director general Mark Thompson as an explicit alignment of choice with PSB: 'MyBBCPlayer … marks a watershed – a major expansion of choice and functionality and a recognition that on-demand is going to become an important – perhaps ultimately the most important – way in which we will put great content in front of the public'.[10] As Mike Flood Page argues of the iPlayer, 'on-demand could justifiably be framed as a natural and legitimate extension of the BBC's mission, as public service broadcasting by another means',[11] legitimising both the BBC's expansion into digital spaces as well as its adoption of the public value test, to demonstrate its value in the digital television marketplace.

If choice had become a prevailing discourse within the BBC by the start of the new millennium, it had also significantly influenced the wider policy sphere. Sonia Livingstone and Peter Lunt trace the nadir of this emphasis on choice in the regulatory shifts during this period that resulted in the hybrid consumer-citizen who replaced the citizen as the body on whose behalf new regulatory body Ofcom policed the television landscape.[12] Whilst we should be careful not to over-emphasise the neoliberal drift in the regulatory shaping of digital television, the UK media policy that shaped the emergence of the digital television landscape clearly emphasised choice and competition in its approach, in turn coinciding with technological developments that have paved the way for a marketplace ripe for new entrants, including from international players such as Netflix, Amazon and more.

In this context, algorithms stand as a logical industrial response to both uncertainty over the ability to maintain audience attention in an era of increased consumer choice and its exacerbation by government policy's focus on deregulation, competition and the market. To return to the anecdote of my own family's viewing patterns, choice can create as many problems as it can solutions. But in an era of neoliberalism, the paternalism inherent in our response is not in vogue as a wider strategy for the management of consumer viewing options. In turn, paternalism has been a key charge leveled at the role of PSB, with the BBC positioned pejoratively as 'Auntie', especially in a digital era in which consumers are better placed to determine their own best interests. But, as Amanda Lotz's theorisation of 'Internet-Distributed Television' as 'portals' implicitly suggests, there remains an underlying sense that such players adopt vast libraries of content to enable choice in lieu of paternalism.

A particularly challenging aspect of theorizing value for cultural goods is that viewers have 'bounded rationality', which means they do not know their own preference for cultural goods. The degree to which viewers often do not know what they want to watch explains the value of libraries that bundle multiple series.[13]

The provision of choice in such a market is, therefore, not simply a case of letting consumers determine their own best viewing interests. Rather, as Hallinan and Striphas argue, the rise of algorithmic culture has profound and fundamental implications for how we understand culture and the factors and forces that shape it. Their study of Netflix leads them to describe the present moment as one increasingly defined by algorithmic culture, in which 'use of computational processes to sort, classify, and hierarchize people, places, objects, and ideas, and also the habits of thought, conduct, and expression that arise in relationship to those processes' is commonplace.[14] As they argue, the ever-increasing sophistication of the recommendation of the Netflix algorithm is equated with customer satisfaction, theoretically creating a 'closed commercial loop in which culture conforms to, more than confronts, its users'.[15] Consumer choice is, therefore, managed not necessarily in the best interests of consumers but rather to accelerate and amplify the value of that virtuous commercial circle. As Uricchio concludes, 'algorithms used as filters, shap[e] our access to the cultural repertoire', but as gatekeepers they help 'determine what will and will not be produced'.[16] Choice, in other words, becomes a proxy for letting platform operators determine what is on offer, with the only measure of success the viewing figures that provide for 'recursive data flows' in which data generated from consumer choices constructs recommendations that shape future use of the site.[17] Thus rather than the 'WH Smith' model posited by Cox, or the portal metaphor proposed by Lotz, consumer choice comes increasingly to look like a hall of mirrors.

Whomever the provider, the goal of algorithms has remained largely the same: to retain, grow and monetise audiences by managing consumer choice and attention. But this focus does surely make them appropriate for a public service landscape if we recall its role in providing us with a 'window on the world'. Far from an outmoded notion about television, this 'window on the world' function was given renewed emphasis in the development of the BBC's digital priorities in contradistinction to that placed on choice. Thus in *The Future Funding of the BBC* (1999) in which choice is espoused as a watchword, the BBC is also charged with providing 'universal access' to the 'information age', whilst in *Building Public Value* (2004) the Corporation had explicitly articulated a policy of acting as a 'trusted guide' to the digital age. These promises are reframed in subsequent policy documents, but guiding viewers remains something that we can take as needing to run in tandem with the abundance of choice offered. The question, then, is what kind of guidance might we expect from a public service algorithm?

iPlayer: Curating and Connecting Choice

Paul Grainge and Cathy Johnson's excellent analysis of the emergence of BBC iPlayer posits the service as increasingly likely to act as a ' "front door" to the BBC … where people might start their viewing journey rather than a site to catch up on missed broadcast content.'[18] As such, they term the iPlayer a 'hybrid space', representative of what it 'means to be a digital broadcaster'. The challenge for the BBC and the iPlayer is not just to be a 'digital broadcaster', but a public service one, requiring it to move beyond the discourses of choice that informed its inception, and the singular emphasis on viewing figures in a market-led approach to algorithms discussed above.

In particular, the last five years have been marked by the growing 'datafication' of the industry, increasingly measuring viewing habits and tailoring and personalising content to allow audiences to become self-schedulers. Thus current director general Tony Hall's 'myBBC revolution', places increased emphasis on the power of data to catch up with commercial VOD providers like Netflix and Amazon.[19] At the same time, however, in a world of big data and the ability to measure everything, it is a peculiarity that television, and PSBs in particular, remain obsessed simply with ratings. Whilst these are increasingly sliced according to demography and geography, it is viewing figures that remain king: determining success and failures, careers and work lives in the industry. But in an era of the datafication of audiences, we must think outside the commercial box and posit some new, and some enduring, principles for how PSBs operate algorithms in a PS context. Otherwise personalisation risks the BBC replicating the echo chambers of familiar pleasures and views described at the outset of this chapter.

As Tony Hall suggests, the BBC should not be 'telling you what customers like you bought, but what citizens like you would love to watch and need to know'.[20] In this vein we might think of different measurements for the success of PS algorithms that can inform the development of iPlayer:

- **Connecting audiences with new content:** measuring not total views, but instead how often an offering outside of the 'norm' of a viewing profile is accessed. For example, if a viewer has *Top Gear* as part of 'my shows', success might be gauged by how often they explore unrelated content, such as a programme on environmentalism and fossil fuel, or *Woman's Hour*.

- **Connecting diverse audiences to shared content**: An enduring principle of PSB has been shared 'national' moments. In an algorithmic culture, PSB should attempt to break citizens out of their echo chambers and measure, and report on, the diversity of audiences watching shared forms of content.
- **Connecting audiences to new experiences and forms**: The world on to which the PSB window can open is more diverse now than simply television programmes. As public service broadcasting becomes public service media (PSM), the duty becomes to connect audiences not to simply more *content* but a variety of PSM forms – for example, VR, gaming, social media, interactive and participatory forms.
- **Connecting audiences with external services and content**: Similarly, we cannot expect to corral viewers into walled gardens of one PSM institution: audiences are always only one click away from a world of entertainment, education and information offered by other providers. There is public value in connecting audiences with shared national treasures and public resources, from the local library to the science museum to the NHS to '.gov.uk' sites. The BBC should use iPlayer as a launchpad for its role as a trusted guide to help licence fee payers navigate the digital age beyond its own borders.

These principles all reframe the notion of digital television as a window on the world, positioning iPlayer as a 'front door' that can broaden viewers' vistas.

But the vision here relies on the BBC not only resisting, at least partially, industry norms and regulatory drives, but also in adapting to the digital age by *curating* as well as connecting. As Amanda Lotz argues, it has become the 'primary task' of on-demand players to curate 'a library of content based on the identity, vision, and strategy that drive its business model'. As Lotz continues, 'curation – although largely untheorised – differs considerably from scheduling, and parallels to the rich insight available about scheduling strategies must now be created for commercial library curation.'[21] There is some evidence of a changing understanding of the role of curation at the BBC, as demonstrated by Grainge and Johnson's recent work, with one senior executive quoted as developing curation tactics that tried 'to recreate serendipitous discovery', much like scheduling had done in the broadcast era.[22] In particular, their work posits a hybrid approach to algorithmic and editorial recommendations. Wider work on algorithms already suggests the importance such thinking has for competitors like Netflix, where 'human intuition for knowing how to talk about and appreciate media content' is combined 'with sophisticated systems for organising categories as uniquely customised for a given subscriber'.[23] Rather than playing 'catch up' again with such commercial competitors, however, the BBC should deploy principles like those articulated here to develop a PS-led approach to algorithms. Fundamentally, deploying such 'human intuition' in the service of PSB rather than commercial logics requires a production culture within the BBC where algorithmic and editorial logics share an understanding of the role and purpose of the Corporation that is measured beyond viewing figures. In short, this demands a strong and imaginative BBC.

Conclusion

In an algorithmic culture, the challenge for the BBC is to find the public service structuring logics of recommendations in order to guide viewer choice. In a digital world, 'inform, educate and entertain' should be appended by 'explore' such that the BBC should once again open up a window on the world. A PSB algorithm would mark the BBC's services out as distinct from the market and connect viewers to a greater breadth not only of the Corporation's amazing output and a diversity of voices and viewpoints, but beyond its walls to other PSM providers. That is what a public service orientation should always be about. One small step that might be taken to iterate and explore this approach is developing a 'serendipity window' on the iPlayer's recommendation panel. At present the iPlayer's screen real estate is crowded with options of more of the same or most popular. Developing a space that pulls in content that might surprise, challenge, confound or even comfort viewers via a serendipity window might place the BBC ahead of the curve in the next battleground of social media as providers realise that people do, after all, want to step out of their echo chambers.

I hope the ideas offered here go some way towards developing the logics of public service algorithms and broadening the vistas of not only my son's, but also my own, preferences, in order to challenge us and to provide the serendipitous experience that ought to be at the heart of public service television. Digital television's window should not become a narrow portal, but instead should continue to broaden our horizons: recent political history tells us how dangerous inhabiting our own echo chambers can be.

Notes

1 William Urrichio, 'Recommended for You', *The Berlin Journal*, 28 (Spring 2015), 8.
2 John Ellis, *Seeing Things: Television in the Age of Uncertainty* (London: I. B. Tauris, 2000).
3 Urrichio, 'Recommended for You', 8.

4 David Puttnam, *A Future for Public Service Television: Content and Platforms in the Digital Age* (London: Goldsmiths, University of London, 2016), 36.

5 Ibid; See also Michael Starks, *The Digital Television Revolution Origins to Outcomes* (London: Palgrave, 2013).

6 Barry Cox, *Free for All? Public Service Television in the Digital Age* (London: Demos, 2004), 28.

7 Home Office, *Broadcasting in the '90s: Competition, Choice and Quality*, White Paper, Cm 517 (London: HMSO, 1988).

8 Department for Culture, Media & Sport, *Review of the Future Funding of the BBC* (London: DCMS, 1999), 136.

9 BBC, *Building Public Value: Renewing the BBC for a Digital World* (London: BBC, 2005), https://downloads.bbc.co.uk/aboutthebbc/policies/pdf/bpv.pdf.

10 Quoted in Mike Flood Page, *The development of BBC on-demand strategy 2003–2007: The Public Value Test and the iPlayer* (Unpublished thesis available at: http://theses.gla.ac.uk/6779/), 146.

11 Ibid., 146.

12 Peter Lunt and Sonia Livingstone, *Media Regulation: Governance and the Interests of Citizens and Consumers* (London: Sage, 2012).

13 Amanda D. Lotz, *Portals: A Treatise on Internet-Distributed Television*, (Ann Arbor, MI: Michigan Publishing, University of Michigan Library, 2017), http://quod.lib.umich.edu/m/maize/mpub9699689/1:3/--portals-a-treatise-on-internet-distributed-television?rgn=div1;view=fulltext.

14 Blake, Hallinan, and Ted Striphas, 'Recommended for you: The Netflix Prize and the Production of Algorithmic Culture', *New Media & Society* 18, 1 (2016), 119.

15 Ibid., 120.

16 Urrichio, 'Recommended for You', 8.

17 David Beer and Roger Burrows, 'Sociology and, of and in Web 2.0: Some Initial Considerations', *Sociological Research Online* 12, 5 (2017), doi. 10.5153/sro.1560.

18 Paul Grainge and Cathy Johnson, 'From Catch-Up TV to Online TV: Digital Broadcasting and the Case of BBC iPlayer', *Screen* forthcoming (2018): n.p.

19 Ibid.

20 John Plunkett and Jane Martinson, 'Tony Hall's vision for the BBC – six things we learned', *The Guardian*, 3 March 2012, www.theguardian.com/media/2015/mar/03/tony-halls-vision-for-the-bbc-six-things-we-learned.

21 Lotz, *Portals*.

22 Grainge and Johnson, 'From Catch-Up TV to Online TV'

23 Nick Marx, 'Industry Lore and Algorithmic Programming on Netflix', *Flow: A Critical Journal of Television and Media Culture* (April 2015), www.flowjournal.org/2015/04/industry-lore-and-algorithmic-programming-on-netflix/.

Part Three

Principles and Purposes of Public Service Television

12

Television and Public Service: A Brief History[1]

Television is in its death throes but has also been reborn; it is a relic of the mass audiences of the 20th century but it has never been more popular or more creative; we are watching more television but television viewing is also declining. Such are the profound contradictions of television in the 21st century.

The television screen remains at the heart of many a British home, and the output of the UK's numerous television companies remains central to British life. Even in the information age of tablets and smartphones, when the idea of broadcasting can seem almost quaint, television remains a powerful – indeed, is arguably still the most powerful – medium for information, education and entertainment.

Television has, in its relatively short history, been connected to major waves of social change. It was one of the main symbols (and accessories) of the consumer boom in the 1950s; it provided a crucial backdrop for many of the struggles that took place in the 1960s; satellite television helped to facilitate the globalisation that occurred from the 1980s while digital television in this century epitomises the abundance of an 'information age'. It has given us new vocabularies and new ways of behaving: we no longer just binge on alcohol or chocolate but on episodes of our favourite TV dramas.

Television also shapes our lives in many different ways. It has a crucial democratic purpose, for example through informing the public about the political process and encouraging us to engage with it, hosting political debate and discussion, investigating and analysing public affairs, and dramatising the most important moments in the UK's political life. Unlike the print and online news media, UK broadcasters are formally required to do all of this impartially. In recent years, television has helped – not without significant controversy – to frame the issues behind the referenda on Scottish independence and EU membership as well as the 2015 general election. Many of the key moments in those campaigns happened on television and much of the reporting that informed the public's decision-making was by television journalists.

Television's highly regulated status has long distinguished it from the UK's notoriously partisan print media, and it is all the more distinctive today amid the cacophony of the internet. Within the existing regulatory framework, television ought to allow for the expression of differences and a respect for opposing views that allows us to work through our conflicts. In a world where increasingly popular social media platforms can act as an echo chamber, it is an important barometer of national and local cultures, forcing us to consider a full range of perspectives and voices.

Television also provides a means of collective experience. It is still largely through television that people can watch major sporting events such as the European football championships and the Olympic Games. This sharing happens on a daily basis too. Television facilitates conversation, both while it is being watched and afterwards. A few shows – *Strictly Come Dancing, X Factor, EastEnders, Coronation Street* – have survived the fragmentation of the multichannel era to remain talking points across the UK. *Sherlock, Downton Abbey*, and *The Great British Bake-Off* have all caught the popular imagination in their different ways. Football fans discuss the matches they have seen live on Sky or BT Sport, or on *Match of the Day*. Much of the discussion in newspapers and magazines, or on Facebook and Twitter, springs from television programmes. At the same time, new voices and platforms have emerged to add to this conversation –from Vice News to YouTube's *danisnotonfire* and *Venus vs Mars*.

Television is a cultural form in its own right, capable of reaching artistic heights, and it is intimately connected with many other cultural forms as part of the wider creative industries and the creative ecology. It also provides major economic benefits. The UK television industry earned revenues of over £13 billion in 2014[2] and is the biggest player in an audiovisual creative sector that employs over 250,000 people, generates more than £10 billion of Gross Value Added and exports over £4 billion of services and products to the rest of the world.[3] Yet none of these possibilities are inevitable nor are they guaranteed to last unless we secure an independent, competitive and creative television landscape here in the UK through appropriate regulatory, technological and creative infrastructure dedicated to this purpose.

The Evolution of Public Service Television

The idea of public service has been integral to the history of broadcasting in the UK, from the foundation of the BBC in the 1920s onwards. The BBC started out as a monopoly provider and a public body acting in the national interest, forged out of the mood of the times, from the specific recommendations of the 1926 Crawford committee, and through the domineering character and singular vision of its first director-general, Lord Reith. In those early days, no broadcasting market was allowed to develop, as it had in the US, and *overt* political interference was generally kept at bay. First incorporated under royal charter in 1927, the BBC was from its earliest days characterised by aspirations towards impartiality and independence, even if in practice these aspirations were not always perfectly fulfilled.[4]

Reith wanted the BBC to be available to everyone across the UK, and he achieved his aim. Monopoly status gave the corporation an almost oracular power as the voice of a nation, a power that to this day it partially retains. As the UK's only broadcaster for more than 30 years, it was synonymous with broadcasting itself and its example influenced everything that followed. Reith may have felt little enthusiasm for television – and he left the BBC not long after the launch of the full television service in 1936 – but his notion that broadcasting should 'inform, educate and entertain' remains the cornerstone of the public service ideal even if this 'holy trinity' has been interpreted in wildly different ways.

When commercial television was launched in the 1950s, the BBC lost its monopoly, but the principle that broadcasting should be public service in character continued into the new era. The ITV network of regional licences set up in 1955 was highly regulated, and required to provide public service programming that was balanced, impartial and high quality in return for the advertising monopoly that made owning a franchise a 'licence to print money'. The regulator held sanctions over scheduling and programmes and could even revoke a licence if necessary. Minority interest programmes were expected to be spread across the schedule, including in peak time, and there were limits on US imports. The regional character of the ITV network was drawn up very deliberately as a way of decentralising the television industry, even if the map was drawn more for the benefit of marketers than with any specific feel for regional identity or local politics.

In the face of this new competition, the BBC had to sharpen up its act: the launch of ITN as a rival news provider to the BBC is credited with many innovations and improvements in broadcast news, for example. It was the BBC too that would be the beneficiary when the development of television was reviewed by the Pilkington report of 1962. Pilkington's scathing criticisms of the output of commercial television led to the BBC being granted the third channel – BBC Two – two years later, and to stronger regulation of the ITV network. Against the backdrop of social liberalisation and under

Hugh Carleton Greene's leadership, the BBC came into its own as a public service television broadcaster in the 1960s.

By the 1970s, the BBC–ITV duopoly was showing its age – its one-size-fits-all approach frustrating for programme makers and failing to reflect the fraying of cultural homogeneity. The time was ripe for a fourth channel, which was the recommendation of the Annan report in 1977. Annan felt that television should serve the various groups and interests in British society and not just aspire to cater for everyone at once. Channel 4 was launched in 1982 along these lines; its addition to the broadcasting landscape expanded the idea of public service to embrace diversity rather than just universality, and allowed for balance across the schedule rather than within programmes. In Wales, the fourth channel was devoted to the Welsh language service S4C.

The Conservative government that had presided over Channel 4's launch was also responsible for the 1990 Broadcasting Act, which significantly changed the nature of commercial TV. ITV licences were to be auctioned off to a highest bidder rather than awarded on merit by the regulator. Elsewhere, a technological revolution was making cable and satellite channels available to anyone who wanted to pay for them. No impediment was placed in the way of this rapidly emerging market, and no requirements were made of these new channels to offer original public service programming (although they were obliged to carry the existing PSB channels). Public service television became the preserve of the four legacy channels, Channel 5 (launched in 1997) and the BBC's new digital services. The Labour government of Tony Blair committed itself to switching off the analog signal by 2012, bringing the digital, multichannel future into focus. By the end of the 20th century, the old public service formula was holding firm but the great technological disruption that so characterises today's marketplace was already under way.

In summary, we can see the history of British public service broadcasting policy in the 20th century as being characterised by a series of very deliberate public interventions into what might otherwise have developed as a straightforward commercial marketplace. The creation of the BBC, the launch of an ITV network required to produce public service programming and the addition of the highly idiosyncratic Channel 4 gave the UK a television ecology animated by quality, breadth of programming and an orientation towards serving the public interest. At each of these three moments, the possibilities of public service television were expanded and British culture enriched as a result.

The 1990 Broadcasting Act and the fair wind given to multichannel services may have ended the supremacy of the public service television ideal. Public service television may now seem like an aberration in an era of apparently limitless con-sumer choice whose discourse is increasingly dominated by economic arguments. Nevertheless, it has survived, through the design of the institutions responsible for it, because of legislative protection, and as a result of its continuing popularity amongst the public. But the goodwill of programme makers and the appreciation of audiences will not by themselves keep it alive in the 21st century. Television does not develop 'naturally' following either a technological or commercial logic. It is worth remember-ing that at all stages, for good or ill, governments of the day have played an instrumen-tal role in shaping the television industry.

Public Service Television Today

Before the multichannel era, all the TV channels were public services in different ways; there were no purely commercial operations. So the trick of providing a mix of programmes that were popular, public service or both was not so hard to pull off and nailing down a definition of what was public service was not an urgent task. Anyone seeking definitions today can find plenty of guidance, if not total enlightenment.

The 2003 Communications Act laid out some of the key features. First, it listed the public service television services as all the BBC's TV services, S4C, every Channel 3 service (which now means ITV in England, Wales and Northern Ireland, and STV in Scotland), Channel 4 and Channel 5.[5] It made it obligatory for these services to be 'broadcast or distributed by means of every appropriate network'.[6] It defined the purposes of public service television broadcasting in terms of programmes that deal with a wide range of subject matters; cater for as many different audiences as practi-cable; are properly balanced; and maintain high general standards of content, qual-ity and professional skill and editorial integrity.[7] It also outlined various genre-based aims for public service television to fulfil, covering cultural activity (drama, comedy, music, films, and other visual and performing arts), news and current affairs, sporting and leisure interests, educational programming, science and religion, as well as pro-grammes for children and young people. It also specified the need for 'programmes that reflect the lives and concerns of different communities and cultural interests and traditions within the United Kingdom, and locally in different parts of the United Kingdom'. Importantly, it did not say which broadcasters should do what, just that the public service channels 'taken together' should produce these outcomes.[8]

The Act required the UK's three commercially funded public service broadcast-ers – the Channel 3 licencees, Channel 4 and Channel 5 – to provide a range of 'high quality and diverse' programming. Channel 4's output must additionally demonstrate innovation, experiment and creativity; appeal to a culturally diverse society; contribute to education; and exhibit a distinctive character.[9] Further detailed requirements in accordance with the act are set out in ITV, Channel 4 and Channel 5's main channel licences (which were agreed in 2004 and renewed in 2015, but have been subject to frequent variations). They are required to broadcast a set number of hours of news and current affairs programming and to fulfil various quotas on production in return for their prominent positions on the electronic programme guide.

Crucially, public service television has been defined more by broadcaster or channel, by (often rather vaguely expressed) principle, and by genre than in terms of individual *programmes*. This is a distinction that is becoming increasingly important in current debates that may seek to restrict the definition of public service to discrete programmes rather than outlets or remits. So while the BBC, ITV, Channel 4 and Channel 5 are public service broadcasters by virtue of the regulatory obligations imposed on them to produce a range of output, a company like Sky, which is responsible for significant news and arts provision, is not described as a public service broadcaster. All of the BBC's output is deemed public service, whereas for the three commercially funded public service broadcaster operators only the main ITV, Channel 4 and Channel 5 channels fall into this category.

Ofcom is required under the Communications Act to review the state of public service broadcasting. Its third and most recent review, published in 2015, said that the system was 'broadly working' but drew attention to changes in the wider marketplace and in consumer behaviour and raised a number of other concerns. For example, it found falling levels of investment in new UK-originated content by the 'PSB channels', with a 44% decline in drama spending.[10] Investment in some genres such as arts and classical music, religion and ethics had 'significantly reduced', while the provision of non-animated children's content outside the BBC was very limited. [...]

Thinking about the television industry as a highly developed and sophisticated ecology – as well as part of a larger creative ecology – allows us to view the challenge of maintaining public service television holistically. But improving and reforming public service television is not a matter of choosing from a menu. There is no point trying to change just one element and hoping that everything else will be fine. It is crucial that we examine today's various challenges alongside each other and come up with solutions that value co-ordination and interaction and secure democratic exchange, diverse representation and meaningful dialogue – in conditions of considerable technological, political and cultural volatility. The challenges that lie ahead are significant but there are also, in our view, some important opportunities.

Notes

1 Edited extract from Chapter 1 of the Puttnam Report, http://futureoftv.org.uk/wp-content/uploads/2016/06/FOTV-Report-Online-SP.pdf.

2 Ofcom, *Communications Market Report 2015* (London: Ofcom, 2015), 147.

3 Creative Industries, 'TV & Film Official Data', no date, www.thecreativeindustries.co.uk/industries/tv-film/tv-film-facts-and-figures/uk-tv-film-government-economic-data.

4 For very different assessments, see, for example, Stuart Hood, *On Television* (London: Pluto, 1997) and Paddy Scannell, 'Public Service Broadcasting and Modern Public Life', *Media, Culture & Society* 11 (1989): 135–66.

5 Communications Act 2003, section 264 (11). It also mentioned the public teletext service.

6 Ibid., section 272 (2).

7 Ibid., section 264 (4).

8 Ibid., section 264 (6).

9 Ibid., section 265.

10 It could also be argued that this fall in spending can be attributed to other reasons including significant production efficiencies and the increase in global investment and co-productions.

13

Principles of Public Service for the 21ˢᵗ Century[1]

Georgina Born

Basic Principles

At the core of previous normative frameworks for public service broadcasting are four interrelated concepts: independence, universality, citizenship and quality. As yet, these norms have not evolved to meet the challenges posed by digital platforms as well as the increasing cultural diversity and stubborn inequalities of modern Britain. The proposition in what follows is that the principles of public service media (PSM), as opposed to public service broadcasting (PSB), have not diminished but *expanded* in the digital era. This chapter explores these principles in relation to PSM as a whole, but is particularly focused on the crucial role in delivering public service played by the BBC and Channel 4 both now and in the future.

Independence

Independence is enshrined in the BBC's current royal charter which says that the BBC 'must be independent in all matters concerning the fulfilment of its Mission and the promotion of the Public Purposes, particularly as regards editorial and creative decisions, the times and manner in which its output and services are supplied, and in the management of its affairs.'[2] It is striking that the charter does not concern itself with the structural conditions that create or impede this independence. But this is inadequate: in future, any such governing document must also concern itself with these conditions. The Broadcasting Research Unit (BRU), reflecting on these issues some 30 years ago, insisted on the need for 'distance from all vested interests, and in particular from those of the government of the day.'[3] A core argument of this chapter is that, particularly in relation to the BBC, independence has been undermined and urgently needs new structural foundations.

Universality

Universality has three important and distinctive meanings:

a) The first is technical and geographical universality: in other words universal access to services, ideally free at the point of use. As the BRU put it, public service broadcasting 'should be available to the whole population.'[4]

b) The second meaning concerns social and cultural universality: as Born and Prosser argue, the provision of services and programming that enhance 'social unity through the creation of a "common culture", as well as those 'that cater for and reflect the interests of the full social and cultural diversity of Britain and its minorities.'[5] Similarly, for the BRU: 'Broadcasters should recognise their special relationship to the sense of national identity', while '[m]inorities, especially disadvantaged minorities, should receive particular provision.'[6]

Crucial to this sense of universality, and at the heart of PSB since its inception, is the relationship between *commonality* and *plurality*: between the creation of a national culture through mass modes of address, and the need to recognise and reflect minorities – from the four nations and all the regions of the UK to the full range of Britain's significant minorities. This relationship remains central to PSM in the digital era; importantly, digital platforms provide opportunities for its expansion.

c) The third meaning refers to universality of genre: as Born and Prosser argue, this centres on the importance of 'the provision of mixed programming, ... the entire range of broadcast genres, thereby meeting a wide range of needs and purposes through the trinity of information, education and entertainment. The aim here is that [PSM] should be truly popular, both as a value in itself ... [and] in order to draw audiences, serendipitously, across different and unforeseen kinds of programming.'[7] Again, this sense remains central to PSM today, but it needs reinvention in digital conditions.

Citizenship

PSM's citizenship purposes have been closely associated with cultivating national identity, social and political community via the public sphere or spheres that provide the grounds for a democratic political culture. This is often linked to PSM's informational role and its intended consequence, the cultivation of rational public debate – functions that can be contrasted with the more individual, consumer mode of address characteristic of commercial media. Recent revisions in the scholarly literature have stressed:

a) The need for citizenship, particularly in multicultural societies, to focus on plurality as much as commonality, on the expression of different identities and the fostering of dialogue between them.

b) The obligation to foster what the philosopher Onora O'Neill calls 'practices of toleration' towards those 'positions and voices that are in danger of being silenced or marginalised',[8] allied to the need to combat political, social and cultural exclusion by ensuring the presence of disadvantaged and excluded groups within mainstream communicative processes.[9]

c) The emergence of the principle of cultural citizenship, such that the space produced by the media is conceived not just as an informational space but also as a cultural space where media are 'involved in the construction of [both] common identities and ... multiple publics'.[10] According to the influential sociologist Stuart Hall, broadcasting has a major role in 're-imagining the nation', not by reimposing an imagined unity but by becoming the 'the "theatre" in which [Britain's] cultural diversity is produced, displayed and represented, and the "forum" in which the terms of its associative life together are negotiated'.[11] Cultural citizenship recognises the key role played by expressive, imaginative and affective media content (entertainment, drama, sports, comedy, arts) in providing frameworks for collective reflection and enjoyment, in addition to the functions fulfilled by news and current affairs in facilitating public knowledge, reflection, and deliberation. Given the role of media cultures in influencing audience tastes and conditioning the wider public culture, then, by analogy with the concern in democratic political theory with the formation of an educated and informed citizenry, cultural citizenship emphasises the importance of the formation of a culturally mature and aware, culturally pluralistic citizenry.[12]

Quality

Accounts of this principle emphasise the *conditions* that promote or impede high quality programming and services. The Broadcasting Research Unit made two points in relation to quality: first, that structural conditions 'should be designed to liberate rather than restrict programme makers' so as to enhance creativity and ambition and therefore quality; and, second, that PSM 'should be structured so as to encourage competition in good programming [and services] rather than competition for numbers [ie ratings]'.[13]

Channel 4's remit has always stressed additional factors enhancing quality: thus the channel must provide 'a broad range of high quality and diverse programming ... which, in particular, demonstrates innovation, experiment and creativity in the form and content of programmes; appeals to the tastes and interests of a culturally diverse society' and 'exhibits a distinctive character'.[14] Notable here is the prominence in Channel 4's remit of both the commitment to universality of genre and the diversity principle central to social and cultural universality and cultural citizenship – both of them conceived as intrinsically significant factors in the quality of PSM output.

An Additional Principle: Diversity

Public service media therefore have a remit both to promote the national commons and to serve minorities, especially disadvantaged and underserved minorities. Given the current insecurities concerning both national and European identities, as well as the rise in divisive racisms, the challenges posed by reflecting and enhancing cultural diversity and pluralism seem more central to PSM than at any time since the mid-twentieth century. A core task for PSM today is therefore to revitalise their offering to multiple social groups, including ethnic and religious minorities, and, in the UK, to address more adequately the distinctive as well as the shared needs of the British population wherever they live. Pressures for increased devolution make this an especially urgent task.

A central contention of this chapter is that, in the digital era, rather than the earlier *two*-way public sphere (commons/minorities), PSM should now employ digital media to shape a *three*-way, multi-platform public sphere. In addition to 'universal' mass or national channels or events, this should take the form of content and services aimed at creating a counterpoint between mass and minority audiences, including services aimed at supporting both *intercultural* and *intracultural* modes of address:[15]

a) *Intercultural* is when *a minority speaks both to the majority and to other minorities,* a core function of pluralist PSM. Here, universal channels and events become the means of exposure to and connection with others' imaginative and expressive worlds via the self-representation of minorities in their own 'voice'. It encompasses 'minority' programming on mainstream channels, including such forms as black and Asian sitcoms, drama and current affairs, community access programming, as well as internet-based content and cross-platform events.

b) *Intracultural* is when *a minority speaks to 'itself'* via services and programming that act as arenas for shared experience and deliberation *by and among minorities about their own cultures, needs and strategies,* enhancing self-expression and self-understanding. Crucially, on PSM this output – whether available on the internet, radio or television – is also always accessible to the majority and to other minorities, who thereby gain understanding of the core minority culture as well as potential pleasure from such encounters.

All three modes of address – universal, intercultural and intracultural – are necessary components of PSM's orchestration, via both mass and niche services and programming, of a democratic communicative pluralism. Clearly, digital platforms have greatly enhanced and will continue to enhance the realisation of this three-way, multi-platform public sphere.

The achievement of pluralism and diversity points towards additional challenges: issues of employment and training, of representation, and of the conditions for production. Guiding principles for addressing these matters are suggested by the political philosopher Anne Phillips when she examines the challenges posed to democratic political practice by marginalised groups. Phillips highlights the importance of the linkage between self-representation and 'presence': rather than conceive of diversity in the terms of diversity merely of belief and opinion, or what she calls the 'politics of ideas', Phillips advocates the 'politics of presence' – the necessity of ensuring the presence within the political process of those most dispossessed from it. As she puts it, '[p]olitical exclusion is increasingly viewed ... in terms that can be met only by political presence.'[16] By analogy, the lack of sufficient diversity in PSM and the wider media culture can be redressed only through a combination of policies aimed at fostering a 'politics of presence' – that is, policies to ensure the due presence of minority and marginalised groups within media production – and greater self-representation among and by such minority and marginalised groups.

Regarding employment and training: several linked steps are necessary:

i) Pluralism and diversity would be enhanced through recruitment drives to encourage more people from minorities to enter the media industries, as well as through training schemes and professional placements; such training, entry and employment policies must be accompanied by measures to support the promotion of minorities to the higher echelons.

ii) These employment and promotion drives should be matched by greater support for minorities to build independent media and production businesses, bringing the linked benefits of increased economic activity and the provision of as yet undersupplied content representing the full spectrum of minority experiences.

iii) Encouraging increased minority presence and activities in the media industries suggests, in turn, the need for policies aimed at providing greater support by the BBC, Channel 4 and other commissioning bodies and distribution channels for SMEs as new entrants to the creative economy: that is, diversification of both scale and source in the production community – issues that have hitherto been attended to primarily by Channel 4.

iv) In light of the weakness of the existing PSM institutions in addressing these issues, we suggest that a new body might be created to advance and coordinate these developments: one model would be the Channel 4 workshops of the 1980s,[17] which had the effect of training and bringing on such world-recognised talents as Isaac Julien and John Akomfrah, themselves influential on filmmaker Steve McQueen.

Regarding representation: such employment, training and industry initiatives are not sufficient, however, unless they are matched by effective support for diverse, innovative and imaginative *representations of minority experience.*[18] A key lesson of recent decades in the UK is that the entry of greater numbers of minority producers and commissioners into the media industries, and specifically PSM, does not in itself lead to stronger representations of the *varieties* of black, Asian, Muslim, queer or disabled experience on screen.

This returns us to the quality principle: for high quality minority representations of this kind require *conditions for production* that support innovation, experimentation, risk-taking and the right to fail – conditions that are still lamentably undersupplied in the PSM ecology, and that we address further in Chapter 23 of this book.[19]

It is striking how key elements of these revitalised diversity and pluralism norms, as they link to quality, are already to be found in the UK in Channel 4's remit. A key question is, then, whether Channel 4's remit now contains core principles that, given their universal importance and undersupply elsewhere, might now be applied more generally to PSM: notably, those concerning diversity, and specifically the conviction in Channel 4's remit that both diversity and quality are intrinsically linked to risk-taking, innovation and experiment in the form and content of programmes. This chapter poses the question: should these principles now be extended to become general foundations for the PSM ecology?

Public Service Principles in the Digital Age: From Institutions to Ecology

Buoyed by the enormous increase in content, platforms and services that has emerged from a less regulated landscape, there has been a concomitant rise in the use of a discourse focused overwhelmingly on the 'market impact' of PSM. But such an approach risks elevating commercial media interests over the public interests served by PSM. Recent economic research reverses this thinking, arguing that publicly-funded interventions can enhance innovation and lead to the creation of new markets, with the potential to fuel wider economic growth.[20] In this context, the following two foundations of PSB in the 20th century, consequent on the above principles, must be reinstated and renewed for PSM in the 21st century in the light of digital conditions:

Public Service Media are Not Synonymous with Market Failure

This principle follows clearly from the underlying relationship between public service media and universality: both universality of genre (mixed programming), and social and cultural universality (i.e. content, events and channels that draw national or mass audiences). Recent governments have overlooked this principle and have attempted to disrupt this orientation by suggesting that public service broadcasters should focus on the provision of content in which commercial providers are likely to under-invest. While it is highly likely that broadcasting, if unregulated, would primarily target the most lucrative and wealthy demographics, PSM should not be seen as vehicles to plug the demographic gaps but, instead, as institutions that challenge the tendency to fragmentation precisely by providing common and overlapping spaces and channels. If PSM are reduced to operating as cultural 'ghettoes' and 'market failure' institutions in a situation of digital abundance, then they are not adequately serving the public and they are likely to decline. Popular and mainstream programming and entertainment that draw high ratings must remain core elements of PSM as they continue to diversify, taking advantage of new platforms and new suppliers.

Public Service Media Refers to an Evolving Digital Media Ecology

PSM is an ecology shaped, as discussed in previous chapters, by institutional design and regulation: it should not be equated with a single institution or channel. In the UK, from the birth of ITV onwards, this ecology has encompassed the commercial public service broadcasters as well as the BBC and, in the future, it will potentially include additional organisations. This definition needs to be reinstated for the digital age. The PSM ecology entails complementarity between the different bodies delivering PSM's public purposes, as well as benign competition to raise standards and stimulate innovation – both central to quality. The PSM ecology optimises the public interest both by creating new markets and by intervening in wider markets.

The current nostrum that PSM's 'market impact' should limit their entry into new and existing markets must therefore be questioned.[21] In contrast, new economic thinking stresses the essential contributions of publicly funded research and development, in technology and culture, to innovation, the creation of new markets and economic growth. We might speak of *distributed innovation* as a property of the PSM ecology through partnerships with start-ups, universities, cultural organisations and so on – through public–public as well as public–private partnerships. This paradigm in the economics of innovation is currently gaining greater visibility and new life. As the leading economist Mariana Mazzucato argues, 'the public sector not only "de-risks" the private sector by sharing its risk, it often "leads the way", courageously taking on risk that the private sector fears.'[22]

Of course, this reframing should not be read as a complete licence for PSM to do everything, everywhere – especially where public resources are limited and commercial provision is highly regarded. As noted in the Puttnam Report, a holistic approach to PSM must consider how changes to one part of the ecology (whether between PSM providers, or between them and commercial providers) are affecting other parts. A more sophisticated approach to market impact would place greater emphasis on the positive and longer term benefits of PSM in creating and sustaining new markets, while also attending to any possible detrimental effects.

Applying Normative Principles to the Funding of Public Service Media

It seems unarguable that the *funding* mechanisms for PSM should follow on from, and be allied to, PSM's *institutional* purposes, values and objectives. It is therefore imperative that the normative principles of PSM, as well as wider good governance principles, should also inform funding.

Universality and Citizenship

As the Broadcasting Research Unit argued back in 1986, it is vital that 'one main instrument of broadcasting [and now of PSM] should be directly funded by the corpus of users'. The BRU insisted on the need for 'a contract between the citizen and the broadcasters that an equally good service ... shall be made available to all for the fee paid.'[23] Ideally, access to PSM services – including those delivered via the internet – should also be free at the point of use in order to maximise this commitment and the universality of these services for citizens.

Independence

Independence is vital in the process of decision-making about setting and distributing the licence fee and other sources of PSM funding, so as to retain a significant measure of autonomy from vested interests. According to the European Broadcasting Union (EBU), funding must not be 'reliant on political favour, thereby promoting public trust in PSM and its role as a truly indispensable service.'[24]

Transparency

Public services, including PSM, should be fully accountable to the public, as enshrined in appropriate good governance models. The funding of PSM, equally, ought to be based on a commitment, as the EBU puts it, to an 'open and clear funding mechanism holding PSM accountable to its audience.'[25]

Redistribution

In accord with the PSM principles of universality and citizenship, new redistributive funding mechanisms should be developed to address structural inequalities and economic disparities both between providers in media markets (as, for example, occurred in the original funding relationship between ITV and Channel 4) and crucially, as we suggest below, between the citizens that compose the audiences for PSM.

Plurality

A healthy PSM ecology is best served by multiple funding sources – and public service providers – in order to minimize, wherever possible, competition for revenue, while maximising competition for innovation and quality. Britain is fortunate to have a television landscape financed by the licence fee, advertising, subscription and even some elements of general taxation (as in the government's small contribution to S4C).

However, it is also important to note specific problems with the existing mechanisms in the light of the normative principles set out earlier:

Subscription favours the better off, discourages universality of genre (mixed programming), and, by fragmenting audiences, damages the attainment of social and cultural universality as well as intercultural modes of address.

Advertising and sponsorship carry risks of commercial influence and of the skewing of provision towards more desirable demographics, thereby providing a disincentive to invest in particular kinds of content oriented to, and representing, less lucrative demographics.

A flat *licence fee* is a regressive payment mechanism in that it is a 'poll tax' which, in relation to the BBC, currently criminalises some of the poorest sections of the British population.

It is therefore important to advance several potential improvements for PSM funding going forward. In relation to the BBC, rather than a flat licence fee, in order to mitigate criminalisation and improve distributive justice, wealth-related payments should be implemented, whether through a revamped and platform-neutral BBC licence fee, general taxation, or a household fee following the German model but based on different tiers, and with substantial exemptions for the low-waged and the unemployed.

In addition, pursuing the principle of distributive justice at the inter-institutional level of funding mechanisms that address the media landscape as an ecology, it would be productive, and is overdue, to examine the use of levies on the profits of the largest digital intermediaries, including ISPs and phone/tablet manufacturers, in order to fund new sources of public service content, as well as to stimulate key genres that are currently at threat due to under-funding, such as children's television and education.

Whatever the merits of these recommendations, governments must in future be urged to base their legislation and policies in relation to PSM funding on the foundational normative principles set out in this chapter. The beneficial effects will be that increased attention is given to curbing inequality of access to the means of citizenship, and that pluralism of funding and provision, linked to greater diversity of services and content, remain at the heart of the PSM ecology in the digital age.

Notes

1 This chapter is an edited and expanded version of Chapter 2 of the Puttnam Report. It draws heavily on Georgina Born's paper, 'Rethinking the Principles of Public Service Media', delivered at the Inquiry's event on the same topic, British Academy, 3 March 2016; and on Georgina Born, 'Mediating the Public Sphere: Digitization, Pluralism, and Communicative Democracy', Chapter 7 in Christian Emden and David Midgely (eds.), *Beyond Habermas: Democracy, Knowledge, and the Public Sphere* (Oxford: Berghahn, 2013), 119–46. Thanks to Des Freedman and Anamik Saha for input.

2 BBC charter, December 2016, 3 (1): http://downloads.bbc.co.uk/bbctrust/assets/files/pdf/about/how_we_govern/charter.pdf.

3 Broadcasting Research Unit, *The Public Service Idea in British Broadcasting: Main Principles* (London: BRU, 1986), 9.

4 Ibid., 1.

5 Georgina Born and Tony Prosser, 'Culture and Consumerism: Citizenship, Public Service Broadcasting and the BBC's Fair Trading Obligations', *Modern Law Review*, 64, 5 (2001), 676.

6 Broadcasting Research Unit, *Public Service Idea*, 7, 5.

7 Born and Prosser, 'Culture and Consumerism', 676.

8 Onora O'Neill, 'Practices of Toleration', in Judith Lichtenberg (ed.), *Democracy and the Mass Media* (Cambridge: Cambridge University Press, 1990), 173; also Born, 'Mediating the Public Sphere', 132–33.

9 By analogy with Anne Phillips, *The Politics of Presence* (Oxford: Oxford University Press, 1995), and see further discussion below; Born, 'Mediating the Public Sphere' 132–33.

10 Born and Prosser, 'Culture and Consumerism', 674.

11 Born, 'Mediating the Public Sphere', 138, citing Stuart Hall, 'Which Public, Whose Service?', in Wilf Stevenson (ed.), *All Our Futures: The Changing Role and Purpose of the BBC* (London: BFI, 1993), 36–37.

12 Born and Prosser, 'Culture and Consumerism'; and for a sustained discussion of arguments concerning the moral, social and political benefits of cultural diversity as a social condition *per se*, see Bhikhu Parekh, *Rethinking Multiculturalism: Cultural Diversity and Political Theory* (London: Macmillan, 2000), Chapter 5, especially 165–78.

13 Broadcasting Research Unit, *Public Service Idea*, 19, 15. On the quality principle and the need to take account of conditions bearing on production, see Born and Prosser, 'Culture and Consumerism', 679–81.

14 Ofcom, *Renewal of the Channel 4 licence*, 25 July, 2013, http://stakeholders.ofcom.org.uk/binaries/consultations/renewal-c4-licence/summary/c4.pdf, 15.

15 For a fuller account of this section, see Born, 'Mediating the Public Sphere', 132–41.

16 Anne Phillips, 'Dealing with Difference: A Politics of Ideas or a Politics of Presence?', in Seyla Benhabib (ed.), *Democracy and Difference* (Princeton, NJ: Princeton University Press, 1996), 141.

17 On the Channel 4 workshops and their innovative conception and output, see Sylvia Harvey, 'Channel Four Television: From Annan to Grade', in Stuart Hood (ed.), *Behind the Screens: The Structure of British Television in the Nineties* (Lawrence & Wishart, 1994); Rod Stoneman, 'Sins of Commission', *Screen*, 33, 2 (1992): 127–44; Hannah Andrews, 'On the Grey Box: Broadcasting Experimental Film and Video on Channel 4's The Eleventh Hour', *Visual Culture in Britain*, 12, 2 (2011): 203–18.

18 On the complexities of diversity initiatives, see Herman Gray, 'Precarious Diversity: Representation and Demography', in Michael Curtin and Kevin Sanson (eds.), *Precarious Creativity* (Berkeley, CA: University of California Press, 2016): 241–53.

19 For a full discussion of these issues, see Anamik Saha, '"Beards, Scarves, Halal Meat, Terrorists, Forced Marriage": Television Industries and the Production of "Race"', *Media, Culture and Society*, 34, 4 (2012): 424–38.

20 See for example Mariana Mazzucato, *The Entrepreneurial State* (London: Anthem, 2013).

21 This was prevalent in the government's thinking in relation to BBC charter review in 2016.

22 Mariana Mazzucato, 'The Future of the BBC: The BBC as Market Shaper and Creator', LSE Media Policy Project, 14 October, 2015, http://marianamazzucato.com/2015/10/16/the-future-of-the-bbc-the-bbc-as-market-shaper-and-creator/.

23 Broadcasting Research Unit, *Public Service Idea*, 12.

24 Richard Burnley, *Public Funding Principles for Public Service Media* (Geneva: European Broadcasting Union, 2016), 3.

25 Ibid.

14

The Purposes of Broadcasting – Revisited

Julian Petley[1]

In 1960 the then Conservative government commissioned a report on the future of broadcasting in the UK. This it did for four reasons. The first was that, because of considerable hostility to 'commercial' television from across the political spectrum, ITV had been introduced in 1954 for only a ten-year 'experimental' period, and this was now drawing to an end. Second, there was mounting criticism that ITV companies were making very considerable profits by lowering programme standards. Third, a second TV channel was to be allotted, and there was a great deal of debate about whether this should go to the BBC or ITV. And finally, the BBC Charter was due to expire in 1962.

The committee which wrote the report was chaired by the industrialist Sir Harry Pilkington, and a particularly notable member was Richard Hoggart – who had recently published *The Uses of Literacy* (1957) and appeared for the defence in the *Lady Chatterley* trial. Its secretary was Dennis Lawrence, a career civil servant in the Post Office.

Much to the fury of the numerous newspapers who had shares in ITV companies, the Committee was highly critical of ITV and decided to award the second channel to the BBC. The Committee in general, and Hoggart in particular, were accused of being, among other things, do-gooders, roundheads, puritans, socialists, authoritarians, paternal, prim, patronising, moralistic, censorious and out of touch with public opinion. Later, Hoggart himself, in his book *An Imagined Life,* described this reaction as 'the usual dreary, underdeveloped litany of fear', and as manifesting an 'Islamic-fundamentalist-like fury'.[2]

The anti-BBC, anti-intellectual and stridently populist tone of the aggrieved reaction against Pilkington carries many a pre-echo of subsequent attacks on the public service broadcasting system and the BBC in particular. The fact that they all come from exactly the same quarter – 'free market' politicians and economists, and a press which is very far indeed from being a disinterested observer of the broadcasting scene – shows just how deep run the roots of the current campaign against not just the BBC, but against public service broadcasting in all its forms. And, by the same token, much of the report's defence of public service broadcasting is as directly relevant to today's battles as to those of 50 years ago.

From this perspective, the most important part of the report is Chapter Three, entitled 'The Purposes of Broadcasting'.[3] It was actually written by Dennis Lawrence, but is highly Hoggartian in spirit.

It is of course important to understand that when the report talks of broadcasting it means public service broadcasting, as embodied in the BBC and ITV, since there was no other form in those days. Now, of course, there is absolutely no shortage of weighty books and articles about public service broadcasting, the public sphere and so on, but in 1960 such serious analyses of broadcasting were rare – and particularly so in official reports. And the report was nothing if not serious about the public purposes and responsibilities of broadcasting.

In its consideration of those purposes and responsibilities, the report argues that 'television is and will be a main factor in influencing the values and moral standards of our society'[4] [...] Inevitably this immediately led to the report being caricatured by its populist critics as calling for broadcasting to play a moralising role in society, whereas what it was in fact doing was simply pointing out that social attitudes, assumptions and values will inevitably be reflected in and to some extent influenced by broadcasting, as well as other forms of modern communication. Consequently, it was important that broadcasters respected the medium and assumed a responsibility for its output, its audience and indeed the wider public and society. An observation that is as valid now as when the report was published.

However, the most significant part of 'The Purposes of Broadcasting' for current debates about the future of the BBC in particular and of public service broadcasting in general is its robust and combative dismissal of the populist approach to television – an approach which thoroughly infused many of the attacks on the report and which has become a hallmark of the many onslaughts on public service broadcasting in the intervening years.

The report notes the argument, familiar even then, that certain programmes are popular with large audiences because they are 'what the public wants', and that 'to provide anything else is to impose on people what someone thinks they ought to like'. But as the report points out:

The public is not an amorphous, uniform mass; however much it is counted and classified under this or that heading, it is composed of individual people; and 'what the public wants' is what individual people want. They share some of their wants and interests with all or most of their fellows; and it is necessary that a service of broadcasting should cater for those wants and interests. There is in short a considerable place for items which all or most enjoy. To say, however, that the only way of giving people what they want is to give them these items is to imply that all individuals are alike. But no two are [...] a service which caters only for majorities can never satisfy all, or even most, of the needs of any individual. It cannot, therefore, satisfy all the needs of the public.[5]

Thus, far from advocating narrowing of the range of television programmes so as to include only those which were 'good' for people, as the report's populist critics suggested that it did, it actually argued, conversely, for a wide range of programmes aimed at a wide range of audiences. [...]

Back in 1989, Rupert Murdoch opened his infamous MacTaggart Lecture at the Edinburgh Television Festival by stating that: 'For fifty years British television has operated on the assumption that the people could not be trusted to watch what they wanted to watch, so that it had to be controlled by like-minded people who knew what was good for us'.[6] But nearly 30 years earlier, the Pilkington report had detonated the rank hypocrisy lurking behind this kind of populist rhetoric, arguing that 'giving the public what it wants' has

the appearance of an appeal to democratic principle but the appearance is deceptive. It is in fact patronising and arrogant, in that it claims to know what the public is, but defines it as no more than the mass audience; and in that it claims to know what it wants but limits its choice to the average of experience. In this sense, we reject it utterly. If there is a sense in which it should be used, it is this: what the public wants and what it has the right to get is freedom to choose from the widest range of programme matter. Anything less than that is deprivation.[7]

Furthermore, far from desiring moral conformism on the part of the broadcasters, the report openly encouraged the expression of dissenting and minority viewpoints, arguing that 'television must pay particular attention to those parts of the range of worthwhile experience which lie beyond the most common; to those parts which some have explored here and there but few everywhere'. Indeed, it stated that television should focus a particular spotlight on what it called society's 'growing pains', because it is at these points that the challenges to existing assumptions and beliefs are made, where the claims to new knowledge and new awareness are stated. If our society is to respond to the challenges and judge the claims, they must be put before it. All broadcasting, and television especially, must be ready and anxious to experiment, to show the new and unusual, to give a hearing to dissent. Here, broadcasting must be most willing to make mistakes; for if it does not, it will make no discoveries.

[...] The Pilkington report played a key role in paving the way for the many invigorating changes that television underwent in the 1960s. In particular, it required the Independent Television Authority to ensure that the ITV companies took their public service obligations far more seriously than they had done hitherto, and the greatly improved programming that resulted caused the BBC to sharpen up its own act considerably. Far from being the near-relation of Mrs Grundy, as painted by the populist press, the report was actually a harbinger of *The Wednesday Play*, *Z Cars*, *World in Action*, *Coronation Street*, *Seven Up!* (and its successors), and many other groundbreaking programmes which had their birth in the decade at the start of which the report was published. However, the caricature of the report as an elitist, moralistic, killjoy charter has been far too useful to the enemies of public service broadcasting – most of whom almost certainly haven't read it – to have been allowed to fade into the obscurity which it deserves.

According to Hoggart, in *Speaking to Each Other*, the Pilkington report was best understood as an argument

about freedom and responsibility within commercialised democracies. It touched on the interrelations between cash, power and the organs for intellectual debate; it had to do with a society which is changing rapidly and doesn't understand its own changes; it had to do with the adequacy of our assumptions and vocabulary to many current social issues.[8]

Today British society, and indeed the world with which it is increasingly deeply and intimately connected, is changing even more rapidly than in the 1960s, and public understanding of those changes is at a woefully low ebb – a situation for which the media, including the public service broadcasters, must take their fair share of blame. Thus we desperately need an analysis of both the strengths and weaknesses of public service broadcasting as it currently exists, as well as a blueprint for its future, which is as profound, challenging, well-informed and intellectually self-confident as was the Pilkington report when it was published in 1962.

Notes

1 Julian Petley is Professor of Screen Media at Brunel University. This is an edited extract from his submissions to the Puttnam Inquiry, http://futureoftv.org.uk/wp-content/uploads/2015/11/Julian-Petley.pdf.

2 Richard Hoggart, *An Imagined Life: Life and Times 1959–91* (London: Chatto & Windus, 1992), 61. I have analysed press reactions to the Pilkington Report in 'Pilkington, Populism and Public Service Broadcasting', *Ethical Space*, 12, 1 (2015), http://communicationethics.net/sub-journals/free_article.php?id=00074.

3 Sir Harry Pilkington, *Report of the Committee on Broadcasting*, Cmnd. 1753 (London: HMSO, 1962).

4 Pilkington, *Report*, 15.

5 Pilkington, *Report*, 16–17.

6 Rupert Murdoch, 'Freedom in Broadcasting', *McTaggart Lecture*, Edinburgh International Television Festival, 1989, www.thetvfestival.com/website/wp-content/uploads/2015/03/GEITF_MacTaggart_1989_Rupert_Murdoch.pdf

7 Pilkington, *Report*, 17–18.

8 Richard Hoggart, *Speaking to Each Other: Volume 1: about society* (London: Chatto & Windus, 1970), 189.

15

Back to the Future: The Uses of Television in the Digital Age

Michael Bailey[1]

Any inquiry into the present-day ecology and future possibilities of British television must necessarily reflect on the historical development of public service broadcasting. Not so as to invoke a nostalgic golden age or a whiggish history of scientific progress; rather, to highlight the democratic purposes that have shaped television as a social technology over the past nine decades. And to also consider how best to re-evaluate that tradition in view of the medium's changing production, distribution and consumption practices in the digital age.

Such an analysis also reveals how and why debates about television are part of a longer inquiry concerning the more extensive relation of culture to society: viz. the articulation between the creative and intellectual capacities of human beings (as expressed in the popular arts, recreation, education, everyday customs and habits of thought), on the one hand, and wider social changes and political forces (industrialisation, urbanisation, enfranchisement, secularisation, welfarism, migration, multiculturalism, neoliberalism), on the other.

It is especially fitting that the Inquiry into the future of public service television, chaired by Lord Puttnam, and based in the Department of Media and Communications at Goldsmiths, University of London, should take its cue from the Pilkington report on broadcasting published in 1962. Apart from being the first (not to mention highly influential) committee of inquiry into television, Pilkington was one of the defining moments in the distinguished career of the late Richard Hoggart, former Warden of Goldsmiths. [...]

Hoggart noted that one of the reasons the inquiry decided not to commission a related audience study was because it did not want the report to 'restrict itself to collecting and ordering objective evidence' or 'to laying out the social alternatives neutrally': interesting though quantitative research may be, it tends to confine itself to 'outlining a great many useful is's'; rarely does such work 'give a single ought'.[2]

The inference of Hoggart's matter-of-fact remarks is that, whilst recognising that *ex cathedra* opinions can be misleading, the committee wanted evidence from people who were not afraid to offer shrewd opinion: 'We were engaged to the best of our ability in a study in social philosophy. We were asking about the nature of good broadcasting in a democracy. We could not enforce our judgements scientifically; we could only say at the end ... "This is so, is it not?" '[3] And of the various metaphysical considerations expressed during the course of the inquiry, foremost was the concern to realise the purposes of broadcasting.

Hence Pilkington's recommendation that broadcasters ought to recognise that they 'were in a constant and sensitive relationship with the moral condition of society', which many critics took to epitomise the report's patrician tone of voice.[4] However, the committee defended this particular clause on the grounds that it was intended to give broadcasters a 'responsibility difficult to define but not easy to shrug off', which necessarily involved them having to steer a course somewhere between the populist Scylla of 'giving the public what it wants' and the autocratic Charybdis of 'giving the public what they ought to have'.[5]

That is to say, Pilkington was not asserting 'a crudely moralistic relationship' in the sense that 'broadcasters had a responsibility for the direct propagation of the Ten Commandments'.[6] Rather, the report was advocating a vocational sense of professionalism that went beyond either a purely commercial or aesthetic definition of broadcasting: a duty to commission programmes that 'bring before us all the widest range of subject matter, the whole scope and variety of human awareness and experience, the best and the worst, the new and the challenging, the old and familiar, the serious and the light [thus] enriching the lives of every one of us'.[7]

Naturally, mediating between these different positions presents all kinds of dilemmas in terms of what broadcasters should prioritise. But Hoggart reminds us that one of the enduring principles established by Pilkington was that 'good broadcasting in a free society ... should not hesitate to reflect "the quarrel of society with itself", even though politicians may not like the result'. What is more, if they shoulder this responsibility, broadcasters will end up becoming 'a sort of yeast in society' in the sense that they 'will be active agents of change', sensitive to new possibilities and unforeseen contingencies.[8]

Hoggart's fondness for positive regulation is well known. But contrary to accusations of him promoting censorship, Hoggart was always clear about the difference between enabling forms of broadcasting policy that say 'Thou shalt' as opposed to prohibitive forms of broadcasting legislation that say 'Thou shalt not': 'Good regulations increase freedom, make for good growth, expand and protect the arena, the living space, for good programming'; 'to windows being opened, not knuckles censoriously rapped'.[9] In other words, unfettered consumerism and uncritical populism are the problem, not sex, bad language and violence.

Commenting on broadcasting policy developments in the early 1990s Hoggart was even more specific. Though never published, *A Broadcasting Charter for Britain* (with Stephen Hearst) remains one of the boldest statements on the duties and rights of listeners, viewers, programme makers and regulators. With typical candour, the manuscript is prefaced with the following statement: 'It would be more fashionable and more generally acceptable to list first – and perhaps only – Rights. But in a democratic society Rights are inextricably bound up with Duties; Duties are the foundation of Rights and so prior to them. No Rights without Duties'.[10]

The public have a duty 'to respect other people's tastes' and 'to look at what is available overall before complaining that there's nothing worth watching'. It is the duty of the programme maker 'to his or her self' (that is to say, to their 'conscience'), 'to do justice to his or her subject' and 'to be creative'. Duties of legislators include the duty 'to create structures and methods of financing for broadcasting' which encourage the production of 'good programmes', 'to enable disparate voices to be heard' and, finally, in Jane Austen's words, 'not to assume it is their duty to "screw people into virtue"'.[11]

Of the many rights listed, the one that best summarises Hoggart's thinking was the declaration that listeners and viewers had a right 'not to be got at, politically, commercially, piously'.[12] That is to say, the public has a right to access the fullest means of information and creative expression in the belief that we can learn to value both our common humanity and our best selves. Only then might we fully comprehend how broadcasting might become truly democratic, comprehensive and socially organic; indeed, the much wider relationship between culture and society generally and, if found wanting, to be in a position to do something about it.

In spite of being sympathetic to the needs of minority communities, Hoggart's claims to represent the broader public interest nevertheless risk excluding those social groups whose cultural tastes and interests are not so easily articulated, much less accommodated, in such prescriptive and general terms. Though he always insisted that 'we should feel members one of another, but also retain all we have of sparky, spikey individuality',[13] there is still a danger that a straightforward Hoggartian analysis could ignore communities of people who have nothing in common with its vision for a common culture.

For example, Hoggart never supported community media initiatives in quite the same way that he supported PSBs such as the BBC or Channel 4. In fact, he actually dismissed the 'small-holding dreams of communications' as at best 'an engaging dream', at worst 'a reversion to parochialism' which will permit 'the ideological toughs and the commercial sharp-shooters' to 'divide and rule'.[14] The subtext of Hoggart's reasoning is that locally oriented media may result in the fragmenting of society into a mass of atomised communities of interest or regional identity, which could put an end to any sense of shared culture and sociality.

Though one can appreciate the cultural and political logic of Hoggart's argument, it is nevertheless his Achilles heel. This is a pity because, whilst broadcasting policy in the United Kingdom has begun to acknowledge the differing needs and wants of a variety of publics, it has been a long time in the making and is still in need of more widespread support. This is partly because of the government's refusal to lift restricted access to the airwaves on the grounds of spectrum scarcity, but also because of broadcasting's tendency towards concentration of ownership, economies of scale, formulaic programming, inward-looking professionalism, and managerial bureaucracy. [...]

There are many other debates concerning public service television and the future of mass communications more generally. The focus of this submission captures only a small fraction of past and current developments. Suffice to say that the above proposals and comments take inspiration from the resurgent positive interest in the Pilkington Inquiry and the related ideas of its best-known committee member, Richard Hoggart. Their criticisms of free-market liberalism and light-touch regulation are as relevant today as they were fifty-odd years ago insofar as they still represent a cogent engagement with the idea of PSB as a primary facilitator of an educated and deliberative democracy.

To quote Hoggart again (writing shortly before his death), 'the arrival of broadcasting in the last century offered the greatest opportunity to create a clear democratic means of communication, one harnessed neither to the profit-making wagon nor to political power'.[15] Furthermore, 'broadcasting can be the biggest and best arena for exposing false democracy and welcoming its opposite', that is a socio-political system which both encourages and is supported by the endless play of free will and a more civil society. And it is for these primary reasons that broadcasting should keep 'going on going on' with 'public service at its heart'[16] – for its sake and ours. [...]

Notes

1 Michael Bailey teaches in the Sociology Department at Essex University. This is an edited extract from his submission to the Inquiry, http://futureoftv.org.uk/wp-content/uploads/2016/05/Michael-Bailey.pdf.

2 Richard Hoggart, *Speaking to Each Other: About Society* (Harmondsworth: Penguin Books, 1970), 189.

3 Richard Hoggart, *An Imagined Life: Life and Times, Volume III, 1959–91* (London: Chatto & Windus, 1992), 62.

4 *Pilkington*, Cmnd 1753, para. 42; though slightly reworded, the clause is emphasised elsewhere in the Report, for example, paras. 83, 89 & 122.

5 Hoggart, *Speaking to Each Other*, 193.

6 Ibid, 194.

7 'The Richard Hoggart Papers' (University of Sheffield Library), 5/9/3.

8 Hoggart, *An Imagined Life*, 66.

9 Ibid, 252.

10 Hoggart Papers, 5/11/75, *A Broadcasting Charter for Britain*; cf. 5/11/73–74 for working versions and notations.

11 Ibid.

12 Ibid.

13 Richard Hoggart, *A Sort of Clowning: Life and Times, Volume II, 1940–59* (London: Chatto & Windus, 1990), 78.

14 Richard Hoggart, *Only Connect: On Culture and Communication* (London: Chatto & Windus, 1972), 89.

15 Richard Hoggart, *Mass Media in a Mass Society: Myth and Reality* (London: Continuum, 2005), 34.

16 Hoggart, *Mass Media*, 138, 111.

16

Television, Quality of Life and the Value of Culture

David Hesmondhalgh[1]

This submission foregrounds two concepts that need to be central in discussions about the future of television in the UK: *quality of life*, and the *value of culture*. It makes four main claims, as follows.

1. Television can Contribute to Quality of Life in Important Ways, But We Should Not Understand That Contribution in Terms of 'Consumer Preferences'

In the last twenty years, much debate about television has been a slanging match between advocates of markets (i.e. of much greater marketisation of the television system) and advocates of public service. Too many contributions on either side treat markets and public service as ends in themselves, sidelining discussion of what the ultimate purposes of the television system ought to be.

One useful and potentially productive way to conceptualise the ultimate goal of any media or cultural system such as television is in terms of its contributions to people's quality of life in the areas reached by that system.

The problem of course is that there are many different ways of understanding quality of life. Mainstream economics, which has exerted considerable influence over television policy (and public policy in general) in recent decades, often conceives of quality of life in terms of 'welfare', understood in terms of *consumers' subjective preferences*.

There are major problems in thinking of quality of life in this way. For people are often mistaken in their appraisals of their own preferences and desired goals, not because they are stupid, but because often they lack information about what kinds of rewards products will provide, and the social consequences of their choices.

In the case of cultural goods,[2] such as television, it is hard to know much at all in advance about what kinds of rewards and pleasures that an individual cultural product might offer. We very often only really know whether we value the experience produced by a cultural product once we have fully tried it. Even trailers, or familiarity with a star name, or source material, can be deceptive. Cultural goods often give greatest reward and pleasure precisely because they surprise, enlighten or delight us by offering a new perspective on the world. The benefits of particular television programmes for individuals, communities and society only become apparent in retrospect. One factor that contributes to this feature of television (shared by many other cultural forms) is that each television programme or series is in a sense a new product, different from all other television programmes and series – sometimes only marginally, but sometimes considerably.

This difficulty for consumers in accurately knowing their cultural preferences is one of the reasons that assertions that media markets necessarily 'give people what they want' are either naïve or made in bad faith.[3] This feature of cultural goods also means that even the more sophisticated expression of that viewpoint, that media markets contribute to people's welfare by efficiently meeting consumers' subjective preferences, is dubious. And it means that to define the kinds of well-being or quality of life that might be enhanced by television in terms of consumers' subjective preferences is mistaken. We need a different conception of quality of life.

2. Television's Contribution to Quality of Life Should Be Thought of in Terms of What it Enables People to *Do* or to *Be*

The 'Capabilities Approach' offers a superior conception of quality of life and its relation to policy. Developed by (among others) two leading neo-Aristotelian thinkers, the US philosopher Martha Nussbaum and the Indian economist Amartya Sen, the Capabilities Approach has served as an important basis for debates about international 'development'.[4] The approach emphasises that 'it is not only what people *have* that is important for their well-being but what they can *do* or *be*'.[5] Capabilities are simply the abilities of people to achieve 'functionings' such as being able to have good health, or move freely from place to place. The focus of the capabilities *approach* therefore is on which functionings should be enabled by public policy, why, and how. It can and should be applied to television (but hasn't been, much).

What kinds of *cultural* functionings might a good television system enable in the population who receive it? Different genres might enable different functionings. In news and current affairs, there is a social need for serious, rigorous and yet accessible information that would allow people to participate meaningfully in democratic life; such a view is of course widely accepted. In 'entertainment' genres (drama, factual entertainment, comedy, talk shows, children's programming) there is a need for a wide range of skilful representations of experiences, so that people can better understand their own emotions, motivations and development. Culture matters for quality of life across a wide range of genres, and there is a danger of elitism in leaving this importance to the market.

Capabilities and functionings can be construed as *needs*, but unlike some other treatments of needs, there is a strong emphasis on freedom. For the capabilities approach does not decree or imply that everyone must achieve functionings whether they want to or not: 'the ability to choose is itself crucial for well-being.'[6] So the approach avoids the paternalism that both left and conservative-libertarian critics would rightly question in television policy, emphasising the very great variability in people's inclinations and practices. But equally it moves beyond the idea that services should be provided on the basis of subjective consumer preferences, by forefronting the need for public deliberation over which cultural functionings a society should enable, and how.

3. Television Markets, if not Well Constructed and Regulated, Are Unlikely to Enhance Cultural Quality of Life Adequately, Because High-Quality Television is a Particular Form of 'Merit Good' and is Therefore Likely to be Under-Produced

The difficulty of knowing in advance what kinds of rewards and pleasures good television might offer (discussed in 1 above) means that good television can be understood as a 'merit good': a product or service of significant social benefit in which individual consumers are likely to under-invest[7] and which therefore markets are likely to under-produce. Other examples would be preventive healthcare or education, in which many individuals and families might under-invest because of a lack of information about the benefits of exercise, diet, or learning, or because of fears that their investments will be worthless. This is why nearly all governments invest in health information and in universal public education, to correct this particular form of market failure.

Even if it is accepted that quality of life (understood in terms of capabilities and functionings) is a good way to think about the goals of cultural systems, and that television can be thought of as a merit good because of certain fundamental features of how it is consumed, some people might object to the preceding paragraph on the basis that there is a false analogy. They might claim that culture should not be treated like health and education, because health and education are simply more important than culture, more a matter of life and death.

Health and education are indeed important but culture matters too, and contributes to quality of life in particular and distinctive ways, as we have seen briefly above. It is only relatively recently that citizens and governments have come to recognise the need for comprehensive social provision of health and education; even neo-liberals would recognise the need for such provision but they just think that the market is best placed to offer it. With the industrialisation of culture since the 1920s, and with generally expanding leisure time and income, culture has become more and more central to people's lives. This process has only been intensified by digitalisation, which allows access to culture to become more mobile, flexible and frequent. Culture *matters* more than ever.

The increasing importance of culture means that it should be considered alongside merit goods in the health and education sectors, as requiring public, democratic provision to prevent under-supply of goods that have a significant effect on people's quality of life. In the realm of culture, consumers will generally over-value in advance the familiar, and underestimate the benefits of the fresh, the innovative and the challenging, because of the difficulties of knowing and understanding unfamiliar experiences. This means that, in marketised systems, while some people will have their cultural needs met, and will expand the range of cultural functionings they can pursue, many will not – especially the poor and less educated, who tend to have less opportunity to take cultural risks on products with which they are unfamiliar. Because culture remains a key marker of social distinction in modern societies, as sociologists from Pierre Bourdieu onwards have amply demonstrated, that will only increase problems of inequality.

4. Digitalisation Does Not Remove the Fundamental Problems Surrounding Cultural Markets and Quality Of Life – It Makes a Public Service 'Common Provider' *More* Important

The above-mentioned digitalisation does not significantly alter the problem of under-production of television as a merit good that enables quality of life. Where digital markets have enabled a number of competing services to offer high quality provision, along the lines of HBO, this is often consumed mainly by white middle-class educated people, even if such services offer products that working-class people might well value once they are exposed to them. So content that crosses class, ethnic and other social divides is likely still to be under-produced; the 'merit good' problem remains. Furthermore, digitalisation, especially the likely proliferation of subscription services (whether consumed via PC, tablet or 'traditional' TV screens) *intensifies the problem of cultural fragmentation*. A version of the current ecology of a public-service oriented BBC, alongside public service oriented commercial providers, must surely remain the prime means by which such cultural fragmentation is countered, by providing trusted sources of varied representations, good explanations, innovative humour, and so on. Only if this public service ecology is generously funded, and positioned as a universal provider, across all major genres, can it serve this purpose, by enabling a wide range of cultural functionings, and providing a real alternative to control of distribution by the big tech oligopolies that now dominate in the realm of digital media, as other sector's oligopolies did in the analog era (again, because of the nature of cultural markets). To undermine or reduce this public service ecology would be to throw away that the potential quality-of-life benefits that investment over 75 years and more has built up.

The BBC and the commercial public service providers have accrued and adapted huge skill and experience in providing programmes that have enhanced life in Britain. The success of the iPlayer and its equivalents in providing a new means of accessing television in the digital age puts the current public service players, especially the BBC, in a strong position to provide widely-shared knowledge and aesthetic experiences, which might redress the massive fragmentation that marketisation and digitalisation have already brought about and are likely to intensify further. The iPlayer needs to be made universally available, and BBC programming – aimed at all citizens irrespective of background and geography – needs sufficient support to ensure that a sufficient number of people continue to watch content via it to justify a universal licence fee.

Notes

1 David Hesmondhalgh is Professor of Media, Music and Culture, University of Leeds. This is an edited extract from his submission to the Puttnam Inquiry, http://futureoftv.org.uk/submissions/hesmondhalgh-david/.
2 I define culture here simply as 'informational and aesthetic-expressive products': a narrow definition, but a common one.

3 For other reasons why commercial media do not give people what they want, see the US legal scholar C. Edwin Baker's comprehensive discussion in his book *Media, Markets and Democracy* (Cambridge: Cambridge University Press, 2002).

4 Martha Nussbaum and Amartya Sen, *The Quality of Life* (Oxford: Oxford University Press, 1993).

5 Andrew Sayer, *Why Things Matter to People* (Cambridge: Cambridge University Press, 2010), 234.

6 Sayer, *Why Things Matter to People*, 234.

7 Des Freedman, *The Politics of Media Policy* (Cambridge: Polity, 2008), 20.

17

Shouting Toward Each Other: Economics, Ideology and Public Service Television Policy

Robert G. Picard[1]

The biggest challenge in determining the future of public service television and the BBC is that there is little debate and informed discussion, but rather a surfeit of partisan viewpoints shouted toward opponents.

We need more objective and reasoned thought as the UK considers what to do about public service television. Proponents of more and less public service television make impassioned, ideologically based arguments that are often deficient in substance and misconstrue the purposes, functions and operations of public service television – thus obscuring the underlying issues and choices that the UK faces.

A sensible debate can only start by recognising that there is nothing sacrosanct about public service television. It is merely a policy tool for achieving desirable social outcomes given the economic characteristics of broadcasting. The fundamental question in the debate is thus whether changes produced by the growth of commercial content provision and contemporary distribution technologies have reduced the necessity and ability of the public service television tool to provide those outcomes.

The fundamental purposes of public service broadcasting are uncomplicated. Its objectives are to provide quality programming that 1) serves the information and entertainment needs of the public, 2) supports national identity and culture, 3) provides service to underserved groups such as children and minorities, and 4) meets the specific local needs of communities and regions.

Public service broadcasting was developed in an age when technical, economic and business conditions made it difficult for other types of operations to effectively serve those social purposes.[2] The UK is now in a national debate about the roles of public service broadcasting in meeting those purposes in the digital age in which other types of operations exist.

In recent decades, critics have vociferously challenged public service television at every opportunity and much of the UK press has been less than forthright about the reasons for their criticisms. Some have criticised it out of self-interests in profits of commercial broadcasting firms. A few critics have opposed the very existence of public service broadcasting on ideological grounds. Most critics, however, perceive value in public service, but argue that its scale and scope reduce effective governance or harm development of the commercial broadcasting sector.[3]

I do not believe that ideology should be absent from the debates over broadcasting. Such discussions are necessarily ideological because they are about choices between reliance on the state or the market. The arguments, however, should be backed by persuasive evidence and made with recognition that the real choice in the current discussion is not one between the state and the market, but rather what is the appropriate balance between them.

To begin with, one must accept the undeniable facts that public service broadcasting has been singularly successful in meeting UK broadcast needs during the past nine decades and since full-scale provision of UK television began following World War II. The UK television market today is lauded by television system observers worldwide for the quality, choice, and social service it provides UK citizens. The UK market is recognised for providing the best public service television and creating a highly successful commercial market with the largest revenues in Europe.

We must also recognise that public service television experienced unparalleled growth until 2010, when its resources began to be reduced and constrained by policy decisions.[4] Nevertheless, the cuts that have already taken place have addressed and are ameliorating many of the criticisms levelled against public service television and it has lower market impact than public broadcasting in many European nations.

Public service television – especially the BBC – has become institutionalised, however, and changes are difficult. It is determinedly supported by those who benefit from the employment it offers and supporters of public service broadcasting who see any criticism as a threat or do not wish to consider other possible tools for achieving the desired social outcomes.

The focus of attention on public service television is appropriate because it is a creation of public policy, funded by policy choices, and because policymakers will have significant influence on its future. As the contemporary debate develops, however, it is important that public service broadcasting not be considered in isolation from developments in commercial broadcasting and the content that is provided by broadcast and digital audiovisual services as a whole. It also needs to be recognised that all broadcasting, cable, and other services have been made possible because of public policy and public investments and that the BBC has played important roles in technological and product developments that benefit commercial providers.

The continued usefulness of public service television as a policy tool must be assessed within the broader perspective of the contributions of the broader broadcasting and digital sectors to national life and should not consider public service television in isolation. For the debate about public service television to contribute to an effective policy solution, bigger questions will need to be addressed:

- What functions should television serve in social and public life? What does UK society need from it?
- What roles will broadcasting – public service and commercial – play in the growing environment of streamed linear and non-linear programming? What will broadcasting contribute to the content environment that non-broadcast providers do and will not?
- What functions and needs does public service television fulfil that are not adequately performed or met by commercial broadcasting and digital streaming? Must public service television only fulfil those functions and needs or should it be allowed to have a broader impact on society?
- To what extent, if any, does public service broadcasting keep the commercial sector from having adequate resources to grow, prosper and contribute to the economic well-being of the UK?
- If the scope and service of public service television are diminished further, what requirements can be placed on commercial broadcasters to meet the social outcomes desired from broadcasting generally or the functions lost by reducing the scale and scope of public service television?

[...] As contemplation of current and future roles of public service television continues, it is useful to consider five salient points:

1. Public service television provides universal access and values all viewers equally, including those who are less valued by commercial firms.
2. Public service television provides social and cultural benefits beyond merely providing content that commercial broadcasters are unwilling to provide. It does more than address market failure to provide some genres of programming and services to minorities.

3. Public service television is distinctly UK-oriented, providing content that serves UK social, cultural and political interests beyond those provided by commercial broadcasters as a whole.

4. Public service television has the potential to interfere with business development of some commercial content providers, constraining the UK economy and denying some additional tax revenue.

5. Funding for public service television has already been significantly diminished, leading to reduced services and impact on the market. The effects of the reduced funding need to be considered during the deliberations taking place about the proper scale and scope of public service television.

Economic issues and policy evidence require that a balance be sought between those who argue for unfettered public service television and those who argue that the market alone can meet the UK's audio-visual content needs. Neither option alone will produce an optimal social outcome. The placement of unwarranted constraints on either public service or commercial provision, however, will reduce the benefits that citizens receive from the UK television market.

Getting the policy choice correct will require thoughtful deliberation and cautious choices, lest the UK risk destroying what is probably the best television system in the world today.

Notes

1 Professor Robert G. Picard is the former research director at the Reuters Institute, University of Oxford. This is an edited extract of his submission to the Puttnam Inquiry, http://futureoftv.org.uk/wp-content/uploads/2015/09/Robert-Picard.pdf.

2 Patrick Barwise and Robert G. Picard, *The Economics of Television in a Digital World: What Economics Tells Us for Future Policy Debates* (Oxford: Reuters Institute for the Study of Journalism, 2012), http://reutersinstitute.politics.ox.ac.uk/sites/default/files/The%20economics%20of%20television%20in%20a%20digital%20world_0.pdf.

3 Robert G. Picard and Paolo Siciliani, eds, *Is there Still a Place for Public Service Television? Effects of the Changing Economics of Broadcasting* (Oxford: Reuters Institute for the Study of Journalism, 2013), http://reutersinstitute.politics.ox.ac.uk/sites/default/files/Is%20There%20Still%20a%20Place%20for%20Public%20Service%20Television_0.pdf.

4 Patrick Barwise and Robert G. Picard, *What If There Were No BBC Television: The Net Impact on UK Viewers* (Oxford: Reuters Institute for the Study of Journalism, 2014), http://reutersinstitute.politics.ox.ac.uk/sites/default/files/What%20if%20there%20were%20no%20BBC%20TV_0.pdf.

18

Everything for Someone: For an Inclusive Definition of Public Service Broadcasting

Brett Mills[1]

Everything for Someone

It is pointless to discuss how public service broadcasting (PSB) can best be delivered unless there is consensus on what constitutes it. Given the need for PSB providers to repeatedly evidence both that their output represents PSB and that PSB represents some kind of social good, there is a necessity for clarity about what it is. In debates about PSB there is rarely little dissent from the view that it can function as a social good; debates instead rest on what kind of social good is appropriate, and the extent to which the 'intervention' of funding is made to 'the market' by that social good. This submission rejects the notion that PSB should function solely, or even primarily, to fill the gaps left by 'market failure'. The provision of social goods is typically seen to be of importance irrespective of whether the market can supply them, and public funding ensures their provision. Just as the existence of bookshops doesn't mean that libraries should only supply the volumes that can't be found in those stores, so PSB providers should not be forced to evidence that they only do what the market can't.

Definitions of PSB throughout the world – typically drawing on the UK, Reithian model – insist that PSB can only function if services are universal. This is conventionally understood as being in terms of access; that all citizens have a right to the material and services offered by PSB, partly because they have paid for it, but primarily because all citizens should have equal access to public services. Debates about PSB, then, often focus on modes of delivery to ensure that access, and, as such, changes in technology such as multi-channel platforms and online services have represented challenges and opportunities to that universal access. While these debates are important, concerns over changes in technology risk crowding out discussion of content.

Universality should not be understood solely in terms of access. Universality must also be considered in terms of content. A universal PSB enables all citizens to see their lives reflected and valued within content, and this is only possible if PSB encompasses as wide a range of genres and programming as possible. It is easy to forget how radical and inclusive the decision that the BBC should 'inform, educate and entertain' was when the Corporation was instituted, with the inclusion of 'entertain' representing a commitment to popular culture that might not have automatically been seen as necessary for PSB. Yet that triad persists and has been implemented in many countries across the world. UNESCO, for example, states that it is '[t]hrough PSB [that] citizens are informed, educated, and also entertained.'[2] In the UK, the notion of entertainment has persisted in PSB definitions, and it has been part of Channel 4's remit since its inception. As such, PSB has been understood as constituting mixed programming whose aim 'was not simply to provide "something for everyone" but, at whatever level, "everything for someone".'[3] Too often, debates about PSB focus on 'something for everyone'; that is, that PSB services reach as large a percentage of the population as it can. However, PSB must also represent 'everything for someone'; that is, that a public service offers all forms of culture desired by citizens. A library that only offers books on certain kinds of topics isn't a public service; and a PSB provider that fails to deliver all components of 'inform, educate and entertain' similarly isn't a public service.

The Problematic Hierarchies within PSB

Despite claims to universality, discourses within which debates about PSB function often hierarchise different kinds of PSB provision. This is evident in both academic and policy material. For example, the Government's 2016 White Paper proposes instituting a 'public service content fund' which would enable broadcasters other than the BBC to deliver 'quality and pluralistic public service content.'[4] While the fund would aim to encourage innovation in content coupled with programming intended to reach a more diverse audience, the White Paper also highlights what it calls 'underserved genres'; these are children's programming, religion and ethics, formal education, and arts and classical music.[5] The White Paper cites Ofcom research as evidencing the 'underserved' nature of these genres. While Ofcom research does indeed demonstrate this, the genres the White Paper lists are not the only ones Ofcom finds to be underserved, as it states that 'There has also been a recent decline in spend on new UK comedy, with spend falling by 30% in real terms since 2008.'[6] The marginalisation of an entertainment genre such as comedy is repeatedly formalised within policy. For example, the Digital Economy Act (2015) requires Channel 4 to produce news, current affairs and film, ignoring other genres.[7] There is a worrying trend of some aspects of PSB being seen as *more* public service than others, with news and current affairs typically hierarchised over entertainment. The BBC's move in 2016 of BBC3 to an online service demonstrates this, given that BBC3 was the largest commissioner of television comedy in the UK by far[8] yet its budget has been significantly reduced. Such hierarchisations are highly problematic yet seem to be becoming normal. Indeed, the Puttnam Report itself refers to '*specific* public service genres, including current affairs, drama, news and sport'[9] as if there are some kinds of services that aren't specifically PSB.

These hierarchisations run counter to the public's views of what PSB is and should be. After all, even today the public, both in the BBC's research and in a recent largescale survey conducted by Ofcom, continue to define public service broadcasting (PSB) not as a narrow set of particular programme categories which the market may fail to provide, but as a broad and integrated system of programmes and services. To them, PSB includes soaps, drama, sport, comedy and natural history just as much as (and in some cases, even more than) the traditional 'public service' categories of current affairs, arts and religion.[10]

The BBC's most recent survey of its audiences found that they 'felt strongly that the BBC's mission to inform, educate and entertain was still highly relevant.'[11] Ofcom's annual survey of audience opinions on the importance of different kinds of programming to PSB only started asking about comedy in 2014. Yet those results show that audiences see comedy as more important to PSB than high-quality drama.[12] That the public might have a quite different view of PSB to policy-makers and academics was evidenced in 2016 by the public's angry response to the BBC's decision to close its recipes website, a decision it quickly changed following an outcry. While content such as recipes has been categorised as 'soft', not in keeping with the 'core' notions of PSB policy-makers insist on, this categorisation clearly does not match the value audiences place upon such material.[13]

Throughout its history the BBC – and the concept of PSB as a whole – has often been criticised as 'elitist and paternalistic.'[14] Yet the inclusion of entertainment in PSB has instead often evidenced a much more inclusive, universal approach to public service which actively responds to how the public defines such services. While this submission has referred to comedy and cookery, it also acknowledges the broader conceptions of PSB and entertainment, and argues for the value of them, including programming such as quiz shows, chat shows, popular factual, panel shows, reality television and so on. To exclude genres or particular kinds of programming from conceptions of PSB is to reinstate elitist, paternalistic notions of culture counter to the ideals of a public service. Similarly, to hierarchise some kinds of programming as more PSB than others is to engage in similarly elitist and paternalistic activity, and to impost a rarefied conception of PSB upon the public whom it is intended to serve. It therefore remains vital that while PSB continues to deliver 'something for everyone', it also offers 'everything for someone'.

Notes

1 Brett Mills is Senior Lecturer in Television and Film Studies School of Art, Media and American Studies, University of East Anglia. This is an edited extract from his submission to the Inquiry which can found at: http://futureoftv.org.uk/submissions/mills-brett/.
2 UNESCO 'Public Service Broadcasting' (UNESCO, 2011). http://portal.unesco.org/ci/en/ev.phpURL_ID=1525&URL_DO=DO_TOPIC&URL_SECTION=201.html (accessed 20 May 2016).
3 Andrew Crisell 'Radio: Public Service, Commercialism and the Paradox of Choice', in Adam Briggs and Paul Cobley (eds) *The Media: An Introduction*, 2nd edition (Harlow: Pearson, 2002), 125.
4 DCMS, *A BBC for the Future: A Broadcaster of Distinction* (London: DCMS, 2016), 71.
5 Ibid, 72.
6 Ofcom, *Public Service Broadcasting in the Internet Age: Ofcom's Third Review of Public Service Broadcasting* (London: Ofcom, 2015a), 12.
7 Channel 4, *Channel 4 Annual Report 2015: Britain's Creative Greenhouse* (London: Channel 4, 2015), 13.
8 BBC, *BBC Annual Report and Accounts 2014/15* (London: BBC, 2016), 80.
9 Future of TV 'How to Submit', *A Future for Public Service Television: Content and Platforms in a Digital World* (2016), http://futureoftv.org.uk/how-to-submit/ (accessed 20 May 2016), my italics.
10 BBC, *Building Public Value: Renewing the BBC for a Digital World* (London: BBC, 2004), 7.
11 BBC, *BBC Annual Report*, 32.
12 Ofcom, *PSB Annual Report 2015: PSB Audience Opinion Annex* (London: Ofcom, 2015b), 10.
13 Mark Sweeney, 'John Whittingdale: I'm not to Blame for BBC Cutting Recipes', *The Guardian* (17 May 2016) www.theguardian.com/media/2016/may/17/john-whittingdale-bbc-recipes-newspapers-bbc-food.
14 Jackie Harrison, *News* (London and New York: Routledge, 2006), 50.

19

Debating 'Distinctiveness': How Useful a Concept is it in Measuring the Value and Impact of the BBC?

Peter Goddard[1]

Defining 'Distinctiveness'

In the 2015 Green Paper produced for BBC Charter Review, the comparison between *Strictly Come Dancing* and *The Voice* represents the most obvious depiction of what 'distinctiveness' seems to mean. In arguing that the BBC should be 'providing distinctive programming across all genre types' including entertainment, the Green Paper contrasts *The Voice*, as a bought-in format similar to ITV's *X-Factor*, with *Strictly* 'which was developed by the BBC in-house and then sold abroad'.[2] Here then, 'distinctiveness' seems to support originality rather than imitativeness, with the added bonus that developing distinctive formats might attract international sales revenues for the BBC. The Green Paper also reports the BBC's own research into whether audiences find its programmes to be 'fresh and new' – perhaps an analogous, if rather vague, description of 'distinctiveness'.[3]

The BBC's own definition is different: 'The test ... should be that every BBC programme aspires to be the best in that genre'.[4] It adds that distinctiveness should also be measured across services rather than just individual programmes and that all BBC services should be distinctive from one another (hence BBC director of content Charlotte Moore's comments about the distinctiveness of BBC One as a service[5]). Here distinctiveness seems to reflect a combination of quality ('the best') and public value. This definition might embrace *The Voice* as well as *Strictly*, as long as both sought to be markedly superior to their commercial competitors.

[...] Despite its prominence in the Charter Review debate, it is interesting to note that the notion of distinctiveness is relatively new in regulatory terms. The words 'distinctive' or 'distinctiveness' do not appear at all in the 2007 Charter and only once (requiring the BBC to enrich 'the cultural life of the UK through creative excellence in distinctive and original content') in the accompanying Agreement.[6] Phil Ramsey notes that 'distinctive' became one of Ofcom's 'PSB purposes and characteristics' only as recently as 2014.[7] It is only in relation to Channel 4 that 'distinctiveness' has a significant history. The 1980 Broadcasting Act required the IBA to 'to give the Fourth Channel a distinctive character of its own' and Channel 4 is still required to exhibit 'a distinctive character,'[8] but legislation has never defined this notion of 'distinctiveness' in further detail. Here, indeed, the implication is that the Channel should be distinctive in relation to the remainder of British television. In view of the radical changes undergone by British broadcasting since Channel 4's foundation, this must mean that its distinctiveness is being measured against an ever-changing target, and the Channel has repeatedly modified its own interpretation of distinctiveness.[9]

Even by this limited range of definitions, then, distinctiveness can be taken to mean original rather than imitative, foregrounding quality and public value, anti-populist (which, note, is not the same as anti-popular), and different from other channels' programming. So, although these might all be worthy aspirations for a public service broadcaster, it appears that distinctiveness is a rather elastic term and at times a contradictory one. These definitional problems show that it would be unwise for 'distinctiveness' to be employed too prescriptively in any future regulatory settlement.

Applying 'Distinctiveness'

When we attempt to apply this loose notion of distinctiveness to the future role of the BBC, several potential issues arise. Most obviously, distinctiveness is subjective, rather like quality (with which it is linked in the Charter Review Green Paper[10]). Most of us probably believe that we can recognise distinctiveness in television when we see it, but how could we prove or measure its existence? Distinctiveness, then, is essentially an ambition rather than a determinate quality, although there may be merit in the BBC being held to its ambition to be distinctive by Ofcom and any other regulatory body to which it is subject.

Another key point is to acknowledge that the BBC has much greater potential to produce distinctive programming because of its publicly-funded, not-for-profit model. Unlike most commercial broadcasters, it has no requirement to produce programming which aims for a commercial return, so meritorious programmes and programme types can be nursed until they find an audience. This has largely been the basis for the globally-significant innovation in British television which supports the UK's thriving production sectors and Britain's remarkable position as the second most successful exporter of programmes and largest exporter of formats in the world.[11] Arguably, many of the BBC's most celebrated programmes of recent years, nationally and internationally, owe their success largely to the ability of this not-for-profit model to develop distinctive programmes. For example, *Strictly* and *Doctor Who* are based on BBC properties originating in the less competitive days of black and white television, *Top Gear* has been grown out of a conventional motoring magazine programme to become something unique, while *The Great British Bake Off* originated as a minority interest series with modest ratings. None represented obvious candidates for commercial success, and the audiences for *Top Gear* and *Bake Off* were generated gradually by the BBC through years of relative invisibility when, presumably, they were supported because of their public value rather than any perception of commercial potential. Today, however, these are among the most lucrative television properties in the world due to programme or format sales or, in the case of *Top Gear*, both. So, paradoxically, the very fact that the BBC represents a different, public service, model for television production, founded substantially on public value rather than merely profit, has been a key factor in its creation of such distinctive, marketable and popular programmes. And, crucially for Britain, the BBC is creating such programmes whilst maintaining a focus on what Robin Foster calls 'UK stories, topics and faces.'[12]

So if the publicly-funded, not-for-profit model itself is a key source for the distinctiveness of the BBC's programmes, what should happen when it generates programmes which become hyper-successful? In Mark Oliver's terms, *Strictly*, *Doctor Who*, *Top Gear* and *Bake Off* are surely passing from being 'breakout hits' to becoming the kinds of 'long running schedule bankers' which his report criticises.[13] In effect, the BBC is being praised for its popular success in creating these distinctive 'breakout hits', but then lays itself open to criticism for a lack of distinctiveness if it continues to commission them. This is where arguments about the BBC's 'market impact' become particularly pernicious. As noted above, one of its key strengths is its ability to create hugely popular and distinctive programmes and formats which commercial broadcasters without public service obligations would probably never consider commissioning. Besides the public value that such programmes create, this represents a strong example of the BBC combatting market failure. But stealing audiences from commercial broadcasters for these same programmes is held by the BBC's critics to be illegitimate in its impact on the profitability of the broadcasting market – something which the BBC should be prevented from doing. This places the BBC in a nonsensical position and defies logic – 'generate popularity from distinctive programming but don't be too popular.' [...] Given the success of such programmes, there are also sound commercial reasons why the BBC's competitors may seek to imitate them, but it would be a perverse disincentive to innovation and distinctiveness if the fact that the BBC has made them seem familiar is used as a reason to criticise the BBC for retaining such programmes in its schedule.

The overriding issue here is the continuation of the BBC's commitment to universality, serving all audiences including those seeking popular, mainstream programming. The BBC's ability to create programmes which are distinctive yet popular comes from the fact that it is a holistic broadcaster which appeals to all, rather than a minority broadcaster like PBS in the USA – often seen as an unpopular alternative.

But while debates about distinctiveness and market impact arise out of the BBC's competitive behaviour, they also embrace the BBC's competitive scheduling. With the rise of on-demand viewing, scheduling no longer has the hold on audiences that it once had.[14] Nevertheless, it seems likely that scheduling will remain significant in two important areas for the foreseeable future – as a 'shop window' for new programming, around which marketing activities can be focused, and at times of the week most associated with shared family viewing, notably Saturday night. It is no coincidence that programmes such as *Doctor Who* and *Strictly*, along with *The Voice*, are part of the BBC's Saturday night schedule. [...]

Notes

1 Peter Goddard is a Senior Lecturer in Media and Communications, Department of Communication and Media, University of Liverpool. This is an edited extract from his submission to the Inquiry at http://futureoftv.org.uk/submissions/goddard-peter/. An expanded version of the submission has been published by the author: ' "Distinctiveness" and the BBC: A New Battleground For Public Service Television?' *Media, Culture & Society*, 2017, DOI: 10.1177/0163443717692787.

2 Department for Culture, Media & Sport, *BBC Charter Review Public Consultation* (London: DCMS, 2015), www.gov.uk/government/consultations/bbc-charter-review-public-consultation, 39.

3 DCMS, *BBC Charter Review Public Consultation*, 38; BBC Annual Report and Accounts 2014–2015 (London: BBC, 2015), www.bbc.co.uk/aboutthebbc/insidethebbc/howwework/reports/ara, 33.

4 BBC, BBC submission to Puttnam Inquiry, http://futureoftv.org.uk/wp-content/uploads/2015/12/BBC-evidence.pdf, 3.

5 Charlotte Moore, speech, 7 May 2016, www.bbc.co.uk/mediacentre/speeches/2016/charlotte-moore-vision.

6 Department for Culture, Media & Sport, *Royal Charter for the Continuance of the British Broadcasting Corporation*, Cm 6925 (London: TSO, 2006); Department for Culture, Media & Sport (2006), *Broadcasting: An Agreement Between Her Majesty's Secretary of State for Culture, Media and Sport and the British Broadcasting Corporation*, Cm 6872 (London: TSO, 2006).

7 Phil Ramsey, 'The Contribution of the UK's Commercial Public Service Broadcasters to The Public Service Television System', submission to Puttnam Inquiry, http://futureoftv.org.uk/wp-content/uploads/2015/11/Philip-Ramsey.pdf, 2; Ofcom, *PSB Annual Report 2015* (London: Ofcom, 2015), www.ofcom.org.uk/tv-radio-and-on-demand/information-for-industry/public-service-broadcasting/public-service-broadcasting-annual-report-2015.

8 Broadcasting Act 1981, 11 (1), http://legislation.data.gov.uk/en/ukpga/1981/68/part/I/1991-10-01/data.htm?wrap=true; Ofcom, 'Channel 4 Licence: attachment to variation number 18', November 30, 2015, www.ofcom.org.uk/__data/assets/pdf_file/0018/40266/channel_4_attachment_to_variation_no._18.pdf.

9 See Suzana Zilic Fiser (2010), 'Social Responsibility and Economic Success of Public Service Broadcasting Channel 4: Distinctiveness with Market Orientation', http://ripeat.org/library/Zilic%20Fiser.pdf.

10 DCMS, *BBC Charter Review Consultation*, 36.

11 Jean K. Chalaby, 'Broadcasting Policy in the Era of Global Value Chain-Oriented Industrialisation', submission to Puttnam Inquiry, http://futureoftv.org.uk/wp-content/uploads/2016/02/Jean-Chalaby.pdf, 4.

12 Robin Foster, 'A Future for Public Service Television', submission to Puttnam Inquiry, http://futureoftv.org.uk/wp-content/uploads/2015/11/Robin-Foster.pdf, 5.

13 Mark Oliver, 'Making BBC1 More Distinctive is not a Threat – it Could be a Benefit', *The Guardian*, 10 March 2016, www.theguardian.com/media/2016/mar/10/bbc1-distinctive-doctor-who-bake-off-strictly; Oliver & Ohlbaum, *BBC Television, Radio and Online Services: An Assessment of Market Impact and Distinctiveness* (London: DCMS, www.gov.uk/government/uploads/system/uploads/attachment_data/file/504012/FINAL_-_BBC_market_impact_assessment.pdf.

14 See Catherine Johnson, 'Video-On-Demand as Public Service Television', submission to Puttnam Inquiry, http://futureoftv.org.uk/wp-content/uploads/2015/11/Catherine-Johnson.pdf.

20

The BBC: A Radical Rethink

Justin Schlosberg[1]

Contrary to widespread expectations and fears, the government's 2015 White Paper on the future of the BBC[2] preserved the BBC's licence fee for the foreseeable future. That came as welcome relief to those who feared a giant sell-off or switch to subscription funding, and an end to the BBC's unique public service mandate as we know it. Though the door remains open to these pathways in the future, public ownership and licence fee funding seemed to have been temporarily secured. A much more worrying development, however, concerned proposed changes to the BBC's governance and a system of appointments that threatened to encroach on the BBC's editorial autonomy.

What's particularly striking about this development is that it pushes in the general direction of growing state control of public service media, spearheaded by countries like Hungary and Poland. A new media law that came into effect in Poland in 2016, for instance, consolidates the executive's power of appointments in public broadcasters. It was one of the first legislative moves of the new government led by the right wing Law and Justice Party. As Reporters without Borders declared: 'This new law, giving the government full powers to appoint and dismiss the heads of the public broadcast media, constitutes a flagrant violation of media freedom and pluralism.'[3]

The BBC White Paper for Charter Renewal proposed a new 'unitary board' of which the majority and most senior members would be appointed by government. For the first time in its history, such an approach threatened to give a direct government appointee overall editorial responsibility for all of the BBC's output.

What's equally striking about this move, is that it flies in the face of what the government has long intimated was at the heart of its charter renewal agenda: to introduce a system of contestable funding to effectively break up the BBC and enable more local and more commercial providers to take a slice of the licence fee. Understandably, that struck fear in the minds of those who rightly believe that the BBC must remain entirely in public hands and entirely not-for-profit.

But defensive arguments against top-slicing tend to oppose any possibility of decentralisation in the BBC's structure and governance, and assume that the BBC's strength lies in its scale and unitary composition. This is assumed to provide a robust defence against both government and market pressures, but there is more reason to think that the exact opposite is the case. A centralised and concentrated BBC is intrinsically more vulnerable to editorial pressures precisely because they can filter down the chain of governors, directors, managers and editors. If a government did seek to shape or control the BBC's agenda, it would have a far more difficult job if it had to contend with a network of editorially autonomous outlets than with a single command and control centre.

Such a network need not involve any degree of privatisation or commercialisation. Indeed, a 'networked' BBC – provided it was structured in the right way – could also be more immune to market pressures that many believe have fostered homogenisation of the BBC's news output and a growing dependency on a commercial-press led agenda.

So what would such a networked structure look like? As it turns out, we don't have to look much further than our own national doorstep for an example. The Nederlandse Publicke Omroep (NPO) in Holland has long been founded on just such a system that distributes airtime and resources among a network of affiliate and member-led broadcasting organisations. Holland was ranked the second freest media system in the world by Reporters without Borders in 2016[4] and although it has faced recent cutbacks and consolidation, the NPO has proved relatively resilient to the pressures of digitisation. Like the BBC, it continues to demonstrate enduring public value, as reflected in the strength of its member-based affiliates and the reach of its online services.

The bulk of channels and airtime assigned to NPO is shared among ten broadcasting associations. Eight of these function as audience cooperatives, with membership bases that reflect the diversity of interests and groups in Dutch society. The remaining two are 'task-based' broadcasters specialising predominantly in news, current affairs and other factual programming. The NPO is charged with administering this network but does not have overall editorial responsibility for output.

With editorial autonomy thus enshrined into its structure, and accountability to audiences cemented by membership-driven governance, the NPO is intrinsically independent in a way that the BBC never has been, from its compromised reporting of the General Strike in 1926 to its infamous capitulation in the face of government flak over the Iraq War in 2003.

If such an alternative sounds unthinkably radical, that only reflects how restricted the terms of public debate over the BBC's future have become. Indeed, the very words 'radical' and 'reform' in the context of the BBC have been so co-opted that they seem to automatically signal cuts or closure rather than any kind of progressive enhancement of the BBC's public service function.

Of course there is always the danger that even consideration of a reconfigured BBC along networked lines – which could take any number of forms – could open a back door route to privatisation or top slicing. But if anything, the White Paper on the future of the BBC took a step in the opposite direction and revealed its true hand: in spite of the rhetoric, a large scale, centralised BBC has always been more consonant with the interests of state-corporate power than it is in conflict, notwithstanding periodic headaches and crises engendered by a pesky journalist.

Of course a much more outspoken critique focuses precisely on the BBC's size and scale which is seen as the major threat to media plurality in the UK. From this perspective, the decline of newspapers threatens to erode any checks on the near monopoly status enjoyed by the BBC. Rather than worrying about the agenda influence of mainstream media in general, these arguments[5] suggest that we should be concerned exclusively with the overarching reach and influence of the BBC.

But how far does the BBC's own news agenda reflect or align with that of its commercial competitors? When scholars at Cardiff University[6] set out to investigate this question during the 2015 UK general election, they found that the BBC's overall issue-agenda appeared to have been consistently led by the predominantly right-wing national newspapers. The extent of this alignment was corroborated by other research conducted at Loughborough University[7] and King's College, London[8] revealing a strong correlation between the range and rank order of issues covered by both television and the press, and one that did not fully accord with public priorities as demonstrated by monthly issue tracking polls.

The important point this raises for the future of the BBC is twofold. First, if commercial press exercise a strong influence over the BBC's political news coverage, it makes little sense to consider it a meaningful counterweight to the BBC's dominance of news consumption. The evidence from the 2015 election suggests that if anything, the BBC amplified an agenda that was set largely by the commercial press. Second, and by the same token, we ought to be equally sceptical of suggestions that the BBC provides a substantive check on the more partisan editorial agenda of the commercial press.

At a time when many public service broadcasters around the world are facing varying degrees of existential crises, public debate is all too often reduced to a choice between preservation or market-based reforms; with the latter usually amounting to cutbacks or closures. What's left off the agenda is the possibility of radical democratic reform aimed at reconstituting the independence and accountability of public service media. The idea that a substantive section of any pluralistic media system needs to be in public hands is one that retains a great deal of force, in spite of the digital transition and corresponding end of channel scarcity. But the way in which public service broadcasters are structured, regulated and governed can have profound implications for independence in relation to both the state and market.

Notes

1 Justin Schlosberg is a lecturer in journalism and media at Birkbeck, University of London. This is an edited extract from his submissions to the Puttnam Inquiry, http://futureoftv.org.uk/submissions/schlosberg-justin/.

2 Department for Media, Culture & Sport, *A BBC for the Future: a Broadcaster of Distinction* (London: DCMS, 2015), www.gov.uk/government/uploads/system/uploads/attachment_data/file/524863/DCMS_A_BBC_for_the_future_linked_rev1.pdf.

3 Reporters without Borders, 'RSF calls for firm EU Stance if Poland does not Abandon New Media Law', January 5, 2016, https://rsf.org/en/news/rsf-calls-firm-eu-stance-if-poland-does-not-abandon-new-media-law.

4 According to 2017 World Press Freedom index, the Netherlands continues to rank highly although it has since dropped to fifth place.

5 See, for example, David Elstein, 'Reflections on the Election: Lessons to be Learned', *Open Democracy*, 20 May 2015, www.opendemocracy.net/ourkingdom/david-elstein/reflections-on-election-lessons-to-be-learned.

6 See Steven Cushion and Richard Sambrook, 'How TV News Let The Tories Fight the Election On Their Own Terms', *The Guardian*, 15 May 2015. www.theguardian.com/media/2015/may/15/tv-news-let-the-tories-fight-the-election-coalition-economy-taxation.

7 See General Election 2015 – Media analysis from Loughborough University Communication Research Centre. Available at http://blog.lboro.ac.uk/general-election/.

8 Martin Moore and Gordon Ramsay, *UK Election 2015: Setting the Agenda*, October 2015, www.kcl.ac.uk/sspp/policy-institute/publications/MST-Election-2015-FINAL.pdf.

21

Ensuring the Future of Public Service Television for the Benefit of Citizens

Voice of the Listener & Viewer[1]

[...] The Social and Cultural Purposes of Television

Many of the arguments which informed the current BBC Charter were developed in *Building Public Value*, the policy document which the BBC published prior to the last Charter renewal. The VLV believes that a vision where the benefits to the citizen are core to the mission of public service television is still valid today:

The BBC's founders believed that broadcasting could make the world a better place. Public intervention would ensure that its astonishing creative power – to enrich individuals with knowledge, culture and information about their world, to build more cohesive communities, to engage the people of the UK and the whole globe in a new conversation about who we are and where we are going – would be put to work to the sole benefit of the public.[2]

VLV believes that public service television should be universally available, available to all free at the point of use and provide something for everyone, including impartial and accurate news and other high quality content for the benefit of the whole of UK society.

Ofcom was established to ensure that the needs of the citizen and consumer are met. These needs may differ but both are equally important. The marketplace will tend to provide for the needs of consumers when there is a benefit to the provider. There may be less financial return from content which benefits the citizen.

The explosion of media platforms in the past decade has led to increased competition for viewers. VLV believes increased competition alongside deregulation of broadcasting has resulted in an increase in the volume of more popular genres and a reduction in the mainstream provision of UK specific content which benefits citizens (for example, UK culturally specific children's content, current affairs, regional and religious content).

Since the introduction of regulation limiting the type of advertising which can be placed around children's programming, the investment of content for children on the commercial PSBs has dropped by 95%.[3]

Despite the proliferation of channels since digital switchover, it is clear that the UK broadcasting sector faces market failure in some key public service genres. 'Market failure' refers to the fact that, whatever the number of providers in the market at any given time, the market may still fail the citizen if there isn't a free to air provision of a wide range of high quality, diverse and informative programming, especially in genres which may not be considered commercially attractive.

The PSB system in the UK has existed up until now as a conscious democratic, social and cultural intervention in the market in order to achieve certain public value purposes: to enrich our lives with high quality engaging content which broadens our horizons, excites us and helps us cohere as a nation.

VLV believes there is currently a balance provided by the existing UK PSBs on free to air television which in most genres provides the public with such content. We would like this system to continue to be supported by a regulatory regime which encourages investment in a range of high quality public service content which is provided for the benefit of citizens as well as consumers. We do not believe that the viewing habits of individual consumers should drive public policy choices which may limit the provision of universal service, and thus the freedom of choice for all viewers. [...]

Notes

1 The Voice of the Listener and Viewer campaigns on behalf of the consumer voice in broadcasting. This is an edited extract from VLV's submission to the Puttnam Inquiry, http://futureoftv.org.uk/submissions/ensuring-the-future-of-public-service-television-for-the-benefit-of-citizens/.

2 BBC, *Building Public Value: Renewing the BBC for a digital world* (London: BBC, 2004), https://downloads.bbc.co.uk/aboutthebbc/policies/pdf/bpv.pdf, 6.

3 Ofcom, *Children's Analysis: PSB Annual Report* (London: Ofcom, 2014), www.ofcom.org.uk/__data/assets/pdf_file/0009/81000/annex_6.ii_childrens_analysis.pdf, 8.

22

The Social and Cultural Purposes of Television Today

Equity[1]

Introduction

Television has a number of social and cultural purposes. Some of these are enshrined in legislation or regulation such as the remit of Channel 4, the regional production obligations to which all public service broadcasters are subject or the public purposes of the BBC, as well as the BBC's overall mission to 'educate, inform and entertain'. Equity supports these principles and duties and has been lobbying for their reform and extension to other commercial broadcasters.

Equity believes that all UK broadcasters should have an obligation to contribute to the UK's cultural diversity through investing in original content production. We have lobbied Ofcom to increase quotas for original and regional drama, comedy, entertainment and children's programmes made in and about the UK, particularly with respect to Channel 3 and 5 licencees.

One of the social purposes of television should be to promote growth and employment opportunities across the creative industries. For some time Equity has been calling on broadcasters to produce more content in the Nations and Regions in order to draw on the skills and talents in these areas and we have urged Ofcom to reconsider its definition of regional/out of London production so as to include on-screen talent.

The creative industries and television and film in particular should provide good jobs and training for the UK's strong base of highly skilled creative workers and performers. Equity has been lobbying, through the Charter Renewal process, for specific references to best practice in employment, training and development both for in-house and independent producers to be included in the BBC's public purposes.

Changing Production, Consumption and Distribution Practices

Equity has worked with all of the major UK broadcasters towards developing new platforms for content delivery and has consistently sought to ensure that content can be made available for use on these platforms when made under Equity collective agreements.

The BBC led the way in terms of establishing the iplayer and Equity has been party to the launch of other such services through the negotiation of agreements with the BBC and other broadcasters for rights clearances. These agreements have contributed to the success of a multiplicity of digital services including All 4, ITV player and Sky Anytime. Crucially, these agreements provide a source of additional remuneration for performers in return for the exploitation of their work.

Most recently Equity has achieved the first agreement outside of the US for the engagement of our members and the reuse of their performances by Netflix. As more new delivery mechanisms and platforms emerge, suitable agreements must be concluded that recognise the rights of performers whose work is exploited across all channels and platforms.

Funding and Regulation of Public Service Platforms

In terms of funding, Equity continues to support the licence fee as the most appropriate funding method for the BBC however we have serious concerns about the most recent settlement, which was again concluded hastily with government without any input from licence fee payers or consultation with those who work for the BBC.

The most pressing concern in terms of future funding of the BBC is the new obligation to provide free licences for the over 75s. We believe this is inappropriate as it confers social policy responsibilities on to the BBC and is likely to lead to a significant shortfall in BBC funding post 2018, despite the government's commitment to end top slicing for broadband rollout and the potential new income arising from the proposed closure of the catch up TV loophole. The projected shortfall in funding could be as much as £350m and this will inevitably lead to large scale job losses, content budget cuts and service closures.

The government's charter renewal consultation during 2015–2016 also considered other funding options for the BBC including a household levy and introducing subscription based services. We believe the latter would undermine the BBC's ability to provide a range of content to all audiences and could lead to the adoption of a much more commercial approach by the BBC. It is also likely that under this model niche services such as radio drama could become underfunded or unaffordable.

BBC Worldwide is an important source of revenue which is re-invested in BBC production. This helps to keep the licence fee as low as possible. It exists to maximise profits for the BBC, but operates under the rules and principles outlined in the BBC's Charter and Agreement. This framework is important as it means that BBCW is independent of government, but supports the BBC's public service mission and is accountable to licence fee payers. There should be no privatisation of any part of the BBC in the coming period and the BBC should instead be free to explore how it can maintain and expand investment in content via all income derived from commercial activities into programme-making.

Equity does not support proposals to divert licence fee funds towards contestable budgets for other broadcasters or producers to create drama or children's content. The BBC's viewers have an expectation that licence fee funding goes predominantly towards the production of high quality programmes for the BBC. Currently this is the case and indeed most UK drama production employing professional performers originates with the BBC and the independent producers who work with the BBC. [...]

The Electronic Programme Guide remains an important element in the funding of all PSBs. Prominence on the EPG remains a key benefit for PSBs and contributes towards their ability to invest in skills and content which provide the vast majority of job opportunities in the audiovisual sector for performers and other creative workers.

Our members are very strongly in favour of a regulatory approach to public service broadcasting that can facilitate the continuation and an expansion in the production of original drama, comedy, entertainment and children's programmes made in and about the UK.

Without regulation which aims to support UK original content production, we fear that certain genres, particularly those for niche audiences, could disappear from our screens at an increasing rate. Ofcom's objective must therefore be to seek to maintain and strengthen existing public service broadcasting commitments, so that UK audiences can continue to get high-quality original TV from a range of providers. [...]

Note

1 Equity is a trade union representing over 42,000 creative workers in the UK. This is an edited extract from its submission to the Puttnam Inquiry, http://futureoftv.org.uk/submissions/equity/.

Part Four

Public Service Television in an On-Demand World

23

Taking the Principles of Public Service Media into the Digital Ecology[1]

Georgina Born

It was suggested in Chapter 13 of this book that the normative principles of public service media (PSM) have become more rather than less relevant, have expanded and have gained a new urgency in the digital era. In what follows, a series of propositions are advanced concerning the ways that such principles find new expression in digital conditions. If the proponents of neoliberal economic thinking argue that the digital economy is best served, and best understood, in terms of the dynamics of competition operating within free markets, then the oligopolistic tendencies that have become pronounced in the last decade, manifest in the dominance of a few key digital intermediaries and in the rapid capacity to establish primacy in new digital markets, disprove such assumptions. This chapter therefore advances the need for public intervention in digital media markets on several levels, each of them important, each founded on and drawing legitimacy from the expanded normative principles set out in Chapter 13.

Given the evidence accruing in support of the wider need for such interventions – for example, as has recently been raised on the grounds of national security[2] – it is remarkable how little public debate has occurred about the desirability and the potential significance of public service interventions in the digital ecology. It is perhaps indicative that where such debates have occurred, it has been less in the global North than in the global South, where the severe and deepening economic, social and cultural disadvantages stemming from dependence on and lock-in by monopolistic commercial media firms based in the North – whether with respect to hardware, software or social media – have, on occasion, become focal policy and legislative concerns.[3]

Despite the success of the BBC iPlayer, DAB and Channel 4's documentary plat-form 4docs, then, it is curious how few sustained innovations exploiting the rich potentials of digital media have come from the core PSM institutions. This chapter lays out several indicative ways in which new research directions, fuelling a new public debate, are urgently needed into the kinds of public service potentials that could be unleashed – in terms of animating digital media-supported cultural, infor-mational, educational and artistic activities – by PSM interventions in the digital space. The founding premise is simple: twenty years into the heyday of the inter-net, society has not yet started this debate, let alone conceived of its parameters and scope.

Digital PSM, Boosting Diversity: Unleashing Low-Budget and Niche Content

A first avenue for PSM intervention centres on boosting the creative potential of participation, user-generated content, low-budget experimental production, niche markets and the 'long tail', as well as platforms designed to host and showcase these activities.[4] The absence of such interventions this far into the digital era suggests a fail-ure of imagination, of sustained R & D, or of institutional commitment – or of all three.

New normative thinking can help to combat this state of affairs, framing new chal-lenges and imagining vital functions to be delivered by PSM. This chapter therefore advances new linked principles: the obligation to *animate participation* and creative practices, and to *curate* and disseminate the results. In related vein, the BBC's direc-tor general Tony Hall spoke in 2015 of *partnership* as a new principle in the digital environment,[5] and this was followed by a focus in the 2016 DCMS White Paper on the BBC on the imperative for the corporation to improve its partnerships with other organisations.[6]

While partnership as a new principle certainly deserves support, it is essential that this commitment should not be limited to the opening up of the BBC, and PSM more generally, to partnering only with established and elite cultural bodies in the UK – such bodies as the Royal Opera House, the British Museum or Tate Modern. Partnership should extend to local engagements with small-scale cultural organisa-tions as well as amateur content producers and platform designers: they too must be invited to participate in the PSM ecology, answering also to the need for greater diver-sity and decentralisation in media and cultural production. It is this imperative that lies, in part, behind the commitment to creating a new fund for digital content provid-ers discussed in Chapter 7 of the Puttnam Report.

The spectrum of PSM-animated and PSM-curated production and services should therefore range from fully professional to amateur and emerging practices: all matter today, and PSM in the digital era is about brokering participation and partnerships across this full spectrum. Emulating the long tail model by utilising the curatorial and distributive powers of public digital platforms will allow PSM to open out, and the productive effect will be to boost its key function of animating the 21st century creative economy.

Together, the new principles of *animation, participation, partnership* and *curation* have synergistic powers. Together, these principles will help to counter the current lack of engagement with the niche possibilities of the digital on the part of PSM. They will stimulate the production both of low budget and experimental content – film, documentary, reality, comedy, ideas- and game-based – and of innovative mainstream output and services, expanding the stream that feeds the discovery and throughput of new creative talent and enabling the curation of this content on public portals. Such exposure will offer the talent behind it higher profile, leveraging their entry into the creative economy as well as into the arts, cultural and educational sectors.

In future, the PSM ecology must involve deep reflection about socially and culturally enriching digital interventions of this kind which are central to the realisation of PSM's diversity principle – for they have the potential vastly to increase the diversity of voices composing the three-way public sphere set out in Chapter 13. At the same time, these digital interventions will encourage the growth of, and consolidate, local, regional and national production hubs, with the added potential to nurture new economic growth. These directions point to the prospect of opening up the category of public service content beyond the PSMs to a host of partners, including, potentially, to all publicly-funded providers of cultural, artistic and intellectual content.

What Do We Want? From Commercial to Public Service Recommendation and Search Algorithms

If a greatly enhanced content production sector is one result of applying public service principles to the digital ecology, a second avenue for PSM intervention concerns the potential radically to reshape existing markets in what have become entirely commercial, and in some cases globally oligopolistic, gatekeeping technologies: search engines and recommendation algorithms. The stakes are high: search engines amount to an infrastructure that powerfully shapes the everyday, incremental and cumulative progression of our information and knowledge practices; while recommendation algorithms amount to a hidden, non-transparent infrastructure that harvests, shapes and directs nothing less than our cultural tastes.

Under the prevailing market impact discourse, obsessed as it is with short-term impacts on competitor revenues and profits, such interventions are almost unthinkable. But the argument advanced in this chapter is that if these interventions derive from the renewed normative principles of PSM set out in Chapter 13 – of independence, universality, citizenship, quality (now extended to include innovation and risk-taking) and diversity – then such interventions in digital markets are fully justified. Indeed, the more pertinent question is why they have been ruled out. In principle, when designing digital interventions in the media ecology, PSM should meet the same criteria as PSB did before it: *they are justified when they complement or raise the game of commercial services.*

A first example of such an intervention consists in the development of what James Bennett has described as public service recommendation algorithms.[7] Such a proposal rests squarely on the PSM norms of universality of genre (mixed programming) and diversity. Current recommendation engines, including the BBC iPlayer, follow a logic of similarity – 'if you liked that, you might also like this' – in this way bringing us more of the same, or only slight variants. But is this really what PSM should do? In the broadcast era, the art of scheduling took audiences through different genres, exposing them to a mixed diet that opened up new experiences and perspectives: from comedy, to news, to drama, to current affairs. In the digital age, in contrast, recommendation engines play safe, corralling and enclosing audience tastes. Bennett asks: 'What if a public service algorithm made ... recommendations from left field, [opening] our horizons? If you liked *Top Gear*, here's an environmental documentary, or *Woman's Hour*. If you liked a music documentary, here's a sitcom. Choice will remain [key]: but it should be genuine choice – to watch more of the same or to explore something new'.

To enable this alternative recommendation logic, PSM should in future take responsibility for researching, designing and hosting algorithms that systematically open up encounters with hitherto unknown culturally, socially or intellectually entertaining and enriching content. But in addition, it might be possible to give users the interactive opportunity to fine-tune the algorithms guiding the content brought to their attention, potentially generating a variable range of settings for individuals, offering them transparent controls over the type and the range of their encounters. Such transparent recommendation algorithms would mark PSM services out as utterly distinct from the dominant commercial offerings; they would work against the commercial enclosure of taste, exposing viewers to a greater breadth of content and diversity of voices and viewpoints, taking them beyond what they currently know – a core principle of PSB.

Two further points can be added in support of this proposition: first, that the results of viewing, listening to and interacting with content guided by commercial taste recommendation engines is to encourage an increasing individuation and nar-rowcasting in reception, with the recommendations invariably operating within a commercial 'walled garden' (e.g. Spotify, Netflix, Amazon). Such narrowcasting works against the commonalities of address proffered by PSM in its universalistic mode. Second, the commercial algorithms favouring such individuation are also intimately involved in the collection, storage and sale of personal data. However, this represents a grave assault on privacy, raising the question of why this dimension of the digital content economy has escaped regulatory intervention. The public service recommen-dation algorithms envisaged here would effectively offer citizens the means to avoid, and alternatives to, the burgeoning commercial market in personal data.

Similar arguments for complementing, reshaping and correcting existing market tendencies can be made for search algorithms. Currently, the rankings resulting from commercial search engines are guided by commercial principles. From the user's point of view this is often unsatisfactory, because the content sought could be two or three pages in, but is trumped by pages that get more traffic. A search engine designed to resist this commercial logic and operate according to different, public service and public interest principles is likely to prove popular. It would get to the desired content directly and quickly regardless of traffic, and it would elevate the online profile of free, non-profit-derived or publicly-funded content. The editorial principles at stake, driv-ing the search algorithm, could again be made transparent and available for ongoing public debate, manifesting the accountability essential for such technologies operat-ing at the core of our future public knowledge ecology.

A complementary intervention would be to regulate commercially dominant search engines: adjusting the competition regime so that obligations are triggered by a certain share of market; regulating Google so that it presents public service options higher up its rankings; or creating 'must carry' rules and obligations that give due prominence to public service content, akin to the current rules for EPGs.[8]

These and other PSM interventions to optimise the design of recommendation and search algorithms in the public interest would require monitoring and analysis of directions being taken by commercial platforms, analysis of society's changing goals and users' evolving needs and wishes, and, as mentioned, the implementation of robust systems of transparency and accountability.

PSM as an Enduring and Accountable Communications Infrastructure: Towards a Digital Public Commons

Similar arguments for PSM interventions in the digital media ecology that complement, correct and raise the game of commercial services can be made at the level of the very foundations of our communicative infrastructure. As Tony Ageh has put it: 'We used to be broadcast beings. We are now internet beings. However with more and higher barriers to entry to the digital realm, we must work hard to ensure that nobody is stripped of the ability to be a citizen of the future.'[9]

Throughout its history the BBC has been not only a broadcaster and programme maker. It has also been a world-leading engineering organisation 'pushing the boundaries on behalf of the population of the UK and the whole of the industry – from manufacturers, to other "competing" broadcasters in both radio and television: sharing technologies to enable broadcasting to go further and faster; introducing standards for pictures, colour, clearer sound, teletext, and HD.'[10] In the digital era, in the face of threatened commercial enclosures of the digital commons, it is vital that the BBC's historic functions of nurturing the next generation of publicly-oriented communications technologies must remain undimmed. For in a future media system that is delivered entirely over Internet Protocols (IP), we will discover that a series of earlier core guarantees delivered by the allocated public broadcasting spectrum – unlimited and secure universal access, free at the point of delivery, to a wide range of reliable and high quality content and services – are at serious risk.

Currently, there is no allocation of IP bandwidth for public access and discourse free of commercial and/or political oversight or imperatives, equivalent to the analog public spectrum. Moreover, the present situation is one in which commercial internet service providers enjoy almost complete control over the internet access of both individuals and public bodies, who are enjoined constantly to invest scarce resources in keeping up to date with rapidly obsolete hardware, software and operating systems. In many cases, obsolescence is built in as the underlying business model such that consumers, public organisations and governments are compelled to pay commercial intermediaries, and to keep on paying, simply to maintain access to ordinary ongoing functioning. Commercial IT operators are at liberty to proceed with these cyclical and unceasing financial demands on citizens and the public sector without due accountability.

In this light, three of the original characteristics of broadcasting that were fundamental to the vibrant democratic nature of the 20th century analog media ecology are now in question:

- Anonymity: the capacity for anyone to watch or listen secure in the knowledge that they could not be overseen and the resulting intrusion on their privacy exploited for commercial or political ends, or to their disadvantage;
- Unmetered consumption: there being no limit to the amount of broadcasting that could be accessed and enjoyed, or where, without additional charge or fees;
- Secure and enduring consumption: the confidence to know that broadcast output could not be taken away, and that access to receive all PSB output was guaranteed.

In future, in order to support the normative principles set out in Chapter 13 and expanded in this one, PSM should provide the foundations for a digital public commons that precludes the repeated creation of barriers to entry either by commercial gatekeepers or, potentially, by politicians. In view of the PSM principles of independence, universality, citizenship, quality and diversity, arguments for a radically new vision of the IP-based future must be advanced – a vision centred on interventions to create public internet spaces devoted to universal access to services free at the point of use. This is a challenge that, given the engineering prowess of the BBC, is entirely feasible and in keeping with the corporation's historic role. As argued earlier, in an era of interactive media, of participation and partnership, it is not only access in terms of reception and use that are at stake, but rights of access to enable, foster and unleash creativity, dialogue, interactivity, collaboration and learning.

In order for PSM to deliver an enduring communications infrastructure equivalent to PSB's analog public space, it therefore needs to envisage, build and sustain the architecture of a PSM-based digital public commons. Such a digital public commons would retain all the elements of its analog precursor, but would also offer additional features and services that were previously impossible or unimaginable:

1. The digital public commons would guarantee access to a protected allocation of internet bandwidth for every citizen, at home and in key public places – just as frequencies within the broadcast spectrum were reserved for PSB.
2. The digital public commons would not require a commercial broadband subscription, but would offer unmetered consumption free at the point of use to all, regardless of status or ability to pay. It would be available anywhere, at any time, to anyone.
3. The digital public commons would offer an ever-growing library of digitised content and assets, and software, owned and provided by our publicly-funded cultural

and educational organisations: PSM, museums, arts organisations, libraries and archives, universities and other educational bodies, and government and public services.

4. The digital public commons would offer innovative products and services that allow people to access, contribute to, interact and communicate with the public cultural, educational and government service sectors.

5. In the digital public commons, citizens would be safe and secure to discover, use, create and share content and data without fear of loss, theft or unintended exploitation of of their creative outputs and endeavours, or of the unintended exposure or exploitation of their personal data.

6. And finally, the digital public commons would not be subject to the threats posed by repeated and unaccountable commercial demands for the upgrading of notionally obsolete hardware, software and operating systems, and the consequent drain both on personal finances and on the public purse. These cycles would be replaced, in the digital public commons, by explicitly justified and well-coordinated evolution of the public digital infrastructure, evolving only under pressure of the public interest.

It is only through such combined aims and amibitions that the digital public commons envisaged in this chapter can match the foundational public benefits delivered in the 20th century by PSB's analog public space – now translated into the digital realm. It is a sign of the faltering of the political, social and intellectual energies and ambitions that guided PSB that such thinking has not taken absolute priority in PSM's, and specifically the BBC's, search for self-reinvention in the face of the challenges posed by digitisation.

Yet, if we imagine that such a tally of objectives is entirely foreign to the core purposes of PSM, then that is mistaken, for these combined aims were, in fact, foreseen by and articulated in a foundational remit written into earlier versions of the BBC's charter, albeit in shorthand. Thus, the sixth public purpose for the BBC set out in the 2006 charter states: '(f) In promoting its other purposes, helping to deliver to the public the benefit of emerging communications technologies and services'.[11] It should be a matter of grave public concern that this purpose was not included in the BBC's 2016 charter.

The proposals advanced in this chapter are designed to contribute to the conceptual work that is necessary, and that is lamentably overdue, in order to develop the capacities of the core PSM institutions to achieve independence, universality, citizenship, quality and diversity *in the digital ecology*. A set of four additional, synergistic principles at the core of the digital media ecology have been identified: animation, participation, partnership and curation. Promoting both universality and diversity in PSM's communicative, informational, cultural and educational offerings both on- and offline requires that citizens are enabled to access, participate, collaborate and create through their engagements with a digital public commons. These are goals that not only resist but roll back what Saskia Sassen has identified as the existing consolidation of 'private appropriations of a "public" space', as well as the profoundly anti-democratic tendencies immanent in the internet's unimpeded commercialisation.[12] Together, the proposals set out in this chapter envisage the transformation of today's citizens away from their reduction to mere consumers and towards their flourishing as 'actors with heightened capacities for [cultural], political and technological activity',[13] through the growth of new sites of public media, cultural, informational and political engagement.The proposals gathered here show ways in which the normative principles of PSM not only can but must find new expression in digital conditions.

Notes

1 This chapter is an edited and expanded extract from Chapter 2 of the Puttnam Report. It draws heavily on Georgina Born's paper, 'Rethinking the Principles of Public Service Media', delivered at the Inquiry's event on the same topic, British Academy, 3 March 2016.

2 A recent case centres on the risks posed by commercial oligopolies to the security of the information infrastructure of, for example, the UK's National Health Service: on the ransomware attack targeting Microsoft computers and operating systems worldwide on 12 May 2017, see, for example, www.telegraph.co.uk/news/2017/05/12/nhs-hit-major-cyber-attack-hackers-demanding-ransom/.

3 See Anita Chan's research on the Peruvian movement dedicated to the adoption of Free/Libre Open Source Software (FLOSS) in government and the public sector in Peru. The movement resulted in 2001 in a Peruvian legislative proposal, itself inspired and supported by the experiences of an earlier Argentinian activist movement, that became influential across Latin America as well as in several European and Asian countries. Anita Chan, 'Coding Free Software, Coding Free States: Free Software Legislation and the Politics of Code in Peru', *Anthropological Quarterly*, 77, 3 (2004): 531–45. Chan details how the Peruvian FLOSS movement promoted quite different goals to the individualistic and libertarian ideals championed by many international proponents of FLOSS.

4 Chris Anderson, *The Long Tail* (London: Random House, 2006).

5 Tony Hall, speech at the Science Museum on the future vision of the BBC, 7 September 2015, www.bbc.co.uk/mediacentre/speeches/2015/tony-hall-distinctive-bbc.

6 Department for Culture, Media & Sport, *A BBC for the Future: A Broadcaster of Distinction* (London: DCMS, 2016), 66.

7 James Bennett, 'Create Public Service Algorithms', openDemocracy, 14 September 2015, www.opendemocracy.net/100ideasforthebbc/blog/2015/09/14/create-public-service-algorithms/. See also Bennett's chapter in this volume.

8 IRIS *plus* 2012–15, *Must-Carry: Renaissance or Reformation*, ed. Susanne Nikoltchev (Strasbourg: European Audiovisual Observatory, 2012), www.obs.coe.int/documents/205595/264629/Must+Carry+Report+%28Dec.+2015%29/bb229779-3fb2-488d-9c0e-d91e7d94b24d.

9 The arguments in this section derive largely from Tony Ageh's paper 'The BBC, the Licence Fee and the Digital Public Space', 3 March 2015, www.opendemocracy.net/ourbeeb/tony-ageh/bbc-licence-fee-and-digital-public-space. Tony Ageh was from 2003 to 2008 Controller, Internet and from 2008 to 2016 Controller of Archive Development at the BBC. As Controller, Internet he was responsible for the creation and delivery of the BBC iPlayer. He subsequently developed more ambitious goals for the BBC's digital activities, as set out in his paper. He left the BBC in early 2016 to take up the post of Chief Digital Officer of the New York Public Library. I am extremely grateful to Tony for his permission to recycle his analysis and arguments in this chapter. I am also indebted to Richard Paterson for making the connection.

10 Ibid.

11 Department for Culture, Media & Sport, *Broadcasting: Copy of Royal Charter for the Continuance of the British Broadcasting Corporation*, Cm 6925 (Norwich: The Stationery Office, 2006), 3.

12 Saskia Sassen, 'Digital Networks and the State: Some Governance Questions', *Theory, Culture and Society*, 17, 4 (2000): 19–33, 20.

13 Chan, 'Coding Free Software', 539–40.

24

Television in a Rapidly Changing World: Content, Platforms and Channels[1]

The Multichannel Revolution

Perhaps the single most striking change in television over the past generation has been the proliferation of channels made possible since the 1980s by the new technologies of cable, satellite and digital compression. The four-channel analog world of the 1980s has given way to a new digital landscape of hundreds of channels and the prospect of an online environment in which linear channels play a less significant role. This explosion of choice was facilitated by governments and regulators but it was consumer-led too; millions of households chose to pay for cable and satellite subscriptions, to adopt the free digital services Freeview and Freesat and to buy the Smart TV sets that 'liberate' them from the tyranny of the electronic programme guide. The process of digital switchover was completed by 2012 without any significant hitches or public resistance.

As a result of this transformation, the analog legacy channels' audience share has halved. In 1988, BBC1, BBC2, ITV and Channel 4 still accounted for 100% of viewing. Ten years later, with Channel 5 now launched as the fifth analog channel, their combined audience share had fallen to 86%. By 2014, they had just 51% of viewing between them.[2] 'Multichannel' services therefore now account for around half of all viewing, bringing new competition for advertising with them. ITV's main channel has been perhaps the most spectacular casualty, its share down from 44% in 1990 to just 15% in 2014.[3]

But the overall impact on the established broadcasters has not been as disastrous as sometimes predicted. They have retained their prominence, thanks to regulation that keeps them at the top of electronic programme guides. ITV may no longer dominate the landscape in the same way, but it remains the UK's most watched commercial channel and retains the commercial clout that comes with that. The old broadcasters have also adapted to the new world by developing new 'families' of channels. Taking those channels into account, the combined audience share of BBC, ITV, Channel 4 and Channel 5 still represents 72% of the total.[4] Of the 20 most viewed channels in 2014, 17 belonged to these four broadcasters, with the five analog legacy channels still the five most popular.[5]

Sky and the Rise of Pay-TV

The only true UK broadcasting powerhouse to arrive on the scene as a result of the multichannel revolution has been Sky. The main satellite TV distributor as well as the operator of a number of channels and a content producer, Sky is a player of real significance. Its reported revenues of £7.8 billion in 2015 were far greater than the BBC's income of £4.8 billion.[6]

So much of Sky's scale and success has been built on the back of its acquisition of sports rights, most importantly those to English Premier League football. It has been the main broadcaster of live Premier League football since the league's creation in 1992. Live football above all else has driven the creation of a pay-TV market in the UK.

Sky's original business model relied on people taking up satellite TV subscriptions to watch content they could not get elsewhere. It grew faster than the cable industry, which was dogged by poor customer service and wasted time and energy on debt-fuelled consolidation and internecine competition before finally coalescing under the Virgin Media brand. Between them, Sky and Virgin now account for just over half of households with digital TV, a proportion that has not changed much in recent years.[7]

As the internet took off, Sky readied itself for the emerging on-demand world, developing the pioneering Sky Plus personal video recorder, moving into broadband provision, and more recently launching the 'over-the-top' service Now TV. Broadband technology has allowed telecoms companies such as BT and TalkTalk to enter the pay-TV market alongside Sky and Virgin. Despite vigorous competition – particularly from BT, which has challenged Sky on the all-important terrain of football rights – Sky remains by far the biggest beast in pay-TV.

Sky's success has not been entirely down to sport – its movie channels, at least initially, helped to drive up subscriber numbers. It has a strong news channel, which is the BBC's main rival, and a well-regarded arts channel. The Sky One entertainment channel has invested strongly in production. In total its channels accounted for 8.2% of viewing in 2014.[8] But Sky has not played a part in the formal provision of public service broadcasting in the UK – nor has it been asked to. In fact, it has often had an antagonistic relationship with the older, more established broadcasters, particularly under the leadership of James Murdoch, who has now returned as the company's chairman. Its relationship with Rupert Murdoch's 21st Century Fox media empire, which currently controls 39% of Sky, lies behind this somewhat feisty anti-incumbent attitude.

Subscription revenues amounted to £6bn or 45% of overall TV industry revenues in 2014.[9] But despite the successful growth of pay-TV, the idea of free-to-air television has not been abandoned. Free, universal access to content is after all one of the cornerstones of the public service television model. Freeview, the terrestrial platform born out of the ashes of the failed ITV Digital, became a powerhouse brand that ultimately made the nationwide switch to digital possible. The very brand names Freeview and Freesat did much to cement the idea that at least some TV should remain free to air.

The Internet and the On-Demand Revolution

Alongside the multichannel revolution and the growth of the pay-TV market, the internet has become a central feature of everyday life and its potential as a mechanism for the delivery of the kind of audiovisual content that has historically been regarded as broadcast material is only starting to be realised. Over the past decade, broadband connections have facilitated the viewing of video content over the internet, while internet-enabled tablets and smartphones have allowed consumers to watch TV 'on the go'.

The statistics are striking: broadband take-up increased from just 31% to 80% between 2005 and 2015. Some 61% of adults now use the internet on their mobile phones, three times as many as in 2009.[10] More than half of adults using the internet say they use it to watch TV or videos, around two thirds of them doing so in the past week.[11]

This rapid adoption of new technology has led to a significant growth in on-demand viewing, both in the home and on the go. Broadcasters have both responded to and driven demand for such viewing, by streaming content as it is broadcast and by launching catch-up services. The most successful of these has been the BBC iPlayer, which has evolved since its launch in 2007 as a simple catch-up service to become a more extensive on-demand platform. Broadcasters have also started to make online-only content as well as putting some programming online first before broadcasting it conventionally at a later date. BBC Three's move online in 2016, while also a money-saving device, was a major step in this direction.

It is worth noting that the habit of watching TV programmes at the viewer's convenience, rather than when broadcast, predates the arrival of on-demand technology: video players have been a part of life for decades and time-shifted viewing through personal video recorders (PVRs) is a significant part of the picture today. DVD box-set viewing, which became a popular way for people to watch TV programmes at their leisure as vast libraries of content both old and new were made available for the first time, has now been superseded by catch-up services and 'over-the-top' online subscription services such as Netflix that have built on an existing appetite for convenient consumption.

The extent to which viewing habits have now shifted away from traditional broadcasting is hard to capture and leads to some strikingly different views about the pace of change. On one measure, only 69% of the total viewing of audiovisual material is through live TV.[12] On another, such linear viewing still accounted for 85% of long-form audiovisual viewing in 2014.[13] But even using the latter methodology, on-demand viewing (which includes catch-up services but not time-shifted viewing) is growing rapidly – from 2% in 2010 to 6% in 2014 – with internet-connected 'smart' TV sets and tablets driving growth.[14] Similarly, even where the precise figures differ, the trend nevertheless remains the same: while Deloitte's *Media Consumer* states that live TV viewing declined from 225 minutes a day in 2010 to 193 minutes in 2014, Thinkbox – using the same BARB source data – shows a slower decline, from 242 minutes to 221 minutes.[15] The key point is that both show that audiences are turning away not from television per se but from linear viewing and towards multi-platform consumption.

This is a widespread trend. Some 57% of adults surveyed in the second half of 2014 said they had accessed at least one on-demand service in the past 12 months, up from 27% in the first half of 2010.[16] The most popular service was the BBC iPlayer, used by 31% of people in 2014.[17] BBC figures show that requests for television programmes through the iPlayer have quadrupled from 722 million in 2009 to 2.87 billion in 2015.[18]

Alongside the catch-up services are the 'over-the-top' subscription services. Dominating this new space are the two US companies Netflix and Amazon, which now have significant ambitions in content production as well as distribution. The rapid success that Netflix in particular has enjoyed since it launched in the UK in 2012 is remarkable. It had nearly 7 million subscribers – some 30% of households – by the beginning of 2017, up from 2.8 million three years earlier, while Amazon Prime had 3.64 million.[19]

But it is not just Netflix and Amazon driving the growth in on-demand viewing. Audiovisual material is now available from myriad sources. Vloggers like PewDiePie with 56 million subscribers and Zoella with more than 12 million subscribers are evidence of the huge appetite for content produced a very long way from the studios of the public service television broadcasters. Newspaper websites are now able to produce video, and cultural institutions can also use the internet to film plays, events or exhibitions. Universities and other institutes of learning make lectures and seminars available online. New entrants in news provision are making a mark – the Vice website targeting a youth demographic, for example, has a digital audience of more than 5 million in the UK.[20] These efforts may not always look like high-quality broadcasting output (though that would be hard to argue in the case of Vice), but they are competing for the time and attention of TV viewers and, according to short-form video specialists Maker Studios, are drastically expanding the very concept of 'content' such that 'consumption can now range from a 6-second Vine to a 10-season Netflix binge marathon'.[21]

These changes appear far more dramatic when the changing consumption patterns of younger people are examined in detail. Reported TV viewing of children between 4 and 15 and adults between 16 and 34 declined by 30% between 2010 and 2015 as compared to the 10% drop across the whole audience.[22] Only 50% of 16–24-year-olds' total audiovisual consumption is through live TV, compared with 69% for all age groups[23] while two-thirds of their TV viewing is live as compared to 86% of those aged above 55.[24] Some 47% of them have an on-demand subscription in the home, against 26% for all age groups.[25] Only 10% of their viewing on Amazon and Netflix services is to BBC or ITV content.[26] They are also increasingly watching short-form content on sites such as YouTube that accounted for 8% of all their audiovisual viewing in 2014.[27] These changing patterns of consumption are not confined to under-25s: there is evidence that 25–34-year-olds and even 35–44-year-olds are also watching material in different ways.[28]

How fast these changes spread remains to be seen, and it is possible that younger people will adopt the habits of older generations as they age, perhaps preferring to watch live TV more as they go out less. But even if this happens – and there are strong reasons to doubt it – it is clear that the formal boundaries between broadcasting and the internet have already effectively collapsed. The trend towards on-demand viewing and the prospects of 'post-network television'[29] point in one direction; it's just a question of how fast the change occurs.

This does not, however, presage the imminent decline of television as a form of popular communication but rather the gradual supplementing of live television with more complex modes of consumption. Indeed, it would be a mistake to equate the appetite for short form video amongst younger audiences with a rejection of long form video when, in reality, those audiences are enjoying both. The increasing popularity of YouTube, as one source of video, 'no doubt poses a challenge for traditional broadcasters' argue Enders Analysis. 'But it is one that concerns the delivery of the content rather than the nature of the content itself – the production of which [comes] ... from a position of experience'.[30] Traditional content providers may have to up their game if they are to keep up with changing consumer preferences but they still retain brand familiarity, access to capital and a track record that suggests they are not likely to disappear anytime soon.

The Arrival of the Americans

As we have seen, the arrival of Netflix and Amazon is potentially of huge significance in disrupting the UK broadcasting sector. They are the biggest names at the moment; others are likely to enter the market, and they are most likely to be US companies. The giants of the technology sector – Google, Microsoft, Facebook, Apple – are all American. Channel 5 is now owned by the US media corporation Viacom, while it is often predicted that ITV will ultimately be bought by a US company. Many of the largest 'independent' production companies in the UK are now US-owned.

It is not parochialism to point this out. The preservation of a vibrant and dynamic British culture and industry, with all its national, regional and local variations, has long been one of the goals of public service television. The protection of UK-originated content and regional news is built into the quotas that are written into the broadcast licences of ITV, Channel 4 and Channel 5, for example.

At the same time, we have to recognise the appeal of much American content. It is many years since the *Financial Times*' television critic, Christopher Dunkley, warned of the dangers of 'wall to wall *Dallas*'.[31] Prime-time schedules are no longer reliant on US series being bought for transmission by UK networks and instead

high-quality long-form television drama has been one of the great cultural phenomena of the past 15 years, from *The Sopranos* to *Breaking Bad*. The availability of DVD
box-sets and the new culture of viewing them at leisure that has developed over the
past 15 years has enabled viewers to sample much more adventurous US-originated
content.

The online subscription services of Netflix and Amazon have followed that pattern. When Netflix and Amazon's customers were asked what programmes they
watched on these services, 49% mentioned US programmes and series, more than
the 37% who mentioned UK material. Some 31% said they watched the original programming now being produced by the distributors such as Netflix's *House of Cards* or
Amazon's *Transparent*.[32] British viewers' exposure to the highest quality US output
has, however, arguably undermined the distinctiveness and primacy of British content and raised questions about whether Britain's creative industry is really matching
the standards reached by the US.

What, Then, is Television Today?

Given all the changes as described above, it is necessary to consider what we even
mean by television today. This Chapter uses the term 'public service television' rather
than the more traditional 'public service broadcasting'. This formulation (which
excludes radio from the discussion) makes sense in a world of increasing on-demand
and time-shifted viewing. Broadcast channels remain with us, and account for a larger
share of viewing than is often appreciated, but the trend is clear: on-demand viewing
is growing and represents the future.

Television does not simply mean broadcast material, or even material that was
broadcast at some point and can also be accessed on an on-demand basis. For it to
have relevance in this rapidly evolving marketplace, the idea must also cover content
that is not necessarily broadcast in the traditional sense but is nevertheless produced
for dissemination to a wide audience, either for free or for payment. Our definition
of television in the UK would cover all professionally produced audiovisual content
intended for a UK audience of significant scale. This means that the services provided
by Netflix and Amazon as well as the digital channels run by Vice and YouTube need
to be part of the discussion.

Moreover, in a world of mixed media and hybrid audiovisual, text and graph-
ics, we must also address the extent to which 'television' should be extended to

include the wider range of online content found across the internet, which might include short form videos and also a range of written material. Here we believe that a pragmatic approach should be taken – in some genres (news, for example), mixed media of this sort are increasingly an essential part of what we think of as public service television. It would be wrong to curtail the expansion of PST into these areas if they are seen by users as the preferred means of accessing content. In others, such as drama, more traditional long form audiovisual content is likely to remain the primary focus.

The Problem of Defining Public Service Television in an On-Demand Age

While we can arrive at a definition of television, pinning down what public service television might mean today is a harder task. In Chapter 1 of the Puttnam Report, we looked at how the Communications Act, BBC charter, and the broadcast licences for ITV, Channel 4 and Channel 5 have given some sort of a definition of public service broadcasting. But it is not a clear-cut or sufficient definition, and it predates the recent changes in technology, the marketplace and consumption habits that we have outlined. By prioritising broadcasters and channels over programmes, it leaves anomalies. How do we define news and arts programmes on Sky, for example? What about original, high-quality drama or documentaries on Netflix or Amazon? What about video items on the *Guardian* website, the National Theatre or the Tate that we discuss in Chapter 25? Are none of these examples of public service television? If they are, do they deserve some form of subsidy too?

So clearly there is a problem defining public service television. The public are likewise not clear about what it is: research commissioned by Ofcom found serious gaps in public understanding. According to Ofcom, spontaneous awareness of public service broadcasting was low, and the public service broadcasters were losing some of their distinctiveness.[33] It also found that viewers were more likely to distinguish between good and bad programmes rather than public service and non-public service broadcasting.[34] Viewers increasingly think in terms of programmes, not providers, which is a problem given our habit of talking about public service broadcasters rather than public service programmes.[35] Yet understanding what public service television is (or is not) in a digital environment will be key if we are to enhance the possibilities for its survival and expansion. In that context, we propose to enlarge the definition of public service television to include all those channels, services and programmes that are subject to regulatory commitments to serve the public interest. PST, we wish to emphasise once more, is not a matter of pure technological or economic compulsion but a purposeful intervention designed to embed public service objectives inside a changing television environment.

Notes

1 Edited extract from Chapter 3 of the Puttnam Report, http://futureoftv.org.uk/wp-content/uploads/2016/06/FOTV-Report-Online-SP.pdf.

2 Ofcom, *Public Service Broadcasting in the Internet Age: Ofcom's Third Review of Public Service Broadcasting* (London: Ofcom, 2015), www.ofcom.org.uk/__data/assets/pdf_file/0025/63475/PSB-statement.pdf, 7.

3 Ofcom, *The Communications Market Report 2015* (London: Ofcom, 2015), www.ofcom.org.uk/__data/assets/pdf_file/0022/20668/cmr_uk_2015.pdf, 192.

4 Ofcom, *PSB in the Internet Age*, 7. Ofcom also reports that the share of viewing accounted for by BBC, ITV, Channel 4 and Channel 5's portfolio channels has risen from 14% in 2008 to 21% in 2014.

5 Ofcom, *CMR 2015*, 204.

6 Sky's revenues are for the UK and Ireland in the year to June 2015 and mostly derive from subscriptions. The BBC's income, quoted for the year to March 2015, is made up of £3.7 billion from the licence fee and £1.1 billion from BBC Worldwide. See Sky and BBC annual reports.

7 Ofcom, *CMR 2015*, 145. In 2014, they accounted for 51% of digital TV households.

8 Ibid., 202.

9 Ibid.,165.

10 Ibid., 340.

11 Ibid., 357.

12 Ofcom, *PSB in the Internet Age*, 19.

13 Ibid., 18.

14 Ibid.

15 Deloitte, *Media Consumer 2015: The Signal and the Noise*, 2015, www.deloitte.co.uk/mediaconsumer/assets/pdf/Deloitte_Media_Consumer_2015.pdf, 4; Thinkbox, *A Year in TV, Annual Review 2015* (London: Thinkbox 2015), 8.

16 Ofcom, *CMR 2015*, 52.

17 Ibid., 53.

18 BBC, *BBC iPlayer Monthly Performance Pack,* January 2016, http://downloads.bbc.co.uk/mediacentre/iplayer/iplayer-performance-jan16.pdf, figures extrapolated from slide 4.

19 Enders Analysis, *Netflix passes 100 million: buy more steak, get less sizzle*, 2017–066, 26 July 2017.

20 Ofcom, *CMR 2015*, 373.

21 Maker Studios, *The Shift Report: The Short-Form Revolution* (London: Maker Studios, 2015).

22 Enders Analysis, *Will the Young of Today ever Turn to Trad TV?* 2016-002, 15 January 2016.

23 Ofcom, *PSB in the Internet Age*, 19.

24 Enders Analysis, *Watching TV and Video in 2025*, 2015–096, 2015.

25 Ofcom, *PSB in the Internet Age*, 20.

26 Ibid., 20.

27 Ibid.

28 Ibid., 21.

29 Amanda Lotz, *The Television Will Be Revolutionized*, 2nd edition (New York: NYU Press, 2014).
30 Enders Analysis, *Does short form video affect long form content?* 12 May 2016.
31 Christopher Dunkley, *Television Today and Tomorrow: Wall to Wall Dallas?* (London: Penguin, 1985).
32 Ofcom, *CMR 2015*, 55.
33 Ipsos MORI, *An Investigation into Changing Audience needs in a Connected World*, 2014, www.ofcom.org.uk/__data/assets/pdf_file/0035/78659/psb-review-ipsos-mori.pdf, 7.
34 Ibid., 41.
35 Ibid., 41.

25

New Sources of Public Service Content[1]

Public service content is no longer confined to the traditional public service broadcasting system. The conventional definition of public service broadcasting, as set out by the 2003 Communications Act and understood by Ofcom, is everything produced by the BBC, and the programming undertaken by the main channels of ITV, Channel 4, and Channel 5 that fulfils the commitments of their broadcast licences.[2]

But there is now much audiovisual material being produced outside these parameters – either broadcast or made available online – that shares many of the traditional features and aims of public service television. Some of this is provided by the many commercial operators that broadcast on multichannel platforms, such as Sky or Discovery, as well as by Local TV services; some of it is offered by the new on-demand services such as Netflix and Amazon; while some of it is being produced online by arts and cultural organisations such as the Tate or the National Theatre, and by many other bodies besides. Here we suggest how some of these new forms of public service content could be strengthened through a specific public intervention.

Public Service Content Outside the Television World[3]

There has been a major shift in recent years in viewing habits, with more and more people watching material on demand, not just through catch-up services such as the BBC iPlayer but also online. Greater broadband speeds have facilitated the viewing of audiovisual material through an internet connection. At the same time, the technical and financial barriers to making such content have fallen. Anyone with a smartphone can make a video. Alongside the amateurs, all sorts of professional organisations have embarked on making content. Video production and programme-making skills are no longer the preserve of professional broadcasters or even of large production studios. Every newspaper, advertiser, campaigning group, agency, corporation and brand is now in the content creation game.

So too are the UK's many and diverse cultural institutions. Ranging from national organisations established in statute to diverse local, regional and charitable establishments, they could prove to be key contributors to a more plural, diverse and dynamic public service media landscape in the future. Many of these institutions, some of which long predate the broadcast era, exist to promote the kind of public service objectives that we have associated with British broadcasting since its emergence in the 1920s – stimulating knowledge and learning, reflecting UK cultural identity, and informing our understanding of the world. Many are active in genres that are currently perceived as at risk or failing in delivery on television – specialist factual, science, arts, children's content. We are not just talking about metropolitan or national organisations; the network of local and regional museums, art galleries and charities is far more widespread, diverse and connected to communities than the outposts of our public service broadcasters.[4]

The technological developments of the past decade or so have given these institutions new digital tools to reach out to the public, and some of them have done remarkable things with audiovisual productions. When Benedict Cumberbatch stepped on to the stage of the Barbican as Hamlet in October 2015, there was a global audience of 225,000 people in 25 countries, courtesy of the National Theatre's NT Live service.[5] Screenings of the play have gone on to make nearly £3 million for NT Live.[6] The Tate now produces its own films and shares them with third parties such as *The Guardian* and the BBC. Its film series TateShots generated 1.9 million views in YouTube in 2014/15. A 'live tour' of its 2014 Matisse exhibition that was broadcast in cinemas worldwide won a Royal Television Society award.[7]

In the past the distinction between television – narrative-driven, entertainment-focused, universally available – and these collection-based institutions, locked into their geographically static buildings, may have seemed absolute. But in the past 20 years the distinction has become far less clear. Take Tate, perhaps the most sophisticated and confident brand in the cultural sphere, with a clear, definable mission: to increase the public's understanding of art. This can be done through galleries and exhibitions, interpretation and education – but for 20 years now, core parts of Tate's intellectual endeavour have been delivered through digital media. Tate has developed a knowledge and skills base that combines editorial and curatorial excellence and digital knowhow to develop what is probably the strongest global cultural brand around contemporary art.

Our cultural institutions, both local and national, have deep specialist knowledge in areas that are core to public service content – whether they be science and technology, ecology and the natural world, cultural identity, history, or dramatic excellence. They also have the editorial knowledge, the assets, the audiences and the expertise to become significant public service content players in the digital world.

What they do not have, by and large, is the money to pursue this destiny. At the moment they operate on relatively modest budgets and are expected to generate much of their own revenue. Even our largest museums and galleries generally have operating revenues of below £100 million. The Tate, for example, had operating revenues of £92 million in 2014/15, of which only about a third was grant-in-aid.[8] As a performance company charging for tickets, the National Theatre generates an even higher proportion of its own revenues: out of its turnover of £118 million, only 15% (just under £18 million), comes from the Arts Council.[9]

None of these organisations has dedicated funding to support digital content creation or engagement beyond the pursuit of their overall public service mission. Whilst initiatives in the 2000s did attempt to support the digitisation of collections and to pilot new services,[10] the galleries, museums and national performing companies have largely had to use their core funding, topped up with bids to the likes of the Heritage Lottery Fund, to develop their digital offerings.

It seems highly likely that these organisations could do much more if they were released into the networked world with a fraction of the resources that we currently provide or safeguard for public service broadcasters. Our cultural institutions have shown they have the creative skills but that they are also in this for the long term. They have core missions that embody a commitment to specific areas of the public realm, with robust corporate governance and detailed statutory frameworks to back them up.[11]

One potential way of getting more from these institutions might be to get them to partner with public service broadcasters. However, the track record of such partnerships up to now has not been good. Cultural institutions talk of projects primarily conducted to broadcasters' priorities and timelines, their resources, knowledge and contacts being exploited, and their brand minimised. Contrast that experience to what the National Theatre has achieved by going it alone with NT Live. Instead of partnering with a broadcaster, the National Theatre has solved the problems of new video production, distribution, rights and business models on its own and is now generating income to return to the core business – £6 million last year, representing 5% of its revenues.[12] Following its own creative and business judgement, it has also become a lead partner and platform provider for other organisations – the record-breaking Cumberbatch *Hamlet* was not a National Theatre production, for example. It is hard to imagine it would have achieved this level of creative and business success if, seven years ago, it had looked to go into partnership for televising plays with the BBC or Channel 4.

Alongside the established cultural institutions, a huge amount of small-scale, grassroots content production is now taking place. While there are some initiatives, for example by Channel 4, to encourage some of this activity, we feel there needs to be a much larger support network and more significant funding to harness the creativity of new or marginalised voices who are squeezed out of the mainstream despite deserving wider attention.

A New Fund for Public Service Content

We believe that the time is ripe for making more of the public service content being developed outside the traditional broadcasting world – both by established institutions and at grassroots level – and to bring it more meaningfully within the sphere of television. The development of this content should not be regarded as a threat to the television model, whether through traditional linear broadcasting or by on-demand platforms, or as giving broadcasters an excuse to opt out of making programming in certain fields.

To take this step will require the injection of public money so that cultural institutions and other bodies from across the UK can bid to use such funds for making television. We suggest the updating of what is now a well-established idea: the creation of some kind of body that would distribute this public money – what has sometimes been called, perhaps unhelpfully, an Arts Council of the Airwaves. Variants of this idea have been proposed before,[13] but it may be that the right moment for it has finally arrived, now that the media landscape has been transformed by ubiquitous broadband, smartphones, and digital switchover.

The government's White Paper on the future of the BBC did in fact bring the idea back into play. The White Paper proposed a 'public service content fund' to operate as a three-year pilot (with grants first made in 2018/19), using money unallocated from the 2010 licence fee settlement. The proposal is somewhat sketchy but it is suggested that the scheme could fund children's programmes or content targeted at under-served audiences such as BAME groups or audiences in the nations and regions.[14]

We believe elements of this proposal make sense. But we do not believe that licence fee income (even if this, for now, is 'old' licence fee money rather than the top-slicing of new income) should be used to fund it: the licence fee should fund the BBC. We also believe that the proposed funding level of £20m a year is inadequate if a new fund is to make a meaningful contribution to the public service television landscape.

We propose a new service for digital innovation: it could be called, for example, the DIG (standing for Digital Innovations Grants). This initiative would be financed by a levy on the revenues of the largest digital intermediaries (notably Google and Facebook) and potentially other sources including the four dominant broadband internet service providers in the UK (BT, Sky, Virgin and Talk Talk). All of these companies derive a huge amount of value from the distribution of existing public service content and we feel that it would be entirely appropriate for them to make at least a small contribution to its continued existence. We estimate that a 1% levy on revenues generated within the UK would raise well in excess of £100 million a year, less than the annual budget of Channel 5 but more than that of BBC Three and BBC Four combined.

In the course of our Inquiry, we heard recommendations to consider levies of this kind. The National Union of Journalists, for example, argued in its submission to us that there was a need to consider new sources of funding including levies and tax breaks to raise additional money for public service content.[15] There is a long history of the use of levies – for example, on recording equipment and blank media – in the European communications industries.[16] More recently, we have seen a £50 million payment by Google to support the French culture industries as well as a new rule that forces video-on-demand operators to invest a proportion of their revenue in French cinema.[17] A report for the thinktank ResPublica suggested a levy on the revenue of large digital news intermediaries to support a fund aimed at sustaining new forms of public interest journalism.[18]

Furthermore, we believe that a levy would be popular with audiences. In a 2015 YouGov poll, commissioned by the Media Reform Coalition, 51% of respondents said that they would support a levy on the revenue of social media and pay TV companies to fund new providers of investigative and local journalism, with only 9% disagreeing.[19] We think that the support would be even higher with a remit to provide a wider array of public service content.

Money awarded by the DIG fund would be disbursed via a new independent public media trust with a clear set of funding criteria, transparent procedures and an accountable system of appointments, as per our proposals for the BBC unitary board. The trust would also recognise the need for meaningful representation from all the nations of the UK.

The DIG would be open to any cultural institutions or bodies that wanted to produce public service audiovisual content and could provide evidence of their creative purpose and expertise. These applicants should not be wholly commercial operations; rather, they should have demonstrable public service objectives and purposes. It should not be for existing commercial broadcasters or production companies to subsidise their content production. The funding could, however, be used to work with partners of any kind, and these might include broadcasters or producers.

In awarding grants, the fund would be mindful of the kind of programming that is not appearing on established channels or is under threat. It could fund local and investigative journalism, for instance, or education, science, history and other specialist factual content. It should look to innovation in form and content, to adopt a phrase from the original remit of Channel 4. In fact, we believe that this intervention could provide something of the energising quality that Channel 4's launch gave the broadcasting world more than 30 years ago. This would be a Channel 4 moment geared to digital convergence and the networked world of today.

It is crucial that all of the content created with DIG funding is made widely available and easily discoverable on all interfaces. Any organisation applying to the DIG would need to provide a distribution and access plan as part of its application for funding, and this would be treated with as much importance as the content of the proposal. We do not believe that DIG content should be tied to a particular platform, while developing a standalone app and brand implies a big overhead in technology and marketing. Applicants for funding may already have their own channels (and brands) with significant audience reach and traction, so DIG funding should not preclude them from strengthening their own public service objectives.

We propose, therefore, that the DIG would create partnerships and framework agreements with the public service broadcasters and other platform owners to promote and distribute DIG-funded content with appropriate branding and acknowledgement. At the heart of this arrangement would be distribution agreements with the BBC and Channel 4 for access to and promotion on the BBC iPlayer and All 4 platforms, which would detail the appropriate editorial presentation and curation of DIG-funded content. The DIG would be expected to make other agreements with other partners that would maximise the prominence, findability and reach of the content it funded.

DIG funding would not be limited solely to linear video content and would include other digital content, applications and mobile and online experiences that met its objectives. Applicants would be expected to use their own digital channels and those of partners to maximise prominence and access to this content.

Qualifying applicants for DIG funding would retain all the intellectual property of their output and retain editorial and contextual control of the content once funded. Applicants would be expected to hold discussions with distribution and funding partners prior to making their application to create both a funding proposal and a distribution and access plan. The DIG would not necessarily be the sole funder, nor would distribution partners be limited to those with which the DIG has a framework agreement. We believe that the work of such a fund would help to transform and revitalise the relevance of public service content for UK audiences.

Notes

1 Edited extract from Chapter 7 of the Puttnam Report, http://futureoftv.org.uk/wp-content/uploads/2016/06/FOTV-Report-Online-SP.pdf.
2 Communications Act 2003, section 264 (11), www.legislation.gov.uk/ukpga/2003/21/pdfs/ukpga_20030021_en.pdf. S4C is also a public service broadcaster.
3 This section draws heavily on Andrew Chitty's paper, 'Beyond Broadcasting: Public Service Content in a Networked World', delivered at the Inquiry's 'Concepts of Public Service' event, British Academy, 3 March 2016.
4 Given their reliance on local authority support, this rich regional landscape of local and regional cultural intuitions is also far more at risk from recent structural changes in public funding, with institutions from large to small at risk of closure or radical reductions in their remit. See *Museum Journal*, 'Cuts Put Regional Museums at Risk', 16 February 2016. Surely this is an argument for diversifying funding rather than continuing to superserve our public service broadcasting system.
5 Rebecca Hawkes, 'Live Broadcast of Benedict Cumberbatch's Hamlet watched by 225,000 people', *Telegraph*, 21 October 2015, www.telegraph.co.uk/theatre/what-to-see/benedict-cumberbatch-hamlet-live/.
6 David Hutchison, 'Benedict Cumberbatch Hamlet takes £3m at NT Live box office', *The Stage*, 9 December 2015, www.thestage.co.uk/news/2015/benedict-cumberbatch-hamlet-takes-3m-at-nt-live-box-office/.
7 *Tate Report 2014/15*, www.tate.org.uk/download/file/fid/56701, 33.
8 Ibid., 89.
9 *National Theatre Annual Review 2014–2015*, http://review.nationaltheatre.org.uk/2014–15.
10 The £50m New Opportunities Fund NOF Digitise programme was launched in 1999 to support the creation of content and the digitisation of collections in the cultural sector. The DCMS's £15m Culture Online programme, which ran from 2002 to 2008, funded new digital services, bringing cultural institutions together with digital media producers. The Treasury has funded discrete initiatives on an invest-to-save basis such as The National Museums Online Learning Project.

11 Primarily the National Heritage Acts of 1980, 1983, 1997 and 2002, and the 1992 Museums and Galleries Act, which set out the governance arrangements, public duties, and the statutory and legal obligations of the museums and galleries applying to artefacts in their collections, and how these may be acquired, loaned and disposed of on behalf of the nation.

12 *National Theatre Annual Review 2014–2015.*

13 In 2004, an independent report commissioned by the Conservative party suggested that a Public Broadcasting Authority (PBA), funded by the Treasury and accountable to Ofcom, should take responsibility for delivery of all public service content, with all broadcasters and producers able to submit bids. *Beyond the Charter: the BBC After 2006* (Broadcasting Policy Group, 2004). Ofcom floated the idea of a Public Service Publisher (PSP) in its 2005 review of public service broadcasting. Andrew Chitty and Anthony Lilley then explored this idea, suggesting a new institution that would ensure the delivery of public service content in the digital age. They proposed a PSP with initial funding of between £50m and £100m, which would be a commissioner of content and could work with a diverse range of suppliers and distribution partners. Ofcom, *A New Approach to Public Service Content in the Digital Media Age – the potential role of the Public Service Publisher,* 2007. The Labour government's 2009 *Digital Britain* report then suggested using a portion of the TV licence fee to allow contestable funding, allocated by an arm's-length body, for public service content where there were gaps in provision. Department for Business, Innovation and Skills, Department for Culture, Media & Sport, *Digital Britain,* 2009.

14 Department for Culture, Media & Sport, *A BBC for the Future: A Broadcaster of Distinction* (London: TSO, 2016), 71–72.

15 National Union of Journalists, submission to the Inquiry.

16 See Institute for Public Policy Research, *Mind the Funding Gap: The Potential of Industry Levies for Continued Funding of Public Service Broadcasting,* 2009, www.ippr.org/files/images/media/files/publication/2011/05/mind_the_funding_gap_1689.pdf?noredirect=1.

17 See IHS Technology, 'France introduces new tax on VoD operators based abroad', September 2014, https://technology.ihs.com/511104/france-introduces-new-tax-on-vod-operators-based-abroad.

18 Justin Schlosberg, *The Mission of Media in an Age of Monopoly,* ResPublica, 2016, www.respublica.org.uk/wp-content/uploads/2016/05/The-Mission-of-Media.pdf.

19 Media Reform Coalition, 'Poll shows strong support for action on media ownership', April 1, 2015, www.mediareform.org.uk/get-involved/poll-shows-strong-support-for-action-on-media-ownership.

26

Designing a New Model of Public Service Television (PST)

Robin Foster[1]

[...] I believe that some of the main building blocks of a new model for Public Service Television (PST) for the future can be identified.

First, although some suggest otherwise, there is still a significant future role for PST. A strong case can be made for a substantial, not just a marginal, intervention in the market. And that intervention should include content across all the purposes of PST identified in this Chapter: information, knowledge, and culture. Without PST investment, there would be fewer UK programmes available, and arguably less editorial innovation and risk taking. Shared experiences should continue to be an important part of PST, via the broadcast of major events but also through the creation of landmark popular programming.

However, reaffirmation of the need for a broad range of public service content should not be seen as underwriting ever-rising funding or as a licence for PS providers to produce just any type of content to attract viewers. While the case for PST's central role in the provision of impartial, independent and in-depth journalism is strong, PST news output will only be of value to audiences if it changes to reflect the opportunities presented by new media to better serve its users. While knowledge building remains a key role, PST must adapt to reflect the new market environment in which it operates, working with the many other expert resources available online. While drama, comedy and entertainment should remain part of the PST mix, there needs to be a renewed search for ambition and distinctiveness – not just across any particular service, but for each piece of content commissioned.

PST's future involvement in some types of content should be scrutinised carefully – for example, questions could be asked about the justification for PST investment in some of the more derivative types of lifestyle and light entertainment programming or online content. And programme volumes in some areas could be reduced, reflecting increased availability of high quality content elsewhere.

While long-form TV programming will remain at the heart of PST, whether on linear channels or (see below) on-demand, the concept of 'television' needs to be broadened to reflect new opportunities presented by digital media. TV news already benefits from the increased convenience and depth offered by online. Having invested in public service newsgathering, it is in the public interest to ensure that audiences can access that resource via a range of different electronic media. Likewise, other genres can be enhanced by an extra online dimension and, in some cases, online will largely replace conventional broadcast TV. PST purposes will endure, but the precise format and nature of content should be flexible enough to change over time to meet audience expectations.

For long-form programming, PST should pro-actively rebalance its portfolio of services away from linear broadcasting channels to on-demand, leading audience behaviour not just responding to it. The advantages of on-demand will include:

- A longer shelf life for programmes which increases the chances of each piece of content being watched
- Improved reach among those audiences who are turning away from linear channels
- Potential to unlock access to the rich and varied programme archive
- Cost-effectiveness as, freed from the demands of a 24 hour schedule, less 'filler' content needs to be made.

Quite soon, the ideal PST portfolio might well consist of one or at most two 'premier' broadcast channels alongside a widening on-demand proposition. The main channels would be the home of live TV and appointment to view programming, while playing a key role in promoting other services and launching new programming.

In parallel, key PST services should be designed to work well with new devices such as smartphones and tablets. It would be anachronistic to restrict PST to conventional broadcast delivery when the audiences who pay for it demand access via new platforms. Universality, in this world, should conceptually encompass platforms which are or seem likely to become mainstream methods of consumption, although the marginal benefits of extending access to such platforms need to be balanced against the costs of so doing.

In this new model, should we focus on the BBC, or encourage a new more plural system, perhaps through some form of contestable funding? Although contestable funding has many attractions, including testing the market for innovation and efficiency, it also faces significant practical problems in implementation, well-rehearsed elsewhere. At a time when PST funding is under pressure, and the commercial market is volatile, it would be counter-productive to tear up the current system completely and start again. A better approach would be to re-cast the way the BBC operates and is held to account, with more internal plurality of commissioning and production, and a greater diversity of programming sources used.

Over the next decade and beyond one might envisage the BBC as a new type of PST institution which is more open, diverse, and devolved in its approach to commissioning, production and distribution, and one which engages more actively and openly with content producers whoever they are – individuals, other institutions or commercial suppliers. Rather than simply commissioning individual programmes or series from external suppliers, this BBC might contract a completely new service from an external provider. Instead of one centralised editorial function for news, a number of independent and diverse news centres might be established to introduce more internal plurality. Local online services could be tendered from other local news sources, rather than set up inside the BBC – and so on.

In parallel with this development, the BBC would be asked to place more emphasis on expert curation of diverse content sources. Audiences increasingly need help to find and navigate their way to interesting content. This is particularly the case for on-demand programming and content on the internet. It is a non-trivial task to do this well, especially in a world where search and sharing are dominated by major US corporations like Google and Facebook, backed by huge investment and R&D budgets. If it is to be of value, this almost certainly requires special executive commitment and substantial new investment to make it happen. Government can help, too, by ensuring that the regulatory framework is updated to secure continuing prominence for PST content on major on-demand gateways (not just the main broadcast EPGs).

Given the risk that audiences increasingly lose touch with PST, another key building block should be to increase the connection between licence payers and the BBC, with the aim of enhancing a sense of real public ownership of PST and its accountability to audiences. At present, the licence fee is in effect a tax paid by anyone owning a TV receiver. In future, it would make more sense to link the payment explicitly to the provision of BBC services, and use the licence fee contract to build a mutually reinforcing relationship between the BBC and its users. Many commercial companies now encourage their customers to join loyalty schemes which provide benefits to users in return for frequent purchases and information given to the company.

Likewise, many charities operate like membership clubs, in which donors are made to feel part of the organisation and have a say in its operations (through annual meetings, voting rights etc.).

There is huge potential for the BBC to borrow the best of these ideas and create a membership or even shareholding scheme for all licence payers, which would ideally help create a closer relationship between the institution and its beneficiaries. Rather than inventing another version of the BBC Trust to 'represent' the licence payer, this would have the effect of directly involving licence payers without an intermediary appointed from among the ranks of the great and the good.

Based on the admittedly impressionistic analysis of the previous section, there seems little evidence that PST in the UK is significantly under-funded at present. In any event, whatever the real funding needs for PST, given the likely economic outlook for the next decade, uncertainties about public support for the licence fee, and the arguments over decriminalisation, it seems unlikely that there will be much potential in future for any significant real increase in the amount of public funding available for PST beyond the current settlement.

For this reason, and also because it is in many ways unhealthy for an institution to rely solely on guaranteed public funding, there is a good case for introducing some elements of voluntary funding into the mix over the next decade. Alongside the core licence fee, users of some of the BBC's peripheral services could be expected to pay for access to those services. For example, it would be possible for access to the iPlayer via mobile devices and PCs to be encrypted, and made available only on payment of a small annual charge. All BBC content would remain universally available, free to air, on the broadcast channels, but added convenience would be available for a modest fee. Alternatively, any BBC membership scheme could have different levels attached to it – again with a comprehensive basic level, but some higher levels for enhanced services.

The trade-off obviously is between creating some financial upside for the BBC, and retaining absolute universality for all. It does not seem unrealistic for such choices to be made in the interest of enhancing overall investment in content while retaining an affordable core fee.

Last but not least, the importance of a competitive UK commercial sector must be recognised. The focus of this Chapter has been on PST provision, and largely on publicly funded provision. However, UK PST has only been so effective to date because it has operated successfully in a wider commercial market (part of which was also regulated). The obligations imposed on the commercial PST sector are now more limited, than before. Existing commercial PSBs like ITV and Five now have a key role to play in helping drive commercial market developments rather than in the delivery of narrowly defined public service goals, although their significance as alternative news providers should not be ignored. More widely, open markets, with their decentralised decision-making, free exchange, scope for trial and error, and speedy ability to exploit technological change, will in future have a key role to play in delivering high quality programming to audiences and in doing so supplementing the effects of PST investment. [...]

Note

1 Robin Foster is an adviser on strategy, policy and regulation in the media and communications sectors and was a member of the Advisory Committee to the Puttnam Inquiry. This is an edited extract from his submission to the Inquiry, http://futureoftv.org.uk/wp-content/uploads/2015/11/Robin-Foster.pdf.

27

Public Service Broadcasting as a Digital Commons

Graham Murdock[1]

A swelling chorus of commentary claims that because in the age of the smartphone and the tablet anyone can access whatever they want, whenever they want, there is no longer any need for publicly funded institutions that offer a comprehensive service. The version of this argument put by Martin le Jeune, former head of public affairs at Sky, is typical.

Judged from the point of view of the *consumer* ... broadcasting is in a very healthy state. There is a good deal more choice for people; they have more ways to access good content ... *If the market is providing more, the state (through direct and indirect intervention) could and should do less.*[2] [italics in the original]

In common with a number of critics he sees public service organisations continuing to have a role but a much more restricted one:

tightly focused on delivering what the market cannot do, or does only to a limited extent. That might indicate a smaller but rather more intellectually distinguished corporation: impartial news and current affairs, factual and documentary programming, children's television, classical music, speech radio – and little more.[3]

This projected future enjoys continuing currency. It features prominently in the consultative green paper issued by the British government in July 2015 inviting responses to a series of questions over the BBC's future role and organisation, in the lead up to the renewal of its governing Charter. Among the key questions put is whether '[g]iven the vast choice that audiences now have there is an argument that the BBC might become more focused on a narrower, core set of services.'[4]

There is no matching question asking for comments on possible directions for further expansion. In a communications environment increasingly organised around digital networks, however, there is a compelling case for extending the BBC's public service remit. There are three reasons for this.

Firstly, successive cuts to public expenditure have seen a major contraction in the public information and cultural facilities previously available in local communities. Public recreational spaces have been sold. Libraries and museums have closed, reduced their opening hours or have only been kept open by volunteers. These cuts render the maintenance of PSB as a comprehensive cultural and informational resource open to all and free at the point of use more essential than ever.

Secondly, this is particularly true of households on low incomes. Despite repeated claims that smartphones and tablets have brought the internet within the reach of 'everyone', research reveals persistent patterns of exclusion by age and class. Recent British figures show that the elderly and the poor are least likely to own a tablet or smartphone. In 2015, 90% of young people aged 16–24 owned a smartphone compared to only 18% of those aged 65 and over.[5] At a time of widening income gaps and cuts in welfare budgets they are also the group most likely to have difficulty meeting the subscription costs for commercial cable and satellite services. For them, PSB is likely to remain their major, and for some, their only, point of access to a diverse range of cultural and information resources which suggests that maintaining a comprehensive, publicly funded service, free at the point of use, remains a policy priority.

Thirdly, users accessing commercially provided 'free' digital facilities now encounter a system where the most popular online activities are dominated by a handful of mega corporations – Google, Facebook, Amazon and Apple – all based outside the UK and generating profits by harvesting and selling users' personal data. In a speech in 2016, the former president of the European parliament, Martin Schulz, characterised the power of these digital giants as totalitarian, arguing that the decisions now being taken behind closed doors were constructing a future without public consultation in which corporate priorities take primacy over the public interest.

Facebook, Google, Alibaba, Amazon: these companies must not be allowed to shape the new world order. They have no mandate to do so! It is and must remain the proper task of the democratically elected representatives of the people to agree on rules and enshrine them in laws.[6]

In the UK the BBC offers the only effective institutional base for a comprehensive alternative to this corporate annexation of the internet. It currently operates one of the UK's most popular web sites, with a unique audience of 40 million, placing it third behind Google (with 46 million) and Facebook (with 41 million). Sky, the only other UK broadcaster to make the top ten, attracted only 28 million.[7]

I have previously argued that PSB institutions should take full advantage of the internet's networking and participatory potentials to become pivotal hubs in the public provision of online resources.[8] In the decade since then, this idea has steadily gained momentum. [...]A variant of this idea has now been incorporated into the Corporation's formal policy proposals outlined into its 2015 manifesto for change, *British, Bold and Creative*. This imagines an 'Ideas Service' with the BBC providing a platform for collaboration that

would bring together what the BBC does across arts, culture, science, history and ideas and add to it work done by many of this country's most respected arts, culture and intellectual institutions ... for curious audiences around the world, the BBC would create and manage an online platform that, working with partners, would provide the gold standard in accuracy, breadth, depth, debate and revelation. It would offer audiences the thrill of discovery and the reassurance of reliability.[9]

This initiative would aggregate content from multiple sources, working across broadcast and on line, providing audiences with content to share, curate and mutate, and encourage participation in citizen science and other collective projects. As the document concedes, however, implementing this vision remains a work in progress, an ambition rather than an accomplishment.

The labels may be different – 'digital commons', 'public space', 'ideas service' – but they are informed by the same basic ambition of deploying the centrality of broadcasting in everyday life to construct and coordinate a public digital network that reinvents the cultural commons for contemporary conditions, grounded in the core commons values of shared access and collaborative activity.

This aim has animated a range of recent BBC initiatives, from the *Listening Project*, in partnership with the British Library, to the collaboration with the LSE on a major survey of contemporary social class and the recent *Global Philosopher* exercise in generating transnational debate on public issues. But a comprehensive effort to build a digital commons that utilises the full range of platforms – broadcast, podcast, website – needs to tackle a series of issues that have not so far been given the attention they require.

Infrastructures

Given the escalating climate crisis it is imperative that any plan to utilise digital systems more extensively addresses the ecological impact of the infrastructures and machineries involved. Cloud storage, for example, consumes very significant amounts of energy.

Proposal: The BBC's purchasing polices for operating equipment and infrastructure should impose stringent requirements on suppliers to meet specified environmental thresholds on the procurement of raw materials and the ecological impact of production practices. Subsequent use should also be subject to strict rules on energy consumption and disposal.

Proposal: Suppliers should also be subject to strict requirements on conditions of work and minimum levels of pay at every stage of the production and distribution process.

Software

Proposal: The BBC should support the open source movement by using non-commercial operating systems and software wherever possible.

Proposal: The Corporation should take the lead in developing a public search engine as an alternative to commercial search engines, allowing users to locate material according to its veracity and social value rather than its popularity.

Participation

A properly inclusive digital commons needs to mobilise participation from the widest possible range of sources. It reaches beyond the major public cultural institutions – libraries, museums, theatres, archives, universities, concert halls – to include the dense networks of voluntary community and freelance initiatives.

Proposal: That the commercial internet companies be charged a fee for their proprietary use of users' personal data and that the money raised be placed in a fund for the production of new digital cultural resources to be added to and accessed through the broadcast commons. These might include: subsidies to local associations wanting to digitalise their archives; grants to teachers developing new educational materials; support for investigative research on key public issues; support for crowdsourced proposals in areas of citizen science.

Internationalisation

Public cultural institutions across the world are now in the process of digitalising their holdings offering an unprecedented opportunity to construct a global online resource. The *Europeana* project, although still in its early stages, demonstrates the immense gains from transnational co-operation. At a time of exacerbated divisions and animosities, facilitating access to materials that illuminate events and situations from within contrasting experiences and perspectives is a priority.

Proposal: The BBC should take the leading role in developing networking arrangements with cultural institutions outside the UK and in enabling users to search for and find relevant material across the full range of available international sources through the public search engine.

[...] For reasons I have outlined, the BBC is the institution best placed to coordinate and build on this ambition, incorporating the 'public' as active contributors as well as consumers.

Notes

1 Graham Murdock is Professor of Culture and Economy at the University of Loughborough. This is an edited extract from his submission to the Puttnam Inquiry, http://futureoftv.org.uk/wp-content/uploads/2016/04/Graham-Murdock.pdf.
2 Martin Le Jeune, *To Inform, Educate and Entertain? British Broadcasting in the Twenty First Century* (London: Centre for Policy Studies, 2009), 3.
3 Le Jeune, *To Inform*, 25.
4 Department for Culture, Media & Sport, *BBC Charter Review: Public Consultation* (London: DCMS, 2015), www.gov.uk/government/uploads/system/uploads/attachment_data/file/445704/BBC_Charter_Review_Consultation_WEB.pdf, 23.
5 Ofcom, *The Communications Market Report* (London: Ofcom, 2015), www.ofcom.org.uk/__data/assets/pdf_file/0022/20668/cmr_uk_2015.pdf, 65.
6 Martin Schulz, keynote speech at #CPDP2016 on 'Technological Totalitarianism, Politics and Democracy', 20 January 2016, www.europarl.europa.eu/former_ep_presidents/president-schulz-2014–2016/en/press-room/keynote_speech_at__cpdp2016_on_technological__totalitarianism__politics_and_democracy.html.
7 Ofcom, *Communications Market Report*, 358.
8 Graham Murdock, 'Building the Digital Commons: Public Broadcasting in the Age of the Internet', in *Cultural Dilemmas in Public Service Broadcasting*, ed. Greg Lowe and Per Jauert, 213–30 (Goteborg: Nordicom, 2005).
9 BBC, *British, Bold, Creative* (London: BBC, 2015), https://downloads.bbc.co.uk/aboutthebbc/reports/pdf/futureofthebbc2015.pdf, 70.

28

'Public Service' in a Globalised Digital Landscape

Ingrid Volkmer[1]

National public service media in Britain as well as in other European countries are in the process of transformation. Current industry debates and scholarly approaches are focusing on the complex practicalities of the convergence, from the 'logic' of broadcasting to the 'logic' of the dynamics of the advanced digital ecology. Of course there are nuances within both discussions; however, overall, it can be argued that both see the transformation as mainly associated with new strategic imperatives such as the production of multi-screen content formats, the embeddedness of interactive content components, the creation of sites for non linear 'catch up' content archives (keyword: 'iPlayer') and linkages with shared platforms of social mediascapes.

It is also argued that in order to maintain a national centrality of public service media, innovative combinations of 'linear' and 'online only' content genres are required to target the emerging divide of generational specific communication practices. While these are important issues in the lens of public service broadcasters, we need to begin the debate of 'bigger' questions to identify the requirements of national public service media in the new discursive scopes of public 'civic' communication. In other words, it is necessary to begin to assess public service as no longer being only in the normative national parameter of territorial 'boundedness', but as a much needed civic space within today's sphere of globalised public communication.

The term 'public service' originates from the time of terrestrial national broadcasting of the early radio days in the 1920s. The national boundedness is to a large extent related to the limitations of antenna reach in the early broadcasting era. In addition, the notion of 'public' service is also rooted in the Habermasian public sphere model of deliberation among citizens as the key component of national public spheres. However, both paradigms are – strictly speaking – no longer sufficient for assessing the public 'service' in today's non-national, non-territorial connected publics. We live in the age of multi-directional discourse spheres of globalised 'threads' that are dis-embedded from territorial 'boundedness'. Furthermore, our communicative world is no longer divided into 'domestic' and 'foreign' communication, or a sphere of inter- or even trans-national communicative 'extensions'. [...] Although nation states will not disappear, national public communicative space is – whether we like it or not – already increasingly disembedded from national territory, seamlessly streamed between servers and screens, shared by peer-to-peer networks, especially among young gener-ations. They no longer understand communicative spheres as territorial but rather as subjective spatial configurations, as their personalised micro-network. We also need to realise that this communication sphere seamlessly amalgamates modern nation states as well as other society types, involving so called 'failed' states and authoritar-ian states. Debates about public service media have to acknowledge these fundamen-tally shifting axes of communication flows – where national public service is only one 'node' among many – to identify a new role for public service communication in such a sphere.

However, it is surprising that the severe paradigmatic consequences of these spheres of civic communication are rarely surfacing in debates of public service media. In fact, globalised public spheres have rarely been addressed in debates of the public remit. For example, even satellite communication, which can be considered as a first important phase of the emerging globalised communication landscape which targeted specifically European countries in the early nineties has – also in scholarly debates – not been incorporated into the paradigmatic debate of public service media in the 1990s. Instead, debates of satellite communication have mainly centred upon the launch of public service satellite channels, such as Euronews and ARTE (a bilateral German/French channel). However, the enlarged communication sphere enabled by satellite footprints has not led to a revision of public service terrains. It is interesting to note that satellite communication remained on the periphery in the assessment of national public spheres despite the fact that thousands of radio and television chan-nels in multiple languages, ranging from Ukrainian, Russian, Portuguese, French, Arabic, to Chinese, Albanian, Croatian, Korean and English, simultaneously acces-sible in Europe as well as North Africa and the Middle East, have created a shared communication universe across these regions. Despite the reality of multi-cultural societies in European countries, and despite such a new dynamic of civic communi-cation, public service media remained to a large degree embedded in a normatively defined national public sphere.

The need to at least question the normative alignment to national imperatives is now even more important. At a time where commercial corporations such as Google and Facebook constitute worldwide monopolies as new types of content provider and producers of new civic communication landscapes, we need to more fundamentally debate new terrains of 'public service'. These multilevel networks no longer operate in the realm of content and 'information' but – one could argue – provide 'public service' knowledge in completely new areas – from 'web search' to virtual libraries to new areas of public service, such as navigation.

In addition, today's transnational terrain public 'reasoning' is situated within, and magnified through, a transnationally available spectrum of choices, loyalties and political alliances. Not only is it possible to engage with digital activism from almost anywhere with internet access but this spectrum has become more 'horizontally' subtle: I can live in Australia, vote in Germany, follow news resources from the US by the minute, watch streaming television from Kenya and engage in 'live' debates about saving the Amazon rainforest with NGOs in South America. These are the new geographies of public 'horizons' which are – and this is important to realise – no longer central to the democratic nation state and also no longer central to other societies! It is a shift towards a subjective axis, determining and selecting engagement in a globalised interdependent public debate of chosen networked formations which has implications for deliberation and legitimacy – again – in a geographically 'horizontal' spectrum. In a way it is the new calibration of 'polis' and 'demos'; my vote contributes to the election outcome in Germany, I take on roles in climate change communities in Australia which are no longer informed by local knowledge or the climate change agenda of national media but rather by subjective public horizons. [...]

Based on this discussion, it is not surprising that the BBC's public remit is still embedded in a bounded conception of the nation. For example, one aim of the BBC's remit is to sustain citizenship and civil society. However, given today's networked structures of communication, citizenship is also perceived as global citizenship, for example, vis-a-vis worldwide risks and conflicts and – in consequence – relates not only to national responsibilities but rather to new responsibilities in a global civil society. The remit's aim to represent 'the UK, its nations, regions and communities' is also challenged as communication no longer relates to 'bounded' content but is embedded in individually produced networks which are no longer 'bounded' but 'fluid'. Collective identity is no longer understood as a 'representation' but is 'shared' within subjectively chosen communities. [...]

In contexts of public service media, we might also have to move away from the media/broadcasting, online/offline duality towards a centrality of civic discourse spheres. [...] Within such a model, public service could provide a civic topography in the larger digital ecology, provide public agency, such as spaces for 'actors', spaces as a critical 'reflector' of discourse or of linking debates as 'interlocutor'.[2] Although this might seem to be a glimpse into the future, it is important to begin the debate.

Notes

1 Ingrid Volkmer is an Associate Professor in the School of Culture and Communication at the University of Melbourne. This is an edited extract from her submission to the Puttnam Inquiry, http://futureoftv. org.uk/wp-content/uploads/2016/05/Ingrid-Volkmer.pdf.
2 Ingrid Volkmer, *The Global Public Sphere. Public Communication in the Age of Reflective Interdependence* (Cambridge: Polity, 2014).

29

Video-on-Demand as Public Service Television

Catherine Johnson[1]

Public service broadcasters (PSBs) are operating in a media landscape in which the increased convergence of the internet and television has blurred the distinctions between broadcast and on-demand TV.[2] In this environment, people can switch easily between live/linear and on-demand viewing within the same interface and access television content through a range of online and internet-connected services. In the UK market there are a number of suppliers providing video-on-demand (VOD) services operating with different financial models (Table 29.1). These services are typically offered across a range of platforms, including PCs and laptops, tablets, smartphones, games consoles, set-top boxes and connected televisions. With increased uptake of connected televisions, tablets, smartphones and superfast broadband, on-demand services are set to become a dominant means for accessing television.

There is a strong argument that as traditional broadcast and internet services merge the case for PSB becomes stronger. In a fully commercial media landscape, economically disadvantaged audiences are under-served because they are less able to pay for services and less attractive to advertisers. By providing free-to-air content that is not determined by ability to pay or attractiveness to advertisers, PSBs ensure that VOD serves all of the UK public with high quality programming that entertains, educates and informs. In particular, UK public service broadcasters provide access to a mixed diet of programming online that includes news and current affairs, genres that are largely absent on subscription VOD services, such as Netflix and Amazon. Despite the increased choice of media providers and services, the UK public increasingly value and trust PSBs, in part because they are mandated by regulation to serve the public's needs over other interests.[3]

Table 29.1

Based on top online VOD services listed in Decipher mediabug – Wave 4 report on claimed use of selected online VOD services in the UK for 2013–2014.[4]

Free	Part of a television subscription package	Direct subscription	Pay-per-view
BBC iPlayer	Sky Go	Netflix	iTunes
ITV Player	Virgin TV Anywhere	Amazon Prime	Google Play
All 4		Now TV	Sky Store
Demand 5			TalkTalk TV Store
UKTV Play			

Increased choice and fragmentation has enabled VOD providers to serve the needs of a wide range of niche audiences, often with large catalogues of content. However, this does not mean that new commercial VODs serve the needs of all audiences equally. For example, in the US Netflix has been criticised for a lack of black TV shows and potential racial bias in its recommendation algorithm.[5] Research indicates that rather than broadening people's diet of opinions, ideas and debate about society, politics and culture, online platforms tends to limit that diet by acting as an 'echo chamber' where individuals find their ideas supported and reinforced.[6] By contrast, PSBs can provide media spaces that encourage encounters with a broad range of ideas, opinions and cultures that are vital for a healthy society and democracy.

PSB has often been criticised for restricting market competition. However, the US case demonstrates that deregulated commercial media markets tend towards conglomeration and there is no evidence to suggest that the internet changes this.[7] New large global corporations have emerged (such as Google, Apple and Amazon) that increasingly control the flow of global digital content and information and seek to limit market competition. These companies not only dominate online search and retail, but also are major players in the provision of VOD.

The new businesses entering the VOD market in the UK and beyond have invested in original content. However, the business model for global providers, such as Netflix, is to produce programming that can be exploited across a range of international markets and there is no guarantee that the global players within the VOD market will invest in UK production.[8] The UK PSBs, on the other hand, remain the primary investors in programming for UK audiences. Beyond programming, UK PSBs can also catalyse technological innovations, develop markets and stimulate demand in new areas where business models are unclear, with Channel 4's and the BBC's lead in the development of VOD a case in point. This PSB investment maintains a world-leading television industry and significantly contributes to the UK's creative industry sector.[9]

If there remains a case for PSB, how might it be funded? VOD makes it possible to pay for PSB via subscriptions. However, introducing a subscription-based model for all public service broadcasting would damage the fundamental principle that PSB should be universally accessible regardless of ability to pay. The option of funding VOD services (such as BBC iPlayer) through subscription (while retaining a publicly funded linear broadcast service for the BBC) places undue costs on the young who make most use of VOD services for accessing television.[10] Indeed, VOD is a crucial tool in delivering public service outcomes to young people.

Since the 1980s there has been increasing competition for advertising revenue which has led to concerns that commercial PSBs will be unable to fulfill their public service remits and remain financially viable. However, there are a number of reasons to suggest that funding through advertising remains a viable option for PSB. First, television content (broadcast and online) remains attractive to advertisers because of the large audiences that it can reach.[11] Second, advertisers and broadcasters are developing technological solutions to prevent ad-skipping and ad-blocking online.[12] Third, the internet enables greater opportunities for viewer data collection which allows targeted, interactive and personalised advertising around on-demand content. Fourth, new relationships are emerging between advertisers and broadcasters to co-create content for broadcast and online with, for example, the rise of programmes funded wholly by single advertisers (currently referred to in the industry as 'advertiser-funded programming').

However, these new developments give rise to a number of concerns. Viewer intolerance of advertising and anxieties about the use of data online might drive them to other media channels for accessing content. In addition, the rise of advertiser-funded programming could erode audience trust in PSBs by undermining (or being perceived to undermine) their creative and editorial integrity. An increase of advertiser-funded programming would also diminish the spaces within which television programmes (both factual and fiction) might engage in useful and necessary critique of the practices and values of advertisers and their clients. In this context, regulation remains important to maintain a balance between ensuring commercial viability and protecting editorial/creative integrity and public service values.

The problems with advertiser and subscription funding demonstrate that there remain strong arguments for publicly funded public service broadcasting. Universal public funding ensures that the benefits of television are available to all and that PSBs are accountable directly to the public they serve. Public funding also spreads the costs of PSB and ensures significant value for money and affordability compared with pay-TV services.[13] Public support for the licence fee has increased and the majority of UK households are willing to pay.[14] Public funding needs to cover all of the services provided by PSBs given the convergence of broadcast and on-demand television. The media landscape in which PSBs are now operating is one in which the boundaries between linear broadcasting and online are continuing to diminish. Public service television for the internet era needs to be understood as a component of a larger networked and connected online infrastructure. In order to be able to provide television programming that serves the public, PSBs have to operate across broadcast television and the internet and need to have the flexibility to respond to new technological developments as they emerge. [...]

In an on-demand environment the organisation and design of interfaces to 'curate' content replaces linear scheduling as the means by which broadcasters shape viewing choices. Ofcom notes that we are witnessing a shift away from the EPG towards advanced search and recommendation through online interfaces provided by a widening number of firms.[15] VOD has the potential to broaden the diversity of content that viewers watch through curation and recommendation. Research suggests that by 2014 42% of people came to BBC iPlayer without a specific programme in mind.[16] While new online providers, such as Netflix, use algorithms to produce recommendations based on existing viewer behaviour (encouraging viewing of more of the same), the BBC has developed online curation focused on enhancing serendipitous discovery of a diverse range of content. Curation can also be used to offer increased access to the rich history of PSB and British culture. All 4 offers free access to box sets of archived programmes, while the BBC has integrated iPlayer into its web pages to create journeys from audiovisual content to curated online articles and third party content. In this way, VOD can be used as the starting point to connect viewers to online content within and beyond PSBs in ways that fulfil the public service remit of informing, educating and entertaining. In this sense, PSBs can act as trusted online hubs to connect audiences to a diversity of content with public service value produced by other organisations, such as galleries, libraries, archives and museums (GLAMS). Within this connected environment, PSBs should be working towards making their online content more shareable and interactive, enabling audiences to freely engage with and spread public service content across the internet.

At present, however, the ability for UK PSBs to provide online access to archival content can be limited by copyright legislation and could be enhanced if the public interest function of extended collective licensing was more clearly articulated.[17] [...]

Notes

1 Catherine Johnson is an Associate Professor in Culture, Film and Media in the Faculty of Arts, University of Nottingham. This is an edited extract from her submission to the Puttnam Inquiry, http://futureoftv. org.uk/wp-content/uploads/2015/11/Catherine-Johnson.pdf.

2 This convergence is by no means smooth or complete, yet with the rise of superfast broadband and roll-out of connected televisions, TV services are increasingly provided through internet-enabled devices.

3 Ofcom's third review of PSB notes that 'the purposes and characteristics of PSB are becoming increasingly important to citizens and consumers, and their satisfaction with the extent to which PSB services deliver on their objectives remains high' (*PSB Annual Report 2015*, 7).

4 Ofcom, *Communications Market Report 2015* (London: Ofcom, 2015), 145. In October 2015 Freeview Play was launched which provides free access to BBC iPlayer, ITV Player, All 4 and Demand 5 on a television set with the purchase of a Freeview Play TV or set-top recorder.

5 Alyssa Rosenberg, 'Why Are So Many Great Black TV Shows Missing From Streaming Services?', *Washington Post*, 27 April 2016. April Joyner, 'Blackfix: How Netflix's Algorithm Exposes Technology's Racial Bias', *Marie Claire*, 29 February, 2016.

6 See for example, Cass R. Sunstein, *Republic.com 2.0* (Princeton, NJ: Princeton University Press, 2009).

7 Stuart Cunningham and Jon Silver, *Screen Distribution and the New King Kongs of the Online World* (London: Palgrave Macmillan, 2013).

8 Patrick Barwise and Robert G. Picard, *What If There Were No BBC Television? The Net Impact on UK Viewers'* (Oxford: Reuters Institute, 2014). http://reutersinstitute.politics.ox.ac.uk/sites/default/files/ What%20if%20there%20were%20no%20BBC%20TV_Executive%20Summary.pdf

9 Ofcom, *Public Service Broadcasting in the Internet Age: Ofcom's Third Review of Public Service Broadcasting* (London: Ofcom, 2015).

10 Ibid.

11 Although the number of ad-supported channels on YouTube increased by 471% between 2014 and 2015, the average views of a YouTube channel have decreased by 56% (*The Long Tail of YouTube*, Open Slate, 15 October 2015, www.openslatedata.com/news/the-long-tail-of-youtube/. By contrast, the five main PSB channels in the UK still account for over half of all TV viewing (Ofcom, *PSB in the Internet Age*, 7) and the proportion of the UK television industry generated by advertising has remained robust despite competition from online (Ofcom, *The Communications Market Report*, 2015, 145–46).

12 For example, Channel 4's new VOD service, All 4, prevents viewers from ad-skipping and from using ad-blocking software.

13 BBC, 'British, Bold Creative: The BBC's submission to the Department for Culture, Media & Sport's Charter Review public consultation' (London: BBC, 2015), 66.

14 Ibid., 67.

15 Ofcom, *PSB in the Internet Age*.

16 Dan Taylor-Watt, 'Introducing the New BBC iPlayer' *BBC*, 11 March 2014, accessed November 2015, www.bbc.co.uk/blogs/internet/entries/3c7c6172-c3e7-34cf-9c04-4b40458c4a6e.

17 For an overview of the current EU reviews of copyright, see Erwin Verbruggen, Réka Markovich, Krisztina Rozgonyi, 'D5.4 Strategic Recommendations to Increase the Amount of Audiovisual Materials on Europeana II', *EUscreenXL Deliverable* (Preprint version), Utrecht, October 2015. Extended collective licensing enables collecting societies to license specific kinds of copyright works across an entire sector and was introduced in the UK on 1 October 2014.

30

Do We Still Need Public Service Television?

Luke Hyams[1]

We still, unquestionably, need public service television. There's a role the public service broadcasters fulfil that neither the independent young creators that we work with at Maker Studios,[2] nor the big media corporations can really fill. There is a sweet spot there in the middle that is so important. We are now at a time when young people really need to get behind the BBC, get behind Channel 4 and re-appropriate young people's vision of public service broadcasting.

In advance of coming here and talking with you today, I spoke to a lot of young people about what public service broadcasting meant to them. And on a lot of occasions, the response I got was: is YouTube a public service broadcaster? Is Instagram or Periscope a public service broadcaster? Because these are platforms that the public have access to, that young people can broadcast through, and where they can share their opinions from in varying levels of creative ways.

There are so many ways in which public service broadcasters do well for under-25 year olds: from incredibly high production values, well thought out dramas and documentaries on Channel 4 to the BBC 6 Music, 1Xtra, the World Service, Radio 4. You know, these are really the things that come up: sports, *Match of the Day*, they come up over and over again. And of course, the objective news perspective. But I think that news is one of the things where there has been a big change for a lot of the young people that I spoke to. They could actually pinpoint it to one night – a Sunday – back in 2011.

When the London riots happened, it was an amazing night. I sat in front of the TV with my laptop open. On the BBC, it was just regular Sunday night news coverage. But on the Twitter feed, there were little videos and photos of Dixons and Foot Locker in Brixton on fire. And it was this constant chat. It was just this incredible disparity.

Obviously the Metropolitan Police didn't want to stoke the fire and get more people to come to the riots and we completely understand that. But the BBC on that night chose to side itself with authority in a way that I think is a bit worrying. For someone who is 18 right now, the BBC offers so many incredible things, so many choices in terms of content, but at the same time, we are living in an age where young people are very sceptical of authority. And for the last five or ten years, the headlines we see are either attacking the BBC or bringing up institutional issues from the BBC's past and I think that that could give a young person quite a clouded image of what the BBC is.

Beyond that, there are serious challenges regarding respecting authority, whether it is with the Tories who make endless cuts or Labour who took us to war or the Murdoch press who had very, very unscrupulous means to get their stories. So I think that what public service broadcasters need to do is to step aside from that. They need to keep the objectivity but really reassess what it actually means to be a public service broadcaster. What can they provide that either the big corporations, or the incredibly authentic young creators who are making stuff themselves, cannot provide?

And I think we need to get this provision *now*. I want to get straight in to the heart of the matter: we need to reappraise who the licence fee is perceived to be for. We are people who have an incredible amount of nostalgia ... you know, our careers have started at the BBC. I remember as a child going past Television Centre like Charlie looking at the Wonka factory and just dreaming to get inside. My whole career started because I sat at home one Saturday night and *Challenge Anneka*[3] built a TV studio training facility for kids a mile away from my house. I'm the luckiest so-and-so who ever lived. So I think that for us, there are lots of reasons why the BBC should continue and we just need to work hard for that next generation, for them to value it in the same way we do. [...]

Notes

1 Luke Hyams is the former Chief Content Officer of Maker Studios, and current Head of Content, Revelmode. This is an edited transcript of his speech given at 'Do we still need public service television?' the inquiry launch event that took place on 25 November 2015 at the Scott Room, The Guardian. The event's participants also included Lord David Puttnam, Film Producer, and the Chair of the Inquiry; Lord Melvyn Bragg, Writer and Broadcaster; Jay Hunt, Chief Creative Officer, Channel 4, and chaired by Jane Martinson (JM), Head of Media, the Guardian. A transcript of the event is available at http://futureoftv. org.uk/events/do-we-still-need-public-service-television/.

2 Maker Studios produces content for on-demand platforms and represents some of the leading content creators on YouTube.

3 BBC reality game show (1989–95).

Part Five

Representing Britain on TV

31

Television and Diversity[1]

'We're just trying to redesign the face of British TV'

(Idris Elba's speech to Parliament, 18 January 2016)

Television is a crucial means through which we come to know ourselves and to learn about the lives of others and public service television, in particular, should provide ample opportunities for dialogue between and within all social groups in the UK. Success for a commercial broadcaster is predicated on reaching the most desirable demographics or on attaining sufficiently high ratings; to the extent that if commercial television does facilitate this dialogue and does address all social groups, it is more of a happy accident. For public service television, on the other hand, adequately communicating with and representing all citizens is not a luxury but an essential part of its remit.

Issues of diversity – based on the recognition that the population consists of multiple and overlapping sets of minorities – are therefore central to the continuing relevance (or impending irrelevance) of any public service media system.

This is far from a new proposition in relation to broadcasting. More than 50 years ago, the Pilkington report insisted that catering for minorities was not an optional add-on or indeed a capitulation to special interests but a vital part of broadcasting's responsibility to serve all citizens. 'Some of our tastes and needs we share with virtually everybody; but most – and they are often those which engage us most intensely – we share with different minorities. A service which caters only for majorities can never satisfy all, or even most, of the needs of any individual.'[2] Some 15 years later, the Annan Committee also agreed that broadcasting could no longer conceive of its audiences as in any way homogeneous; contemporary culture, it argued, 'is now multi-racial and pluralist: that is to say, people adhere to different views of the nature and purpose of life and expect their own views to be exposed in some form or other. The structure of broadcasting must reflect this variety.'[3] Broadcasting, it famously asserted, should be 'opened up'[4] in order both to promote the most diverse range of experiences and perspectives and to more effectively communicate with a changing population.

If television in the 21st century is to retain legitimacy and relevance, then it has little option but to recognise the desire of all social groups to be listened to and to be properly represented. This is especially the case when, for example, devolution, inequality, immigration and the establishment in law of 'protected characteristics' – such as age, disability, gender, race, sex, sexual orientation and religion – have further weakened the idea of the UK as a 'singular' space in which we all face the same challenges and share the same dreams. Public service television has somehow simultaneously to recognise our common interests and to serve the needs of different minority and under-represented groups.

This means that diversity, as it applies to television, needs to take on board issues of voice, representation *and* opportunity. It needs, in other words, to provide a means by which all social groups are able to speak, to be portrayed respectfully and accurately, to have equal employment prospects and, finally, to have access to a wide range of content.

The US academic Phil Napoli has identified three dimensions of broadcast diversity that connect to these capacities: source, content and exposure diversity.[5] We dealt with one element of *source diversity* in the previous part of this book where we examined the prospects for new suppliers of public service content in a digital age; we will consider another crucial area of source diversity later in this chapter where we confront the fact that television continues to be an industry dominated by white middle-class men. We discuss *content diversity* both in relation to the need to support the broadest range of television genres (in the next part) and, later in this chapter, in relation to how minority groups are represented on television as well as how they themselves perceive this representation. *Exposure diversity* – in other words, 'the degree to which audiences are actually exposing themselves to a diversity of information products and sources'[6] – is particularly difficult to measure and to mandate but our belief is that if audiences are presented with a television environment that is more open and receptive to the labour, lifestyles and languages of minority groups, then they are far more likely to seek out this material and to cultivate more promiscuous consumption habits. Public service television, we believe, has a crucial role in delivering both surprises and certainties to a curious (and diverse) population.

Are You Being Served?

Many viewers appear to be content with the quality of television in general. Ofcom reports that audience satisfaction with the delivery of public service broadcasting has risen from 69% of respondents in 2008 to 79% in 2014[7] and, while half of all adults believe that programme quality has stayed the same in the last year, the gap between those who think it has improved (17%) in relation to those who believe that things have got worse (30%) has more than halved in the last ten years.[8] Research carried out for the BBC Trust found that the public's 'overall impression' of the BBC has increased since 2008 earning an average score of 7.4 on a scale of 1–10 with 60% of respondents claiming that the BBC offers them 'quite a bit', 'a lot' or 'everything I need'.[9]

The problem is that satisfaction levels are not shared equally by all the population and that some groups – notably ethnic, regional, national and faith-based minorities – have expressed significant dissatisfaction with how they are represented or with the range of programmes relevant to their interests. So, for example, the wealthiest audiences are more than 50% more likely to praise the BBC's performance than those in the poorest households while English viewers are significantly more positive than Scottish ones.[10] Just 44% of Christian and 47% of non-Christian audiences agree that the BBC adequately represents their faith while only 41% of non-white audiences and a mere 32% of black audiences are happy with how the BBC represents them.[11] Just consider the implications for the BBC that less than one-third of black audiences report that they are satisfied with Britain's main public service broadcaster. In fact while public service television channels (including their portfolio channels) account for some 73% of the viewing of white audiences, the figure drops to a mere 53% for black, Asian and minority ethnic (BAME) audiences.[12] Overall satisfaction levels may look impressive but there are serious fissures behind the glossy headline figures.

This unevenness in satisfaction levels spills over into Ofcom's figures for audience perceptions of both visibility and portrayal of a range of social and geographical communities across all public service television channels. For example, while 42% of viewers in Northern Ireland think that there are too few people from Northern Ireland on TV, a mere 4% of Londoners think there are too few Londoners on TV; while only 6% of Londoners think they are shown in a bad light, some 20% of those from the North of England think they are represented negatively; similarly, while a mere 8% of men aged 55 and above think there are too few of them on TV, the number rises to 27% of women who think that there should be more older women on our screens. Finally, while there is a broad consensus among both the general viewing population and those viewers with disabilities that there are too few disabled people on TV there is no such agreement about the representation of black ethnic groups where 16% of all PSB viewers feel they are portrayed negatively in contrast with the 51% of black respondents who felt they were shown either 'fairly' or 'very' negatively.[13]

It is true that all minority groups are naturally more likely to want both to increase their visibility and to draw attention to the frequency and scale of negative representations. Who, after all, wants to feel either marginalised or caricatured? The more important point, however, is that if sections of a viewing public that is meant to be at the heart of public service broadcasting do not see themselves on screen or do not recognise the representations that do exist as valid, then broadcasters have a credibility problem they need to address. As the equality campaign Creative Access put it to us, the media 'cannot reflect society if society is not reflected in the media' and they warned of the consequences for broadcasting if it does not 'represent visually the society that pays its bills'.[14] The slogan 'No Taxation without Representation' may have originated in the run-up to the American Revolution in the 18th century but 21st-century broadcasters have much to fear if they neglect its message. This is all the more crucial in a situation in which there are more platforms and channels to choose from and where, as the actor Idris Elba put it in his call for broadcasters to embrace diversity, 'if young people don't see themselves on TV, they just switch off the TV, and log on. End of ...'[15]

We are not at all suggesting that public service television is a monocultural space or that broadcasters have totally failed to recognise the identity claims as well as the demographic and social shifts that are changing the face of the UK. Channel 4's heavy investment in and promotion of the Paralympics and the BBC's commissioning of a range of programmes concerning transgender issues is evidence of such recognition. What we are arguing is that 'opening up' television – to a full range of voices, cultures, narratives and identities – is an ongoing process and that public service television needs constantly to renew itself. If it fails to keep pace with changing tastes and attitudes, then it will undermine both its popularity and its legitimacy.

Indeed, as long as different social groups are not adequately addressed and as long as they are ignored, stereotyped or patronised, then struggles over visibility and representation will continue. One topic that has generated a significant amount of debate in recent years is the representation of working class lives in reality television,[16] a genre that has, formally speaking, allowed 'ordinary people' to enter a television world in which their presence, until then, had been largely confined to soap operas, 'kitchen sink dramas' and Alan Clarke productions from the 1970s. Factual entertainment is relatively cheap to produce, popular with audiences and has the added attraction of dramatising the experiences of ordinary viewers for ordinary viewers. It has won hearts and minds with programmes like *The Great British Bake-Off* but it has also antagonised whole sections of the population with, for example, what has been described as 'poverty porn'[17] – programmes (usually with the word 'benefits' in the title) which explore the 'reality' of life for some of the poorest in society. In his lecture to the Royal Television Society, the writer Owen Jones condemned the 'malignant programming' that 'either consciously or unwittingly, suggest that now – in 2013 – on British television, it's open season on millions of working-class people ...'[18] Professor Bev Skeggs, a sociologist who has studied reality television, put it to us that this is 'social work television, the moral television that tells people how to behave as better mothers (though very rarely better fathers interestingly) and how to look after children'.[19]

Of course, broadcasters themselves insist that television programmes that can help to stimulate a discussion about, for example, how to cope with poverty in 'austerity Britain' are invaluable and responsible. This was precisely the argument provided by the producers of Channel 4's *Benefit Street* in 2014 where the claim by the channel's head of documentaries that there is no more 'important job for programme makers than to record what life is like on the receiving end of the latest tranche of benefit cuts'[20] was countered by accusations that the programme simply 'demonised the poor and unemployed'.[21] The fact that death threats were issued to local residents, that Ofcom received nearly 1,000 complaints and that a petition condemning the programme gathered more than 50,000 signatures suggests that the perceived dangers of misrepresentation remain very real.

Yet while we have plenty of data on what audiences think of television content, regulators are not required to collect data on the actual on-screen representation of different social groups. Instead we have occasional pieces of industry and academic research that attempt to monitor specific areas of content. For example, the Cultural Diversity Network carried out research in 2009 and 2014 that found that women, disabled people, lesbian, gay and bisexual and BAME individuals were all significantly under-represented on television in relation to their proportion of the UK population.[22] Professor Lis Howell's annual 'Expert Women' project examines the representation of women experts on television news bulletins. Its most recent findings in November 2015[23] showed that there were five men to every woman on ITV's *News at Ten*, a ratio of three to one on Sky News with Channel 5 News coming out on top with a ratio of 1.6 men to every woman.

A similar study in 2014 led by Professor Howell in association with *Broadcast* magazine about the ratio of white to black, Asian and visible ethnic minority (BAVEM) contributors revealed a far more mixed picture: while the ratio of white people to eth-nic minorities in the UK is approximately six to one, researchers found that ITV per-formed worst with a ratio of over seven to one in its programmes while both Channel 4 and the BBC had ratios of 4.3 to one with Sky, a non-PSB channel, performing especially well with a ratio of three to one. The study, however, also identified a 'diversity gap' in relation to specific genres like topical, factual and entertainment leading Howell to conclude that a major problem lies in drama (apart from soaps) and 'in factual enter-tainment programming where BAVEM's are almost invisible.'[24] Unfortunately, the research was not followed up and, without a commitment from either broadcasters or regulators to commission such research, detailed data on representation – both quan-titative and qualitative – is likely to remain scarce and impressionistic.[25]

Of course better data about representation and even increased visibility of minor-ity groups will not, by itself, necessarily lead to more favourable representations. However, without a comprehensive record of who is being portrayed and in what cir-cumstances, it will be even more difficult to attain a more diverse on-screen television landscape.

Diversity Strategies

UK television, therefore, does not yet look like the audience it is supposed to serve. This is also true in terms of the composition of the television workforce that remains, some 15 years after the former director-general Greg Dyke's comment that the BBC was 'hideously white'[26]: it is disproportionately white, male, over-35, London-based and privately educated. This is accentuated at top levels where women occupy 39% of management positions while BAME individuals occupy a mere 4% of executive posi-tions, well below their respective proportion of the population (of 13%).[27] This is not quite as bad as the situation in the UK film industry where Directors UK found that women directed a mere 13.6% of films made between 2005 and 2014, leading them to conclude that 'there has not been any meaningful improvement in the representation of female directors.'[28] There is, however, no room for complacency in relation to televi-sion and a real need for concrete measures to address the situation.

Lenny Henry certainly touched a nerve in his celebrated BAFTA lecture in 2014 where he argued for action to address the fact that BAME individuals make up only 5.4% of the creative industries (precisely the same figure as in 2000) and that, while the sector has grown overall, fewer BAME people are working in it.[29] Data from Directors UK suggested that 1.5% of television programmes were made by BAME directors while, of the 6,000 directors on its database, a mere 214 (3.5%) were from BAME backgrounds.[30]

In response to this deficit, diversity has become a key buzzword inside the television industry with all broadcasters publishing 'diversity strategies' that relate to their plans to develop more 'inclusive' hiring and representational practices. For example, the BBC has recently published its latest Diversity and Inclusion Strategy, Channel 4 introduced its 360° Diversity Charter in 2015 while ITV has a Social Partnership strategy that it aims to embed throughout its programming.[31] While all these initiatives are to be welcomed as a sign that broadcasters have accepted that they have to improve their performance in relation to diversity, they are not without their own problems.

First, there is the definitional issue. We have already argued that diversity in television needs to be understood with reference to voice, representation and opportunity and that, therefore, it cannot be restricted to the portrayal of a specific social group. However, there is a danger that diversity becomes a 'catch-all' phrase that refers to a blissful state of 'inclusion' rather than a commitment to tackle previous patterns of 'exclusion'. When the cover of the BBC's strategy document insists that 'Diversity includes everyone' – with a photograph of *Bake-Off* winner Nadiya Hussain along with Paul Hollywood and Mary Berry – the implication is that diversity is all about the creation of a 'happy family' as opposed to the commitment to challenge the structures and ideas that have undermined prospects for inclusion and equality.

Even Channel 4, which, was launched with a remit to target minority audiences and which regularly attracts high levels of BAME viewers to its news bulletins, is keen to shift diversity onto less contentious ground.

Diversity is not about the colour of someone's skin; it goes way beyond that. Diversity is about being all-inclusive, regardless of culture, nationality, religious persuasion, physical and mental ability, sexual orientation, race, age, background and addressing social mobility.[32]

The problem is, however, that diversity *is* about skin colour, gender, sexual orientation, class and other characteristics, and therefore about how specific marginalised groups have not been sufficiently well integrated into the television workforce and television programming. So, for example, when Idris Elba stood up in front of parliamentarians in 2016 to insist – quite rightly we believe – that diversity is 'more than just skin colour' and is mainly about 'diversity of thought',[33] the fact remains that he was asked to deliver the speech precisely because of a growing concern that opportunities for BAME participation in the TV industry remain very limited. Race, as well as other forms of 'difference', cannot be so easily 'erased' from diversity talk.

Indeed, Sara Ahmed, who has written widely on diversity and public policy, argues that there remains a 'sticky' association between race and diversity. While, in reality, it is not so easy to move 'beyond' race, the language of diversity is 'often used as a shorthand for inclusion'[34] – a way of recognising difference but freeing it from negative associations concerning actual forms of discrimination. Diversity, she insists, can then be used to avoid confrontation and simply to highlight the contributions and achievements of different groups without asking more fundamental questions of why these achievements were marginalised in the first place.

Television historians such as Sarita Malik remind us that diversity policy was not always like this. When Channel 4 first started, it operated as a 'multicultural public sphere' with a series of programmes that engaged directly with 'questions of representation and racial stereotyping'.[35] Malik identifies a change in programme strategy after the closure of its Multicultural Programmes Department in 2002 as part of a more general shift in broadcasting from a 'politicised' policy of multiculturalism to a more consumerist emphasis on cultural, and now creative, diversity. What we are now left with is the possibility of a 'depoliticized, raceless "diversity" consensus'.[36]

The implication here is that broadcasters are using justified complaints about a lack of representation to pursue commercial strategies to appeal to diverse audiences without fundamentally changing commissioning and funding structures. The cultural theorist Anamik Saha describes this as the 'mainstreaming' of cultural diversity which 'while no doubt increasing the visibility of blacks and Asians on prime-time television, had actually has little impact on the quality of representations'.[37] So while BAME individuals may be increasingly visible on TV, the quality of their representations has not fundamentally changed and we are still stuck, all too often, with a repertoire limited to 'terrorism, violence, conflict and carnival' or, in terms of how Muslims are portrayed, to 'beards, scarves, halal meat, terrorists, forced marriage'.[38]

This connects to the second potential problem with broadcasters' diversity strategies, especially with regard to employment: the reliance on targets, the provision of small pockets of funding and training and what the Campaign for Broadcasting Equality described to us as 'Flash-in-the-Pan initiatives which are announced with a great flourish but which fail to deliver structural change'.[39] Let us be clear: additional money for diversity is a positive sign but the BBC's £3.5 million spend in 2014 that was dedicated to increasing diverse employment constituted less than 0.1% of the BBC's overall budget. Similarly, targets are entirely welcome and a very useful focus for organisations seeking to highlight the need for change but they are rarely successful by themselves, can be easily manipulated and are painfully slow in their realisation.

The fact that there have been, according to Lenny Henry, some 29 target-led diversity initiatives adopted by the BBC in the last 15 years, bears witness to this.[40]

The BBC has now launched its 30th such initiative promising to ensure that, by 2020, half of its workforce *and* its screen time will be composed of women, 8% of disabled and LGBT people and 15% of BAME individuals.[41] Channel 4 have announced similar targets (actually more ambitious in terms of BAME figures) and have announced 'commissioning diversity guidelines' which require independent production companies to demonstrate their commitment to diversity both on- and off-screen.[42]

It is not clear to us, however, how these targets, no matter how necessary they are, will overcome the structural barriers that have undermined diverse employment in television up to this point: the employment networks that favour friends and contacts, the reliance on unpaid interns and the reluctance of commissioners to take risks. Small steps in the right direction will do little to counter the pressures pushing in an opposite direction. So, for example, while there are a number of training schemes aimed at entry level positions, this can simply reinforce the notion that it's the talent that is the problem and not the institutions themselves. 'Training schemes and initiatives,' argues Simone Pennant of diversity campaigners the TV Collective, 'inadvertently create the perception that the reason why Black, Asian and ethnic minority talent are leaving the industry or not striving in their careers is because they are "not good enough" for existing roles.'[43] According to Lenny Henry: 'When there aren't enough programmes from Scotland we don't give the Scots more training. We place more commissioners up there to find good Scottish programme makers to make decent programmes. Let's do the same to ensure BAME representation.'[44]

We believe that Lenny Henry is right to argue that 'systemic failures' have led to a lack of diversity in the industry and we believe, therefore, that 'systemic' solutions are required alongside the provision of targets and training schemes.

This takes us back to the importance of the principle of quality that we discussed earlier in this book: that high quality minority representations require *conditions* that support innovation, experiment, risk-taking and the right to fail, conditions that arguably undersupplied in the current PSM ecology.

So firstly, we need to tackle the blockages at commissioning level. Idris Elba, for example, warned in his speech to Parliament that, all too often,

Commissioners look at diverse talent, and all they see is risk. Black actors are seen as a commercial risk. Women directors are seen as a commercial risk. Disabled directors aren't even seen at all. In general, if broadcasters want to stay in the game, their commissioners must take more risk with diverse talent.

We need to change the culture of commissioning and to provide incentives for commissioners to take risks. This might be enhanced if the Equality Act 2010 were to be amended so that commissioning and editorial policy would then be covered by public service equality duties.[45] There is also a need to create new and more diverse commissioning structures at the same time as placing new obligations on existing commissioners to break from a 'risk-averse' mindset by working with a broader base of talent. As one BAFTA member warned us: 'There's so little risk taking … that we risk stifling a whole new generation of makers and audiences.'[46]

Secondly, public service broadcasters who after all have a specific remit to serve multiple audiences, should be required to use a range of instruments to improve minority employment and representation. As the founder of the Campaign for Broadcasting Equality told us, 'there need be no conflict between ring fenced funds, quotas, targets and other measures to promote diversity. They are complementary.'[47] In particular, given the worryingly high levels of dissatisfaction of BAME viewers, together with the under-representation of BAME talent in the industry itself, we believe that public service broadcasters should be required to increase their investment in BAME productions through significantly enhanced – and ideally ringfenced – 'diversity funds' along the lines that Lenny Henry has called for[48] in order to secure conditions for a more representative workforce (at all levels) and prospects for more representative content.

We recognise that television alone cannot be expected to solve issues of under-representation given the inequality we see in relation to access to other services like health, education, employment and housing. But television certainly has a role to play both in addressing these issues and in involving minority audiences in the dialogue that will be necessary if we are to live together and to act collectively to overcome all forms of discrimination. For that to happen, appropriate targets and quotas need to be complemented by sufficient resources if aspiration is to turn into reality.

Notes

1 Edited extract from Chapter 8 of the Puttnam Report, http://futureoftv.org.uk/wp-content/uploads/2016/06/FOTV-Report-Online-SP.pdf.
2 Sir Harry Pilkington, *Report of the Committee on Broadcasting 1960* (London: HMSO, 1962), 16.
3 Lord Annan, *Report of the Committee on the Future of Broadcasting* (London: HMSO, 1977), 30
4 Annan, *Report*, 16.
5 Phil Napoli, 'Deconstructing the Diversity Principle', *Journal of Communication* 49, 4 (1999): 7–34.
6 Napoli, 'Deconstructing', 25.
7 Ofcom, *Public Service Broadcasting in the Internet Age* (London: Ofcom, 2015), 7.

8 Ofcom, *UK Audience Attitudes towards Broadcast Media* (London: Ofcom 2016), http://stakeholders. ofcom.org.uk/binaries/research/tv-research/attitudes-to-media/UK-audience-attitudes-towards-broadcast_media-2016-summary.pdf, 6.

9 NatCen, *Purpose Remit Survey UK Report*, BBC Trust, 2015, http://natcen.ac.uk/media/1012994/The-BBC-Purpose-Remit-Survey-UK-Report-Autumn-2014.pdf, 6, 7.

10 NatCen, *Purpose Remit Survey Report*.

11 NatCen, *Purpose Remit Survey Report*, 31, 33.

12 Channel 4, *Annual Report 2015*, http://annualreport.channel4.com/downloads/Channel-4-annual-report-2015.pdf, 36.

13 Ofcom, *PSB Diversity Research Summary*, June 2015, https://www.ofcom.org.uk/__data/assets/pdf_file/0023/59333/psb_diversity_report.pdf, 7,9, 15, 20, 34.

14 Creative Access, submission to the Inquiry.

15 Idris Elba's keynote speech to Parliament on diversity in the media, 18 January 2016, www.channel4. com/info/press/news/idris-elba-s-keynote-speech-to-parliament-on-diversity-in-the-media.

16 Reality television, as the format expert Jean Chalaby reminds us, 'is a broad church, with many strands in constant evolution, and therefore does not lend itself easily to grand statements'. It includes a variety of categories including observational documentaries, factual entertainment, reality competitions, talent competitions and constructed reality. See Jean Chalaby, *The Format Age: Television's Entertainment Revolution* (Cambridge, Polity, 2016), 43–44.

17 See 'Who Benefits? Poverty Porn' at the Edinburgh International Television Festival, 23 August 2013, www.youtube.com/watch?v=6haFqSt4mwQ&list=PL0vfD77BRT94IAZjPbgXjOAu9HF0WIFxy&index=7.

18 Owen Jones, 'Totally Shameless: How TV Portrays the Working Class', 25 November 2013, https://rts.org.uk/article/rts-huw-wheldon-memorial-lecture.

19 Comment at 'Are you being Heard?', Inquiry event at Goldsmiths, 23 March 2016.

20 Nick Mirsky, 'Benefits Street Struck a Nerve – Exposing how Vital a Documentary it is', *The Guardian*, 10 January 2014.

21 John Plunkett, 'Benefits Street to be Investigated by Ofcom following viewers' complaints', *The Guardian*, 25 February 2014. In the end, Ofcom found that *Benefits Street* was not in breach of its rules, www.bbc.co.uk/news/entertainment-arts-28086213.

22 Creative Diversity Network, *Diversity Monitoring: the Top TV Programmes*, August 2014, http://creativediversitynetwork.com/wp-content/uploads/2014/08/CDN-diversity-portrayal-pilot-2014.pdf.

23 Hannah Gannagé-Stewart, 'ITV News at Ten Slammed for "shocking" lack of Expert Women', *Broadcast*, 25 November 2015.

24 Robin Parker, 'Entertainment Shows Fail Diversity Test', *Broadcast*, 21 August 2014. It is also worth noting that newer white ethnic minorities, for example Polish and other Eastern Europeans, are not captured in this data.

25 Project Diamond, an industry-wide diversity monitoring system, was launched in 2015 and aims to collect data on the backgrounds of both on- and off-screen talent. It is, however, voluntary and therefore unlikely to provide the comprehensive picture that is required. See http://creativediversitynetwork.com/news/diamond-is-coming-are-you-ready/.

26 Amelia Hill, 'Dyke: BBC is Hideously White', *The Guardian*, 7 January 2001.

27 Creative Skillset, *2015 Employment Survey*, March 2016, http://creativeskillset.org/about_us/research/creative_skillset_employment_survey_2015.

28 Directors UK, *Out of the Picture: A Study of Gender Inequality Amongst Film Directors in The UK Film Industry*, May 2016, www.directors.uk.com/news/cut-out-of-the-picture, 7.

29 Tara Conlan, 'Lenny Henry Calls for Law to Boost Low Numbers of Black People in TV Industry', *The Guardian*, 18 March, 2014.

30 Directors UK, *UK Television: Adjusting the Colour Balance*, 2015, www.directors.uk.com/news/uk-television-adjusting-the-colour-balance.

31 See the House of Commons Library briefing paper on *Diversity in Broadcasting*, No. 7553, April 12, 2016 for an overview of diversity strategies.

32 Channel 4 – Equality Objectives, March 2012, www.channel4.com/media/documents/corporate/C4_Equality_Objectives_2012.pdf.

33 Idris Elba, speech to Parliament on Diversity in the Media, January 18, 2016, www.channel4.com/info/press/news/idris-elba-s-keynote-speech-to-parliament-on-diversity-in-the-media.

34 Sara Ahmed, *On Being Included: Racism and Diversity in Institutional Life* (London: Duke University Press), 14.

35 Sarita Malik, "Creative Diversity': UK Public Service Broadcasting After Multiculturalism', *Popular Communication* 11, 3 (2013), 10.

36 Malik, 'Creative Diversity', 17.

37 Anamik Saha, "Beards, scarves, halal meat, terrorists, forced marriage': television industries and the produce of 'race', *Media, Culture and Society* 34, 4 (2012), 430.

38 Cited in Saha, 'Beards', 425, 435.

39 Campaign for Broadcasting Equality, submission to the Inquiry.

40 BBC News, 'Lenny Henry Criticises BBC Chief's Diversity Plans', 24 June 2014, www.bbc.co.uk/news/entertainment-arts-27992392.

41 BBC, *Diversity and Inclusion Strategy*, April 2016, http://downloads.bbc.co.uk/diversity/pdf/diversity-and-inclusion-strategy-2016.pdf, 22.

42 Channel 4, 360° Diversity Charter –One Year On, 2016, www.channel4.com/media/documents/press/news/24114_C4%20Diversity%20Report_FINAL.pdf.

43 Simone Pennant, TV Collective, submission to the Inquiry.

44 Lenny Henry, comments at 'Are you Being Heard?' Inquiry event, Goldsmiths, University of London, 22 March 2016.

45 As recommended to us by Simon Albury of the Campaign for Broadcasting Equality in his submission.

46 Survey of BAFTA members undertaken for the Inquiry, March 2016. See Appendix 2 of the Puttnam Report.

47 Simon Albury, Campaign for Broadcasting Equality, submission to the Inquiry.

48 For example, in his demand for catalyst funding for BAME output that takes its cue from the quotas drawn up for content production in the nations and regions (Inquiry event at Goldsmiths, University of London, 22 March 2016).

32

Public Service Television in the Nations and Regions[1]

Public service broadcasting has previously been described as 'social cement'[2] in relation to the role it plays in bringing together and solidifying the various communities
of the UK. At a time when the UK's constitutional shape is changing and when devolutionary pressures are increasing, what kind of role should television play both in
maintaining the cohesiveness of the UK and in reflecting and giving voice to these
hugely important shifts?

This is not, of course, an entirely new question. Back in 1951, in the very early
days of television, Lord Beveridge chaired a committee on the future of broadcasting
in which he spoke of the need for 'greater broadcasting autonomy' for the constituent
countries of the UK. This was rejected by the government of the day that nevertheless
acknowledged their 'distinctive national characteristics, which are not only valuable
for their own sake, but are essential elements in the pattern of British life and culture.
It applies in only lesser degree to the English regions which also have a rich and diver-
sified contribution to make and should be given full opportunities for making it'[3]

Some 65 years later, with the emergence of devolved governments and assemblies
as well as 'city-regional machinery' in places like Manchester, Leeds and Birmingham,
there has been a clear shift to what Tony Travers at the London School of Economics
calls a 'quasi-federal UK'[4] Significant powers have been devolved to the administrations in Scotland, Wales and Northern Ireland and some additional powers transferred to municipalities in England. According to the Royal Society of Edinburgh, with
the 'passage of the Cities and Local Government Bill in 2016 some 55% of the population will be experiencing a form of decentralised decision-making.'[5]

Yet Whitehall and Westminster continue to exert a decisive influence on major
areas of everyday life. For example, the UK remains one of the most fiscally centralised

of all major western countries with only a tiny proportion of tax raised locally. So while there has been devolution of power and resources in some policy areas, there has not been a similar shift in relation to fiscal policy, defence, pensions, competition law and foreign policy that are matters 'reserved' for the Westminster parliament.

Furthermore, England continues to dominate the UK not just politically but also in terms of population and wealth. It has 84% of the population and 86% of GDP although these headline figures gloss over some significant differences. While the South East's share of GDP has risen from 38.6% to over 45% of the total in the last 50 years, the share held by the North West and North East has declined by a quarter: from 16.8% to 12.7% of GDP. According to Travers, 'despite the substantial redistribution of resources from place to place, significant territorial inequality has persisted'.[6]

This chapter will explore the extent to which these 'territorial inequalities' are relevant to the UK television system and discuss the kinds of action that broadcasters have taken to address the situation. Given that television policy remains a 'reserved' matter for the Westminster parliament, with devolved administrations having little control over the shape and content of television, the chapter also seeks to consider whether the present arrangements are fit for purpose or whether, in the light of changing constitutional arrangements, they need to be updated and a new approach developed that more adequately serves all the population of the UK.

Television's Role Across the UK

Unlike their multichannel counterparts, public service broadcasters are required to cater to all the geographical constituencies of the UK and, according to Ofcom,[7] they do this in several ways.

First, they make programmes either produced or set in different parts of the UK to transmit to all UK audiences. Recent 'network' programmes have included *The Fall*, produced in Northern Ireland, *Doctor Who*, which is made in Wales, *Broadchurch* made in Dorset and *Happy Valley* and *Last Tango in Halifax* produced by the Manchester-based RED production company. The intention here is both to represent parts of the UK to the whole of the UK as well as to redistribute TV budgets outside of a London base that has long performed the same role for British television as Hollywood studios have for US television.

PSBs also produce news and current affairs programmes in and for Scotland, Wales and Northern Ireland and the English regions as well as a small range of non-news programmes. This refers to the crucial 'intracultural' form of address in which a community speaks to itself in order to get to grips with shared experiences and problems. The BBC and Channel 3 licence holders are required to produce a specific amount of each genre broken down into news, current affairs and non-news (although ITV is no longer required to produce standalone non-news programmes in its regional English output).

Finally, there are services aimed at minority language speakers: for example, S4C provides Welsh-language television for the more than half a million people who speak Welsh while BBC Alba provides programming for Gaelic speakers in Scotland.

Research carried out for Ofcom as well as the BBC Trust[8] shows that that there is especially strong demand for material produced in and for 'the nations' – as Scotland, Wales and Northern Ireland are referred to – and the English regions. Although there are very different political and cultural contexts that pertain to the 'nations', as distinct from the 'regions', they are key spaces in which communities are able to find out about issues that directly pertain to their lives and their identities. As the managing director of UTV told us, audiences for its *Live at 6* news bulletin are often bigger than those for *Coronation Street* while Ofcom research suggests that 'the importance people place on their Nation or region being portrayed fairly to the rest of the UK has increased across the UK since 2008'.[9]

The concern that we wish to highlight is the growing gap between expectations and performance. This gap is likely to grow given the increased demands of audiences together with current pressures on public service broadcasters to cut budgets and to secure 'value for money' which, if narrowly interpreted, could lead to a further reduction in 'minority' services.

For example, despite the fact that we have had a Scottish parliament and assemblies in Wales and Northern Ireland since 1999 and despite the increased infrastructural investment linked to the creation of both a 'Northern Powerhouse' and a 'Midlands Engine', investment in television for the 'nations and regions' has not kept pace with these developments. Non-network output in 1999 reached 17,891 hours (that is first-run original output produced for the 'nations and regions' by the BBC, ITV and S4C; by 2014, 15 years *after* devolution, it had fallen to 13,814 hours (and that includes programming by BBC Alba), a decline of nearly 23%.[10] The main reason for this is the reduced obligation for Channel 3 licence holders to provide such programming though there have also been significant declines in BBC output – in Wales, for example, the BBC's English language television output has dropped by 27% since 2006/7.[11]

If we focus only on the period between 2009 and 2014, the picture appears to be more stable with an overall 7% increase in hours. However this headline figure disguises a 9% fall in Wales, a 3% decline in Northern Ireland and a small fall in the English regions. The picture is affected by the very welcome 57% increase in hours in Scotland but, even here, there were very specific explanatory factors, notably the increase in resources provided to cover the 2014 Commonwealth Games and the independence referendum as well as the distorting impact of STV's low-budget, overnight programme, *The Nightshift*, that ran from 2010 to 2015.[12]

Spending on programmes produced for the 'nations and regions' has also declined markedly in the past few years: from £404 million in 1998 to £277 million in 2014, a drop of just under one-third in real terms. This is due to the significant decrease in Channel 3 spend which has overshadowed a small increase in BBC investment.[13]

The most worrying declines have been in the English regions and in Wales with spending down by 11% and 16% respectively. It could be argued that the situation in Wales has been improved by the contribution of S4C to the Welsh cultural economy although its own creative capacity has been squeezed by a highly uncertain economic picture. It suffered a 24% cut to its core funding in 2010 when the bulk of its source of income was transferred from the government to the BBC, while BBC Wales' contribution to the channel is also set to decline. According to the Institute of Welsh Affairs, these reductions threaten the ability of Welsh broadcasters to tell the full range of stories in the widest possible range of forms: 'pluralism needs to be viewed not just in terms of the number of providers, but also in terms of the range, form, purpose and tone of programmes and the voices they carry.'[14] Rhys Evans of BBC Wales told us at our event in Cardiff that 'a fully developed national television service should go beyond news and sport and should help create and define a wider culture. We need to be entertained as well as informed.'[15]

A similar picture affects the prospects for BBC Alba where small pockets of funding from the Scottish government and the BBC allow for a mere 1.7 hours of original material per day with a 73% repeat rate overall. Despite its popularity with Gaelic viewers, its director of development and partnership, Iseabail Mactaggart, told us that insufficient funding 'creates really serious audience deficits' that need urgently to be addressed.[16]

Years of declining output and spend have, therefore, hindered the ability of broadcasters to more effectively cater to national and regional audiences and, in the case of some communities, have done little to dispel the idea that a centralised UK television system could *ever* adequately recognise their distinct needs and identities. The TV producer Tony Garnett, who has a distinguished record with the BBC, now talks of a 'Central London Broadcasting Corporation' that 'steals from the rest of the country by taking its money and spending it on itself'. Instead of truly reflecting the diverse lives of its population, the BBC – the main, but not the sole, target of his criticism – 'reflects distorted slivers of privileged life, for the international market; then it goes downmarket to caricature everyone else in soaps'.[17]

At a time when more viewers are associating themselves with a 'sub-national' UK identity, how should policymakers and television executives react and what steps should be taken to best meet the needs of viewers from across the UK? We first examine the emergence and impact of the 'nations and regions' strategy and then consider some alternatives.

Going 'Beyond the M25': The Emergence of a 'Nations and Regions' Strategy

Simply put, fundamental shifts in the UK's political tectonic plates, and an indefensible imbalance in investment in the UK creative economy provided the key motivations for developing a 'nations and regions' strategy – especially for the BBC and Channel 4, organisations without the regional structure that ITV at least used to have. The licence fee is collected in every corner of the UK yet for most of its history, the vast majority of spending took place where only a minority lived. In 1992, 80% of BBC network television programmes were made in London and the South East which then had 25% of the UK population[18] and which are areas that are not culturally, politically and socially representative of the entire UK.

Demands for a more decentralised service also reflect the realities of everyday lives, many of which continue to be lived locally despite increasing patterns of mobility and migration. According to research carried out for TSB in late 2015, people live on average 60 miles away from their childhood home with some 60% of people continuing to live in the same area where they were born. 'Even in an age of easy, cheap travel, instant global communication and the chance to experience life across the world, a significant proportion of Brits remain firmly connected to their origins'.[19] As people grow older, have children, buy homes and plan their recreational time, so their appetite for local information and expression grows. The celebrated phrase, 'think global, act local' reflects the significance of supra- and sub-national spheres of interest and the idea that, paraphrasing Daniel Bell, the nation state is too small for the big problems in life and too big for the small problems.

So there was real pressure in the late 1990s on the BBC – as the 'national' broad-caster – to address its deep-seated metropolitan bias and to shift some production from London to other parts of the UK. The generous licence fee settlement granted in 2000, shortly following John Birt's term in office as director general, had very clear 'out of London' requirements which were then supported by the new DG, Greg Dyke. Once the argument had been accepted inside the BBC, Channel 4, which already had a strong pedigree in culturally representative programming, was left exposed and immediately followed suit.

There had already been a BBC 'regional directorate' throughout the 1980s and 1990s. Scotland had lobbied especially hard against being seen as a 'region' and so in 1999, Mark Thompson was appointed as director of national and regional broadcast-ing followed in 2000 by a new director of nations and regions. Stuart Cosgrove was given the same title at Channel 4 not long afterwards.

The 2004 *Building Public Value* initiative and subsequent charter review process emphasised the BBC's commitment to meet the needs of an increasingly diverse and fragmented UK. The BBC promised to strengthen its programming for the devolved nations, to step up its local services, both in the nations and in the English regions and to develop its network of 'Open Centres' and 'digital buses' where less well-off people could access online technologies for no additional cost, seven days a week.[20] Whole departments and channels were to leave the London base with Salford announced as the main destination.

However, the main focus of this strategy was on increasing *network* output in Scotland, Wales and Northern Ireland with only a very limited expansion of local ser-vices in the English regions including the launch of a local television pilot that was subsequently refused permission by the BBC Trust following heavy lobbying by the newspaper industry. In 2008, Jana Bennett, the director of BBC Vision, unveiled pro-posals that she described as a 'radical shift in the whole set up of broadcasting'[21]: a promise that spend on network programming in the nations would go up from 6% of total spend in 2007 to 17% by 2016, representing their share of the overall UK pop-ulation, and that 'out of London' spend overall would rise to 50% by 2016 (still sig-nificantly below its share of the population). For the first time in many years, the gravitational field in British broadcasting was due to change – a situation that would be further cemented by the requirement imposed on Channel 4 in 2014 to allocate 9% of its budget to 'out of London' productions by 2020.

This strategy, it could be argued, had an inescapable logic and an underlying sense of fairness. 'Sustainability' was seen as a key objective of the BBC's approach in which just four new or enhanced centres of network production, one in each nation and the new Media City in Salford, would be established. Thus real concentrations of craft and talent could be created and developed.

There have been undoubted successes. The targets for 2016 have been met and indeed have been exceeded: as of 2014, the 'out of London' spend was over 53% while Scotland, Wales and Northern Ireland accounted for over 18% of total spend.[22] Cardiff Bay has built quite an industry around *Doctor Who*, *Torchwood* and *Sherlock*; in Northern Ireland, strengthened BBC foundations (along with a significant contribution from Northern Ireland Screen and the Northern Ireland government) have enabled the creation of *Game of Thrones* (albeit for HBO) and much more including the network series *The Fall*; *Question Time* is now produced out of Scotland which has also excelled at Saturday night National Lottery programmes like *In It to Win It* and *Break the Safe*.

And therein lies a major problem with the existing nations and regions strategy for network programming: that it may have shifted elements of production out of the capital but there is little guarantee that this will lead to rich and complex representations of the nations themselves. 'While drama production has been a beacon of success in Wales', argue Cardiff University's Sian Powell and Caitriona Noonan, 'this drama rarely reflects life in Wales and Wales is solely a location for filming rather than part of the narrative setting'.[23] Angela Graham of the Institute of Welsh Affairs told us that 'it's ironic that BBC Cymru Wales is enjoying such great and welcome success when its domestic output is tragically low'.[24] It has made *War and Peace*, *Casualty* and *Doctor Who* but it lacks the resources to dramatise experiences that more directly speak to people from Cardiff to Caenarfon. *Doctor Who* may be about many things but it is not, at least overtly, about the people of Wales.

There is also the problem, as with Scotland in particular, that a 'tick box' approach to 'out of London' programming may not necessarily lead to the emergence of a sustainable production infrastructure. Production has indeed been shifted but often by temporarily transferring labour and resources during the programme run: the so-called 'lift and shift' strategy. Additionally, commissioning, finance and most national channels remain within the magic circle that surrounds W1A – a pattern that is replicated by the vast majority of big, successful, independent production companies.

So despite the positive impact of increased network spend across the UK, it can be argued that the balance of power has not fundamentally shifted. Key positions – including those of director general, director of television, director of England and director of BBC Studios – are all still based in London; network production in the nations is now under the creative leadership of genre heads based in London while the main conurbations of England, with their massive populations, are not directly represented at the BBC's most senior management table in London. Meanwhile, funding pressures remain intense both on the nations as well on the BBC's output across the English regions. Given all these developments, one could make the argument that power is now actually more centralised inside the capital than it was previously.

At least some of this has been acknowledged by the broadcasters themselves which explains why many are stepping up their commitments, particularly with regard to Scotland, Wales and Northern Ireland. Tony Hall, the BBC's director general, for example, accepts that not enough has been done to provide programming and governance structures that adequately reflect the demand for a 'louder' voice from the nations. 'Audiences have told us ... that they think we need to do more to capture distinctive stories from across the UK and share them across the country, as well as doing more to reflect the changing nature of the UK and support democracy and culture.'[25] He now promises to complement the quotas for network content with, for example, new drama commissioning editors in each nation, dedicated 'splash' pages for its news websites and the iPlayer, and increased support for English-language programming in the nations. In February 2017, Tony Hall announced the creation of a new channel, BBC Scotland, backed up by a £20 million investment into Scottish content.[26]

We welcome these commitments but we note that they do not signify a meaningful shift in power away from W1A: decisions about the nations will continue to be taken in London while the new drama commissioners will still report to the overall controller of commissioning in London. We believe that a new approach is now needed: one that accepts both that a centralised structure and culture can never adequately represent all citizens and that a changing political settlement will require a robust response from broadcasters.

In reality, despite some who thought that *any* significant shift of production out of London might weaken the BBC as a whole, the 'nations and regions' strategy was developed not to undermine the BBC's role as a 'national broadcaster' but precisely to rescue it. As Greg Dyke forcefully argued back in 2005, such changes were necessary 'if the BBC really wants to be the national broadcaster and not what it is today, a broadcaster aimed disproportionately at the South of England middle classes. This bias will only change if more broadcasters live away from the South East and more BBC programming commissioning is done away from London.'[27] For some critics, however, the existing 'nations and regions' strategy was only ever 'a response from institutions reluctant to devolve real power, which construct this offering as a means to retain control in London.'[28] At a time when, as we have already argued, more and more decisions are being taken by directly elected assemblies and parliaments as well as by mayors, local crime commissioners and unitary authorities in the English regions, we feel that a more full-blooded engagement with decentralisation is not simply advisable but necessary if the BBC in particular is to retain loyalty from viewers across the UK.

A 'Devolved' Approach to UK Television

At its most basic level, a devolved television system would simply allow distinct communities to decide what stories to tell and how to tell them. The present arrangements, based on centralised budgetary, commissioning and editorial control, all too often prevent them from doing this. This lack of autonomy has stirred up some lively debates on the possible devolution of television policy. The Institute of Welsh Affairs, for example, argues that responsibility for broadcasting 'should be shared between the UK government and the devolved administrations'[29] while the academic Robert Beveridge put it to us that Scotland should have full control over its media policy.[30] In a high-profile speech at the Edinburgh International Television Festival in August 2015, the Scottish first minister, Nicola Sturgeon, called for a 'federal' BBC,[31] a demand that was rebuffed in the UK government's 2016 White Paper but one that we think is likely to resurface in any future referendum debate and that merits very serious discussion. While there is little point in pre-empting constitutional change, there is also little point in refusing to acknowledge significant shifts in the public's appetite for increased autonomy.

In the meantime, as Robert Beveridge told us, 'we need to establish new and better ways of working within which to secure the Scottish public interest within the evolving constitutional settlement.'[32] Following this logic, devolved administrations are energetically making the case for further decentralisation. The Scottish government, for example, has asked for the ability to spend the £323 million raised by Scottish licence fee payers on content and services of its own choosing including, of course, content produced centrally.[33] This form of 'budgetary control over commissioning', it argues, could even be achieved within the terms of the existing charter and ought to be seen as a fairly basic democratic principle. The Welsh assembly is recommending that commissioners for the nations and regions should be based in those areas and provided with greater control of network funding, 'as a means of increasing the range and diversity of output, both locally and for the network.'[34]

There appear, however, to be few spaces in UK-wide policy circles in which to argue for these sorts of policies without being dismissed as either 'nationalist' or 'parochial'. This is particularly the case in Scotland where, as we have already noted, the BBC already receives the lowest performance ratings in the UK. We ought to recognise the strength of the Scottish government's mandate to secure more control over the country's future but we also need to disentangle what are sometimes still seen as 'partisan' nationalist politics from the wider opinions of the Scottish public – not every demand for more autonomy is necessarily a full endorsement of Scottish National Party policy.

So while we welcome the quotas for network spend for and the creation of new commissioners in the nations, we believe that real commissioning power should follow shifts in production. Too many decision makers continue to walk the same metropolitan (and sometimes suburban) streets and eat in the same restaurants to truly appreciate, and hence reflect, a fast changing UK. For this to happen, commissioners need to be in charge of budgets that should be devolved with them. There is no particular reason why drama commissioning could not be based in Cardiff, comedy commissioning in Glasgow and children's commissioning in Belfast. If, as it is mooted, Tony Hall is set to restructure the BBC around new divisions focused on education, information and entertainment,[35] then a new opportunity arises to devolve power via commissioning budgets.

We also welcome the government's commitment in its 2015 BBC White Paper to main minority language television services. Indeed, such programming may be more important than ever in a multichannel age and it can hardly be accused of lacking 'distinctiveness'. The arguments for S4C were originally made in 1982 when the UK had just three channels. 'How much stronger', asked S4C's Huw Jones at our meeting in Cardiff, 'are those arguments today when the English language offering for the viewer consists of more than 500 channels, while in Welsh we still only have the one?'[36] We note the fact that the government intends to review S4C in 2017 but we are mindful that the government's commitment to language programming has to be backed up with secure, long-term funding. Given the particular purposes they serve in relation to national heritage, cultural diversity and education, we feel that they should be at least partially funded by ringfenced money – either from central government or another source – and not left to survive on whatever the BBC can find from its (declining) budgets.[37]

A devolved strategy would also recognise what is possible in other countries. The significant success of Danish drama is the result of imaginative government intervention and the support of the industry – soft power achieved in subtle ways in 'smaller' states. As the former controller of BBC Scotland John McCormick told us, devolved structures are common in other European countries. 'While comparable audiences in Ireland and Catalonia are each served by half a dozen or more TV channels located in their territory, the German länder have one by right under federal law and the Dutch provinces have one. In Scotland, apart from BBC Alba we still have the twin TV channel opt out model established in the earlier part of the premiership of Harold Macmillan.'[38]

As well as a new and more vigorous strategy for the devolved nations, we also need a far stronger remit for the English regions with specific responsibility for diverse ethnic and faith-based representation. English regions – with the notable exception of the North West – have failed to benefit from the existing 'nations and regions' strategy and, indeed, Bristol has had its drama base 'lifted and shifted' to Cardiff. Ethnically and socially diverse areas like the East and West Midlands and Yorkshire, which are home to far more licence payers than those in Wales, Scotland and Northern Ireland, enjoy little or no network television production and are underrepresented in most genres. The announcement by the BBC to locate several departments in Birmingham, including its centre of excellence for skills, recruitment and talent development, Diversity Unit and HR functions is very welcome and, in part, a response to the energetic campaign run in the city to secure improved broadcast representation. The BBC's agreement to move its online channel, BBC Three, to Birmingham by 2018 is more evidence of a willingness to reflect demands for greater investment in infrastructure outside of London.

We are not arguing that these devolutionary changes should be at the expense of core PSB services for the UK where demand remains strong across the nations and regions. Indeed, some of the highest viewing figures for network content are in Wales; that fact does not preclude the need, at the same time, for more Welsh content. As John McCormick of the Royal Society of Edinburgh put it in relation to Scotland, 'it's important to find a way of articulating the need for adequate Scottish public service broadcasting without losing sight of the value of existing provision from London, from which we all benefit enormously. And the desirability of not harming it.'[39] Our point is that public service television – and this is not restricted to the BBC alone – will be strengthened if it is restructured on a more democratic and accountable basis that recognises both the demand for UK-wide content as well as a growing appetite for output that fits the changing political configuration of the UK in the 21st century.

Notes

1 Edited extract from Chapter 9 of the Puttnam Report, http://futureoftv.org.uk/wp-content/uploads/2016/06/FOTV-Report-Online-SP.pdf.
2 Paddy Scannell, 'Public Service Broadcasting: The History of a Concept', in *Understanding Television*, eds. Andrew Goodwin and Garry Whannell (London: Routledge, 1990), 14.
3 Cited in Robert Beveridge's submission to the Inquiry.
4 Tony Travers, 'Devolving Funding and Taxation in the UK: A Unique Challenge', *National Institute Economic Review*, 233, August 2015, R5.
5 Royal Society of Edinburgh, Response to Scottish Parliament Education and Culture Committee report on BBC Charter Renewal, Advice Paper 15–21(A), November 2015, 6.
6 Travers, 'Devolving Funding', R6.
7 Ofcom, *Public Service Broadcasting in The Internet Age: The Nations of the UK And their Regions* (London: Ofcom 2015), http://stakeholders.ofcom.org.uk/binaries/consultations/psb-review-3/statement/PSBR_natreg.pdf, 5.
8 For example, see the BBC Trust's Purpose Remit Survey reports at www.bbc.co.uk/bbctrust/our_work/audiences/previous_prs_reports.html and Ofcom's *The Nations of the UK And their Regions*.
9 Ofcom, *The Nations of the UK And their Regions*, 6.
10 Data from relevant Ofcom *Communication Market Reports*.
11 Institute of Welsh Affairs, *IWA Wales Media Audit 2015, Executive Summary*,http://futureoftv.org.uk/wp-content/uploads/2016/04/IWA_Executive-Summary.pdf 2.
12 Ofcom, *Communications Market Report: Scotland* (London: Ofcom, 2015), 46–47.
13 Ofcom, *The Nations of the UK And their Regions*, 11.
14 *IWA Wales Media Audit 2015*, 2.
15 Comments at Inquiry event at Cardiff University, 6 April 2016.
16 Comments at Inquiry event at the Royal Society of Edinburgh, 13 April 2016.
17 Tony Garnett, 'The BBC should Explore the World beyond London', *The Guardian*, 17 April 2016, www.theguardian.com/commentisfree/2016/apr/17/bbc-explore-world-beyond-london.
18 John Birt, *The Harder Path* (London: Time Warner, 2003), 312.
19 TSB, *Homebirds*, 2015, www.tsb.co.uk/tsb-home-reports-homebirds.pdf.
20 BBC, *Building Public Value: Renewing the BBC for a Digital World*, BBC, 2004, https://downloads.bbc.co.uk/aboutthebbc/policies/pdf/bpv.pdf, 75–77.
21 BBC, 'Jana Bennett Unveils Major TV Production Shift outside London', press release, 15 October 2008, www.bbc.co.uk/pressoffice/pressreleases/stories/2008/10_october/15/bennett.shtml.
22 *BBC Annual Report and Accounts 2014/15*, http://downloads.bbc.co.uk/annualreport/pdf/2014–15/bbc-annualreport-201415.pdf, 82.
23 Sian Powell and Caitriona Noonan, submission to the Inquiry.
24 Comments at Inquiry event, Cardiff University, 6 April 2016.
25 Tony Hall, 'The BBC in the Devolved Nations: Progress Update', letter, 12 May 2016, http://downloads.bbc.co.uk/mediacentre/nations-progress-update.pdf.
26 BBC, 'Biggest BBC Investment in Scotland in Twenty Years', press release, 22 February 2017, www.bbc.co.uk/mediacentre/latestnews/2017/scotland-investment.
27 Greg Dyke, 'On Broadcasting', *Independent*, 3 January 2005, http://downloads.bbc.co.uk/mediacentre/nations-progress-update.pdf.
28 Neil Blain and David Hutchison, 'A Cause Still Unwon: The Struggle to Represent Scotland', in *The Media in Scotland*, eds. Neil Blain and David Hutchison (Edinburgh: Edinburgh University Press, 2008), 14.

29 *IWA Wales Media Audit 2015*, 5.

30 See, for example, Robert Beveridge's submission to the Inquiry.

31 Kate Devlin, 'Nicola Sturgeon calls for New Scots Channels in BBC Revolution', *The Herald*, 27 August 2015.

32 Robert Beveridge, submission to the Inquiry.

33 Scottish Government, *Policy Paper on BBC Charter Renewal*, February 2016, /www.gov.scot/Resource/ 0049/00494159.pdf, 8.

34 National Assembly for Wales, *Inquiry into the BBC Charter Review*, March 2016, http://senedd. assembly.wales/documents/s49465/Report%20on%20the%20BBC%20Charter%20Review%20 March%202016%20Conclusions%20and%20recommendations.pdf, 3.

35 Steve Hewlett, 'Tony Hall's Grand Reorganisation of the BBC "is playing with fire"', *The Guardian*, 1 May 2016, www.theguardian.com/media/2016/may/01/bbc-reorganisation-tony-hall-inform-educate-entertain.

36 Comments at Inquiry event, Cardiff University, 6 April 2016.

37 See the submission to the Inquiry from Teledwyr Annibynnol Cymru, the association of Welsh independent producers, that makes an eloquent case for increased funding of S4C.

38 Comments at Inquiry event, Royal Society of Edinburgh, 13 April 2016.

39 Comments at Inquiry event, Royal Society of Edinburgh, 13 April 2016.

33

Are You Being Heard?

Lenny Henry[1]

I made a speech at BAFTA recently, where I spoke about my shock at a Skillset census that revealed that between 2006 and 2012, the number of BAMEs (Black, Asian and minority ethnic people) working in the UK TV Industry has declined by 30.9%.[2] Skillset's figures clearly showed that BAME representation in the Creative industries in 2012 stood at 5.4% – its lowest point since they began taking the census. This is an appalling figure, especially when you consider that London, arguably the UK's biggest creative hub, is 40% BAME. Interestingly, figures from Directors UK in 2015 show that 98.5% of directors in the industry are white.[3]

Back then everybody said something needed to be done. Government ministers said something needed to be done, the BBC said something needed to be done, Channel 4 said something needed to be done and Sky announced their 20% BAME targets. I was invited to talk at a Parliamentary Select Committee. Diversity was going to be addressed. Life was great.

But this is where I differed from many of the big TV companies and broadcasters. They seemed to think more training initiatives were the easy fix – not training courses for those in positions of power on how they could be more diverse and inclusive in their employment practices and commissioning – but instead further training for the BAME talent base!

They set up tons of BAME training schemes, management training, youth training, even trainee commissioners. Now, I am not arguing against training, far from it. Nor am I suggesting that these initiatives lack merit or the best of intentions.

My concern is this: that when the only tangible solution on the table to create significant and sustainable change is training, it can be argued that, inadvertently, the perception being perpetuated of the BAME creative community – the reason why BAME people are leaving the industry, the reason why our numbers are at their lowest in years – is because we're not good enough.

Please don't misunderstand, I am not arguing against training; training initiatives are what ensure a strong, capable workforce. What I am saying is by all means create training to improve the skill base in the creative industry, but what good is training as a tool to improve inclusion and diversity in the industry if the very systems that have inadvertently created the problem fail to address their systemic failure?

So here's my revolutionary thought; why don't we change the system?

[...] I think everyone in the television industry today would agree that ensuring that diversity in front of the camera, diversity behind the camera and a diversity of programmes and voices that speak to all the nations, regions and communities must be our ultimate goal if we are going to truly serve our viewing audiences now and most importantly in the future!

I think making sure programmes of all different genres being made by a diverse production team is just as important as making sure programmes are made by Scottish and Welsh production teams.

If there weren't enough news and current affairs programmes being made, we wouldn't blame the journalists and give them more training to make better programmes. We give a current affairs commissioner a budget and a number of hours on TV and she or he has to find programme makers to make those programmes.

When there aren't enough programmes from Scotland we don't give the Scots more training. We place more commissioners up there to find good Scottish programme makers to make good programmes.

Let's do the same to ensure BAME representation.

Let's create a number of commissioners and give them real power (and that means money) to find productions made by diverse teams to make great programmes.

And let's not ghettoise these diverse programme makers by saying they can only make programmes about black or Asian issues. Just like Scotland can make *Eggheads* and it's a Scottish programme, I want BAME professionals to have access to make programmes across the TV landscape, from high-end period dramas to *Panorama*.

[...] We have a once in a decade chance to change history, to make diversity a celebration of our nation rather than a problem!

Let's make history and let's all make sure we change the face of the television industry forever.

Notes

1 Sir Lenny Henry is a comedian, actor, writer, and television presenter. He is one of the founders of Comic Relief and a prominent campaigner for the increase of ethnic diversity in media. This speech was given on March 22, 2016 at the Inquiry event 'Are You Being Heard?' held at Goldsmiths, University of London. Other panelists included Dawn Foster, writer on politics, social affairs and economics and Bev Skeggs, Professor of Sociology, Goldsmiths, University of London. The event was chaired by Pat Younge, MD of Sugar Films, former BBC Chief Creative Officer and a member of the Inquiry's Advisory Committee. To access the speech in full or to watch the recording of the event, go to http://futureoftv.org.uk/events/are-you-being-heard-representing-britain-on-tv/.

2 Creative Skillset, *Employment Census of the Creative Media Industries 2012,* https://creativeskillset.org/assets/0000/5070/2012_Employment_Census_of_the_Creative_Media_Industries.pdf, 17.

3 Directors UK, *UK Television: Adjusting the Colour Balance,* November 2015, www.directors.uk.com/news/uk-television-adjusting-the-colour-balance.

34

Skills and Training Investment Vital to the Success of Public Service Broadcasting

Creative Skillset[1]

[...] The UK has one of the most vibrant TV production sectors in the world, with an enviable track record of producing innovative and high quality content across genres. Our PSB system has been a driving force: the PSBs, between them, are responsible for some 80% of total investment in UK original non-news content.[2] Independent producers are responsible for around 60% of total commissioned hours on the five main PSB channels.[3] This has been bolstered by the recent introduction of tax credits in PSB related industries.[4] According to the report by Olsberg and Nordicity, the High End TV Tax relief created some 16,800 jobs in the UK and generated some £852m for the UK economy.[5]

Given the rapid pace of change around technology, audience behaviour and business models, the demands placed on those working in production are constantly evolving. The creation of content and its distribution on multiplatform devices is integral to delivering PSB – and this relies on relevant talent and skills. PSB needs to serve both the media consumer and the citizen. A healthy and sustainable media market needs high quality, home-grown, innovative content, effective distribution and plurality. Future PSB models need to support and invest in our industries' workforce in order to keep content relevant and our creative industries competitive.

As new platforms and formats emerge, old divides are blurred. A holistic and collaborative approach across not just PSBs but all screen-based industries is increasingly vital to ensure that the Creative Industries' talent base can compete globally. This requires upskilling and re-skilling with an integrated view and a systematic approach to tackling barriers to entry and enabling progression within an ever more casualised workforce.

Skills Challenges in the Creative Industries

Barriers to entry – and the ability to attract, maintain and sustain a highly skilled, creative and productive workforce – are critical factors affecting the growth of the PSB workforce. There is currently a high proportion of graduates entering the Creative Industries, but a workforce from a wide range of backgrounds, with a rich mix of skills is vital to creativity and employability. Factors that might influence industry entrants' social and educational backgrounds include the systemic culture of those wishing to gain industry skills having to undertake unpaid 'work experience': in a 2014 survey of the creative media workforce, 82% of those surveyed who undertook work experience did so unpaid.[6] Similarly, there is still a lack of open recruitment practices: 71% of the creative media workforce in the same survey reported that they heard about their current job through informal routes.

This arguably affects the current state of diversity in the PSB workforce. Positive progress has been made: currently, overall representation of those from BAME groups in the TV industry stands at 9%[7] – which, if compared to BAME representation in the overall UK workforce (10%[8]), and compared with similar exercises, shows improvement. There is still more to be done, however, and we believe an ideal starting point includes better monitoring at a more granular level to help PSBs. There is broad commitment across the Television sector to 'Diamond', the diversity monitoring system that will provide detailed and consistent reporting in a way that has not been possible before and is coordinated by the Creative Diversity Network.

The government's apprenticeships levy[9] could, if implemented effectively, provide a timely opportunity for industry to help diversify and supply a cohort of new entrants to PSB. As both a graduate and non-graduate route, new apprenticeships could – with industry backing – be a powerful driver for greater creative industry workforce diversity via paid, job-ready entrants.

Creative Skillset has worked with employer groups and training providers UK-wide to deliver 'trailblazer' Apprenticeship standards. We have developed content for roles including broadcast production junior, content producer and props technician – and, at a more advanced level, an apprenticeship in outside broadcast. We know from our work with partners including Channel 4 and the BBC that the appetite for apprenticeships is growing, and we see a positive opportunity for PSB and its supply chain to benefit by working with us to create the right content and conditions to ensure that investment delivers the best value for levy-paying employers.

Within the current workforce, skills demands have been driven by growth in PSB 'tax relief' sectors (children's, TV, high-end TV, film, animation), spurring a demand for rapid access to the latest skills and talent. However, in a 2015 survey some 22 out of 24 high-end TV companies reported difficulties in crewing for their latest production.[10] These sectors share similar skills requirements, so there is a greater need to sustain a more flexible workforce that can move as demand requires between sectors – sharing know-how, injecting energy and stimulating innovation and transformation.

Coupled with this, PSB industries, and TV in particular, comprise a high proportion of SMEs, with freelancers forming around 40% of the workforce. Many freelancers work in production-related roles: for example, 67% of camera staff and 60% of post-production staff are freelance.[11] Perhaps unsurprisingly, freelancers are more likely to report a training need (57%) compared with permanent employees (45%); and 74% claim to experience barriers to training and professional development compared with permanent employees (55%).[12] There is a risk of market failure in skills provision since our casualised workforce will, on the whole, not have employers investing in their training. The shift toward 'portfolio' careers is also likely to increase the proportion of freelancers in the PSB industries.

PSB broadcasters are investing in upskilling their workforce through measures such as the high-end TV levy, contributing over £4million to training and skills for their workforce since 2013. This has resulted in several successful schemes delivered in collaboration with Creative Skillset. An example includes the 'Stepping Up' Programme, which aims to facilitate high-end producers' progression into the role of TV Drama Producer across scripted, factual and other broadcast media. This initiative supports ten production placements, ultimately supporting talent progression in a critical PSB genre while, at the same time, increasing the wider PSB talent pool. Despite this progress, a stronger and more coherent industry-wide approach is needed to tackle investment in upskilling and mobilising the PSB workforce. [...]

Notes

1 Creative Skillset is the creative industries' key skills partner, working with and for industry to enable relevant skills provision that improves productivity, employability and creativity across sectors including film, television, animation, games, visual effects, radio, publishing, advertising and marketing communications. This is an edited extract from their submission to the Inquiry, http://futureoftv.org.uk/wp-content/uploads/2016/05/Creative-Skillset.pdf

2 Oliver & Olhbaum Associates Ltd, *TV Producer Consolidation, Globalisation and Vertical Integration: A Report for Pact* (London: Oliver & Olhbaum Associates Ltd., 2015).

3 Ibid.

4 A high-end television programme is defined to mean a drama (which includes comedy) or documentary production that is intended for broadcast and has expenditure per hour of slot length of not less than £1 million. See BFI, *British High-End Television Certification* (London: BFI, July 2015).

www.bfi.org.uk/sites/bfi.org.uk/files/downloads/bfi-british-high-end-television-certification-cultural-test-guidance-notes-2015-07.pdf.

5 Olsberg SPI and Nordicity, *The Economic Contribution of the UK's Film, High-end TV, Video Games and Animation Programme Sectors* (2015). www.o-spi.co.uk/wp-content/uploads/2015/02/SPI-Economic-Contribution-Study-2015-02-24.pdf. Commissioned by the BFC, BFI, Ukie, Pact and Pinewood, details the impact of the Film, High-end Television, Animation and Video Games Tax Reliefs on the economy, infrastructure and job creation.

6 Creative Skillset, *The Creative Media Workforce Survey* (London: Creative Skillset, 2014). https://creativeskillset.org/assets/0001/0465/Creative_Skillset_Creative_Media_Workforce_Survey_2014.pdf.

7 Creative Skillset, *Creative Industries Workforce Survey* (London: Creative Skillset, Autumn 2015).

8 Office for National Statistics, *Labour Force Survey, October-December 2015* (London: Office for National Statistics, 2015). www.ons.gov.uk/employmentandlabourmarket/peopleinwork/employmentandemployeetypes/bulletins/uklabourmarket/ december2015.

9 Government announced an Apprenticeship levy in Summer Budget 2015. The levy is to apply to all UK employers in both the private and public sectors and is payable on annual pay bills of more than £3 million. Employers can spend their levy funds on training their apprentice against an approved standard or framework. This includes either existing staff or new recruits as long as the training meets an approved standard or framework and the individual meets the apprentice eligibility criteria.

10 Creative Skillset, *The Full Picture: The Demand for Skills in UK TV Production*. (London: Creative Skillset, March 2015). https://creativeskillset.org/assets/0001/8052/The_Full_Picture_-_The_Demand_for_Skills_in_UK_TV_Production.pdf.

11 Creative Skillset, *Creative Industries Workforce Survey*, 2015.

12 Creative Skillset, *Media Workforce Survey*, 2014.

35

The Media Cannot Reflect Society if Society is Not Reflected in the Media

Creative Access[1]

[...] Our evidence is focused specifically on the question of skills and the pool of talent from which our creative sector currently recruits. There is a marked lack of ethnic diversity in the creative sector's recruitment processes, which remains a significant problem throughout the media and creative industries.

The Scale of the Problem

Media does not represent visually the society that pays its bills. There is a significant under-representation of people from black, Asian and other minority ethnic backgrounds (BAME) working in the media and creative industries.

- 60% of media workforce is graduate level;[2]
- 23% of UK undergraduates are BAME;[3]
- Yet 6% of creative sector workforce is BAME;[4]
- In television, the figure ranges from 9.5% in terrestrial broadcast to 5% in independent production;
- At senior levels the figures are much lower, circa 3%;
- Almost one third of all UK creative sector jobs are in London,[5] where 30% of the working population is BAME.

UK media is missing out on an enormous pool of talent. This is despite many years of efforts by individuals and organisations designed to improve ethnic diversity within the industry.

Diversity is Economically Important for the Creative Industries

If this problem is not tackled, in the long run it is the creative sector that will lose out: in not recruiting black and Asian workers, it is limiting its labour resource and it will be unable to understand and sell back to a significant proportion of the UK population that is non white. There is a vast pool of talent out there and the creative sector is currently failing to tap into it.

It is widely recognised that diversity enhances the creativity within an organisation.[6] From a creative company or organisation's perspective it makes both commercial and ethical sense: if you want your production to link to your audiences then your audiences have to be part of the production.

Why the Problem Persists

From Creative Access' research among media companies, colleges and universities and among hundreds of applicants looking to find ways into creative positions, it is clear that there are many reasons why access to the creative industries for young people from ethnic minority backgrounds is poor.

These include:

- Lack of awareness among BAME young of the opportunities available, especially those with no friends or family working in the sector (lacking the necessary social capital);
- The appearance of a closed shop based on networks of personal contacts, mentors and role models;
- A history of 'expenses only' internship placements for extended periods;
- Closed recruitment networks within the media;
- Limited knowledge on the part of school and college career services of the opportunities available in the creative sector.

What is particularly clear is that there is no single place for BAME young people to look for training and employment opportunities in the creative industries and that is where Creative Access comes in. In just over three years, Creative Access has placed over 500 BAME interns with 214 media partners across 13 creative industry sub-sectors. [...]

What More Can Be Done

Those holding the purse strings should use their leverage to promote greater diversity and to extend schemes such as the BFI and Channel 4 diversity charters to all broadcasters and to secure public funding for creative endeavour. This would ensure that companies applying for funding or commissions would be required to take practical steps on diversity in order to succeed in business. [...]

Notes

1 Creative Access is a charity established in 2012 which aims to provide opportunities for paid, year-long internships in the creative industries for talented young people from black and Asian backgrounds, with a view to improving their chances of securing full-time jobs and, in the longer term, increasing diversity and addressing the current imbalance in the sector. This is an edited extract from their submission to the Inquiry, http://futureoftv.org.uk/wp-content/uploads/2016/04/Creative-Access.pdf.

2 Creative Skillset, 'Workforce Survey Calls for Fairer Access to Creative Media Industries', *Creative Skillset,* 19 May 2015. http://creativeskillset.org/latest/press_office/3412_workforce_survey_calls_for_fairer_access_to_creative_media_industries.

3 See HESA, *Data and Analysis,* 2015/16. www.hesa.ac.uk/data-and-analysis

4 Creative Skillset, *Employment Census of the Creative Media Industries* (London: Creative Skillset, 2012). http://creativeskillset.org/assets/0000/5070/2012_Employment_Census_of_the_Creative_Media_Industries.pdf (This figure represents a decline from 6.7% in the previous – 2009 – census).

5 Ibid. See also DCMS, *Creative Industires: Focus on Employment* (London: DCMS, June 2015). www.gov.uk/government/uploads/system/uploads/attachment_data/file/439714/Annex_C_-_Creative_Industries_Focus_on_Employment_2015.pdf.

6 See for example this 2015 study by Vivian Hunt, Dennis Layton and Sara Prince, *Why Diversity Matters* (McKinsey & Company, 2015). www.mckinsey.com/business-functions/organization/our-insights/why-diversity-matters.

36

Does Television Represent Us?

Ken Loach[1]

[...] Does television represent us? No, absolutely not. Does it do justice to the nuances and the subtleties and the intricacies of people's lives and their concerns and their worries? No, absolutely not. It never has. It has marginally done better at some times than others. But I want to say that really broadcasting is about control. Broadcasting is about ensuring that the main tendencies of the state stay in place. Tony Benn said once: 'Britain doesn't need the KGB, it's got the BBC'. And there's a lot of truth in that. It's about control.

Back in the day, there were investigative programmes, there were dramas, there was the voice of the individual writer, which did give some variety, which did give some acknowledgement of the diversity of the life that we should reflect. The voice of the individual writer is very rare now. You have one in Liverpool, Jimmy McGovern, who's a terrific writer, fine writer, he writes brilliantly. But most writing in drama now is formulaic. The talent is there, but people are put into a situation where the formula transcends the writer because television is about making commodities, it's not about making communications. And in the making of commodities you have to shape it and fix it so that it will sell. And then you refine it so it has a shelf life, and you sell it for as long as you can, and then you drop it. Writing individual communication is much more subtle, much more personal, much more driven by what people have to say. Broadcasting is now driven by commodities, not communication. It is rarely driven by unearthing something you should know about, which we do want to know about, it's driven by questions that the powers that be are happy for you to deal with.

There's one story that reveals the role of the BBC when it was at its foundation in the early 1920s. Soon after it had been established there was a general strike in 1926, and it was a major event. Churchill, who was in the government, wanted to deal with the BBC as an agent of government. He wanted to control it. He wanted to use it as a propaganda machine. Baldwin, who was Prime Minister, said, 'no, no, no, you're very crude. What is much more convincing is if people believe the BBC is independent they will take what it says as important. If they see it as a government propaganda sheet they will ignore it'.

So what happened was, Lord Reith, who was the man in charge at the time, moved into a government office. He wrote the news the government wanted the people to hear. He even considered banning the Archbishop of Canterbury from speaking because it was thought he might be too sympathetic to the strikers. He put out government propaganda, but the people believed it because they believed the BBC was independent. The BBC has never been independent from that day to this, and that's why it doesn't represent us, because the people have interests that the BBC will not represent.

Because you have to think: who writes the news? Someone goes through all the things that come in on the teleprompter or whatever, and somebody says 'well, that's important, we'll put that in, and we'll adjoin it to that, and this is how we'll frame it'. And then you'll have the current affairs programmes. Somebody decides the editorial line. Somebody says 'we'll ask this question, we won't ask that one, that's the language we'll use, that's the subtext of our questioning'. And the BBC is the master of this because the BBC, like the British ruling classes, is urbane, sophisticated, nuanced, very subtle and knows how to appear to be fair minded while actually getting you by the goolies. And that's the subtlety of the BBC, and they tell you what to think without you realising it, and that's why it doesn't represent us. And it goes to the heart of who they are.

They've always had political pressure from the 1920s onwards, but of course it's been more intense. When I joined the BBC in the 1960s it was very class conscious, but there was a space for a few ruffians from the Midlands in our case, and from other parts of the country to come in and do stuff. That is largely closed up now. Broadcasting deals with people at the lower end of society. Benefit scroungers, poverty porn, fascist TV really. Setting people up to be diminished, demeaned, loathed, derided. You could make a list of what the BBC believes in. The BBC believes in monarchy. And how they believe in monarchy. They believe in organised religion. Most of the people in the country are probably not religious at all. When was the last time you heard a Humanist or Secularist on Thought for the Day? Don't exist. No other thoughts. No other view of what you might call one's spiritual imagination. Only organised religion gets a hearing.

Most of all they believe in the market. That is the political correctness that the BBC espouses. Don't challenge the market. Politicians are testing them, are you business friendly? That's the test. If you're not business friendly then obviously you're unspeakable. The free market equals freedom in the eyes of the establishment and has reflected through television and broadcasting. Because the BBC represents the state, not the government, so the BBC will be dismissive of the far right but give it huge coverage because they're fascinated by it. So you'll see Farage on and you'll see Trump on wall to wall. They're fascinated by the far right. Did you see Bernie Sanders as much as you saw Donald Trump? No, of course not. They hate Corbyn and the Labour party now. Absolutely hate it. [...]

The BBC's political programmes are a joke. Who thinks watching Andrew Neil with that curious haircut is ever going to be about politics? About how we live together, about how we teach our children, about how we look after each other when we're ill, how we get work. Is that about politics? Andrew Neil and a few deadbeats from Westminster. Is it hell! It's nothing to do with it. And even if they patronise us and put it in a studio somewhere outside London you'll have the same bunch of deadbeats boring us to death again. That's not about politics. Politics is about how we live, it's how we survive, it's how we treat each other.

I just want to say a couple of other things really quickly. One thing is about the huge exploitation in the broadcasting industry. It's run on people trying to get their CVs and working for nothing. It's run on trainees being forced to do overtime without payment. There's huge exploitation. Any inquiry into broadcasting must take that into account, and the BBC must stop commissioning programmes on budgets that they know will require them to exploit their workforce. [...]

The micromanagement is something else that's never mentioned when they talk about their business. When I began [my career on television], [...] there were one or two people at the top and a lot of people making programmes. Now there's lots of people telling other people what to do and the person making it. We've just done an interview for Sky. The cameraman was the recordist, he was the spark, he was driving the van, he was working the communication on top of the van. The director was holding the mic and was also the sound recordist. That is rubbish. It's not professional. We need to stop the micromanagement from the top and give proper budgets for people to make proper programmes.

Finally, there's a huge fear of privatisation, but we have to defend public broadcasting, we have to make it genuinely accountable. It has to be based in the regions with proper budgets, and then those programmes can be broadcast nationally so that we speak to each other. We want competition in ideas, we want no government control, no appointees from the government telling people what they should be organising and making. It happened under all parties, whether it was Alastair Campbell or Bernard Ingham,[2] and it's certainly happening now. It must be independent and it must be democratic, but we must defend public service broadcasting, and my god, we've got to make it better.

Notes

1 Ken Loach is a television and film director and social campaigner. This is an edited extract from his speech at the inquiry event, *Does Television Represent Us?*, hosted by Writing on the Wall festival on May 4, 2016 at the Black-E, Liverpool. The event was chaired by Lord Puttnam, and other participants included screenwriter and producer Phil Redmond CBE, Chair of Hansard Society Ruth Fox, and producer and CEO of Nine Lives Media, Cat Lewis. For the full transcript and recording of the event, go to http://futureoftv.org.uk/events/537-2/.

2 Respectively, the former director of communications under Tony Blair and the former chief press secretary for Margaret Thatcher.

37

Public Service Television in Wales

Caitriona Noonan and Sian Powell[1]

[...] Our research and our engagement with the local television sector tells us there is certainly evidence of success and a well-placed sense of optimism in Wales. The international visibility of the Welsh-language drama *Y Gwyll/Hinterland* (S4C 2013–) and investments in the new BBC drama studios in Roath Lock and by Pinewood Studios are testament to the growing confidence and capacity of the television production sector, much of which began with the reimagining of *Doctor Who* in 2005.

However, mixed with that renewed confidence is an awareness that further interventions, resources and accountability are needed if these successes are to be fully leveraged by local industry and audiences. [...]

The Crucial Role of Public Service Broadcasting and its Broadcasters in Wales

Public service broadcasting plays a crucial role in enhancing citizens' understanding of their culture, history and political system. The process of political devolution in the UK has made this event more important but also more complex. [...] In Wales particularly, there is a limited range of news sources about devolved politics. So, for example, a 2016 survey in Wales found that many readers relied on news produced in England, or UK-wide news, which has limited information about Welsh affairs.[2] Following the 2014 Scottish Referendum, further power transfer to the National Assembly for Wales is being discussed and so consideration should be made to the way in which our media portrays the differences between the nations of the UK.[3] Therefore, while the provision of an effective PSB service is crucial throughout the UK, we believe that it is central to the future of a well-informed citizenry and publicly accountable government in Wales.

One of the biggest changes in television provision in Wales relates to English-language programming. There has been a significant decrease within both the BBC and ITV in terms of both output and spend.[4] The consequence of this has been a narrowing of programmes and genres in Wales. For instance, there is little content produced specifically for Welsh audiences in the genres of arts, children's, and comedy.

While drama production has been a beacon of success in Wales, this drama rarely reflects life in Wales and Wales is solely a location for filming rather than part of the narrative setting. This is a major disappointment to both audiences and local industry who believe that Wales and Welsh life deserves/needs to be represented both to itself in its opt-out service and to the wider UK audience on network television.[5] Without such representation it is difficult to see how the BBC can fulfill at least one of its public purposes 'to reflect the many communities that exist in the UK' (BBC 2016). Successful content like *Happy Valley* (BBC 2014–) and *Last Tango in Halifax* (BBC 2013–) demonstrate that there is an appetite amongst audiences for content which is specific to a locale and based outside of London.

In order to achieve the kind of content which represents local communities, we would like to see a sustainable change within the BBC which encourages network commissioners to engage more proactively with the nations, along with a commitment to Welsh-specific content especially on network news and drama. Research by Noonan[6] suggests that in order for decentralisation of broadcasting services to be successful, three ingredients are necessary: financial resources, local decision-making and cultural commitment to change. We are confident that the provision of each of these in Wales will enhance the creative resources of the BBC going forward.

The Future of S4C and Welsh-Language Provision

S4C's contribution to the continuation and survival of the Welsh language has been well documented. With so much choice in terms of English language television channels, the importance of the one and only Welsh language television channel, S4C, should be highlighted at every possible opportunity. The role of broadcasting is to reflect audiences and their communities, and in Wales this means both in Welsh and English. Therefore it is vitally important that the Welsh language is a visible and vibrant part of the television system in the UK. Welsh language broadcasting offers Welsh speakers and learners the opportunity to hear Welsh being spoken both formally and informally within a range of contexts and on a day to day basis.

Any efficiency savings regarding PSB should acknowledge the wider significance of broadcasting in Wales.

The ongoing cultural and social impact of S4C is often overlooked when metrics concentrate exclusively on economic value. That is not to say that S4C does not offer value for money or economic support to the Welsh creative economy. Indeed its economic significance is evidenced by the diverse and highly skilled jobs it supports directly[7] and through the independent television companies and external partners it works with closely. It is impossible to compare the funding of S4C with other media organisations: S4C doesn't exist within a multi-channel context, it is the only Welsh language channel and, therefore, its unique contribution to the UK's creative and social identity must be taken into account. [...]

Notes

1 Dr Caitriona Noonan is lecturer in Media Communication in the School of Journalism, Media and Cultural Studies (JOMEC), Cardiff University. Dr Sian Powell was assistant researcher at JOMEC and is Director of Hedyn Communications. This is an edited extract from their submission to the Inquiry, http://futureoftv.org.uk/wp-content/uploads/2016/05/Sian-Powell-and-Caitriona Noonan.pdf.

2 Steven Cushion and Roger Scully, 'British Media is Failing to Give Voters the Full Picture Ahead of Elections', *The Conversation,* 4 April 2016. http://theconversation.com/british-media-is-failing-to-give-voters-the-full-picture-ahead-of-elections-57020.

3 Sian Powell, 'Wales, Devolution and the Scottish Independence' *Scotland's Referendum and the Media: National and International Perspectives,* eds. Neil Blain, David Hutchinson and Gerry Hassan (Edinburgh: University of Edinburgh Press, 2016).

4 IWA, *Media Audit 2015.*

5 Ruth McElroy, Caitriona Noonan, 'Television Drama Production in Small Nations: Mobilities in a Changing Ecology', *Journal of Popular Television* 4, 1(2016), 109–27. Accessed 24 May 2017, doi: 10.1386/jptv.4.1.109_1.

6 Caitriona Noonan, 'The BBC and Decentralisation: The Pilgrimage to Manchester', *International Journal of Cultural Policy* 18, 4 (2012), 363–77. Accessed 24 May 2017. doi:10.1080/10286632.2011.598516.

7 There were 129 according to S4C. See S4C, *Annual Report 2015* (Cardiff: S4C, 2015). www.s4c.cymru/abouts4c/annualreport/acrobats/s4c-annual-report-2015.pdf.

38

Public Service Broadcasting: A View from Scotland

Robert Beveridge[1]

[...] Scotland has given much to the values, structures and content of public service broadcasting. One need only think of the four Johns: John Logie Baird (1888–1946), inventor; John Reith (1889–1971), founding Director General of BBC; John Grierson (1898–1972), GPO Film Unit and Documentary Theorist; and John Gray (1918–2006), producer and Director of *West Highland*, the last of the lyrical documentaries and co-founder of Edinburgh TV Festival.

However, the relationship between broadcasting in Scotland and broadcasting on a UK basis is analogous to that between the UK film industry and Hollywood. In Lord Puttnam's apposite and acute analysis of the latter, it could be said to be an 'Undeclared War'.[2]

At the very least, it can be defined as neglect: sometimes benign; sometimes malign but in many cases, and over many decades, a refusal to enable the culture(s) and identities of the nation of Scotland to find full expression.

Media and broadcasting policy is an articulation of the balance of powers in a given State. They generally reflect specific historical and political circumstances and are always, even by default, an expression of a settlement between various interest groups and stakeholders in the context of the prevailing zeitgeist. It was no surprise therefore when the Scotland Act 1999 ensured that powers over broadcasting remained reserved to Westminster.

It was no surprise therefore that when Ofcom was established, there was no place at the top table for the voice of the nations, including Scotland. This policy decision – to have partners on the main board of Ofcom rather than representatives – struck at the heart of democratic accountability, however imperfect, and threw away decades of representation for Scotland, via named members, on the regulators who had been merged into Ofcom.

It was no surprise therefore that when the Controller of BBC Scotland, John McCormick, proposed changes to the BBC Scotland news offer that this was rejected by the then BBC Governors, following what can only be described as a conspiracy between the then Director General John Birt and the New Labour government, thus compromising the independence of the BBC in the interests of the British state.

BBC Scotland had proposed, in preparation for devolution, that the news offering from BBC Scotland be enhanced by the introduction of what became known as the 'Scottish Six' i.e. the national (Scottish), national (UK) and international news edited and broadcast from Glasgow.

This was a step too far for the British state but was clearly not in the interests of licence fee payers in Scotland nor of the BBC which has continued throughout this charter to exhibit worrying and low levels of audience approval in Scotland.[3]

It is not a surprise therefore that the concerns of citizens and consumers in Scotland extend beyond the BBC into Channel 3 and Channels 4 and 5. In November 2014, I enquired into the steps that have been taken by Ofcom to ensure that its licencees take full account of the evidence on problems in reporting – with due impartiality, accuracy and balance in news and current affairs – coverage of the four nations. In particular, I asked how were the lessons of the King report addressed by Ofcom and what steps have been taken to monitor and report upon these issues?[4]

Ofcom's reply was far from satisfactory and basically stated that they already had guidelines and would continue to monitor their licencees and judge complaints on a case by case basis. They apparently chose not to draw the attention of their licencees – affecting for example ITN and Channel 4 News – to research which was and remains relevant in terms of the balance, impartiality and accuracy of reporting stories in the complex context of a changing UK. [...]

The examples provided are but a few of the ways in which Scotland's broadcast structures, cultures and operations have had a mixed experience in the field of broadcasting regulation and policy. [...] However, it would be churlish to state that there has been no progress and it is important to draw attention to the establishment of BBC Alba and the enormous success, evident in many indicators, of the quality and appreciation for broadcasting in the Gaelic language.

Indeed, MG Alba and BBC Alba provides a salutary lesson as to what can be achieved from minimal investment and one can only hope that further investment and statutory or charter security for this indigenous language broadcasting can be secured. Further investment over even a few million pounds would reap substantial economic, cultural and democratic benefits. [...]

Recommendations

The Scottish Parliament should be fully responsible for media policy and media regulation in and for Scotland, including BBC Scotland. This would require an amendment to the Scotland Act so that powers over broadcasting are no longer designated as reserved. [...]

The scale and scope of BBC Alba should be increased. The BBC must increase its programme contribution to BBC Alba from the current 230 hours per annum to 520 hours to match that of S4C.

Having the headquarters of a channel with funds and commissioning power based in Scotland would transform the Scottish broadcasting and creative industry sectors. Incidentally, it might also help the creative industries in the North of England and Northern Ireland. At a stroke, this would increase production and thus economic impact and investment.

BBC Scotland needs to have control of its own scheduling and to adopt an opt-in rather than an opt-out policy towards programming, thus taking account of the distinct and distinctive nature of the Scottish television and media market and patterns of consumption. There is no reason why BBC Scotland cannot have some of the same independence as STV in relation to London centre. Should in-house production quotas and terms of trade allow greater competition and what impact could this have on the Scottish broadcasting industry? [...]

The BBC should be required to achieve approval of at least 60% in terms of licence fee payer satisfaction with performance in relation to the Corporation's duty to represent and serve all of the nations of the UK – the fourth of the BBC's five public purposes. [...] Failure to meet this target should result in reductions in performance related pay for both the Director General and the Director, Scotland as a minimum. This recommendation comes after a decade, if not decades, of underperformance in Scotland and failures to devise and implement policies which enable the BBC to have reduced the purpose gap – despite substantial evidence of the problem and exhortations from the Audience Council of Scotland and others to do so.

The real issue is what levers exist to ensure that there is appropriate executive action when the purposes are not fully achieved. The evidence has been clear for decades. The licence fee payer in Scotland deserves better.

The BBC tries to support distinctive Scottish content but – to take one example which is also a value for money issue – *River City* is not shown on the BBC network across the UK. Given the substantial investment in this series – over many years – one has to wonder quite why it has not been shown in peak time or even at another time. It is not necessarily my choice of programming but that is neither here nor there. Why is *River City* not on the BBC One UK network and what are the implications of this?

This is not a new problem: and it affects how much can be spent on distinctive Scottish content. In Scotland we are exposed to numerous adaptions of Jane Austen but have yet to make a television series of the renowned, popular and high quality Scotland Street novels by Alexander McCall Smith although the BBC in London managed the *Number One Ladies Detective Agency*. [...]

The BBC can also enhance support and development of talent and skills in Scotland. The BBC is the UK's best training operation in television and radio but the proposal above to establish a federal BBC would bring about enhanced support and development on talent and skills in and for Scotland.

There needs to be more investment and more programming in Edinburgh, the capital of Scotland. At present, Edinburgh must be the only capital city of its stature and status in the world to have such a poor broadcasting infrastructure. At a minimum, BBC Scotland needs to establish a multimedia studio to do more than radio phone-ins and Parliamentary reporting. The BBC's proposal for the creation of an interactive digital service for each of the Nations of the UK could provide an opportunity for Edinburgh to be better served. At present the commercial broadcaster STV does a better job than the BBC of reporting Edinburgh. [...]

Notes

1 Robert Beveridge is Professor at the University of Sassari, Sardinia. This is an edited extract of his submission to the Inquiry, http://futureoftv.org.uk/wp-content/uploads/2016/02/Robert-Beveridge.pdf.

2 David Puttnam, *The Undeclared War: Struggle for Control of the World's Film Industries* (London: Harper Collins, 1997).

3 See the BBC Trust's research into purpose gaps, NatCen, *Purpose Remit Survey UK Report* (BBC, Winter 2012–2013), http://downloads.bbc.co.uk/bbctrust/assets/files/pdf/review_report_research/ara2012_13/prs_reports/uk.pdf.

4 BBC Trust, *Impartiality Report: BBC Network News and Current Affairs Coverage of the Four UK Nations* (London: BBC Trust, June 2008), http://downloads.bbc.co.uk/bbctrust/assets/files/pdf/review_report_research/impartiality/uk_nations_impartiality.pdf.

Part Six

Content Diversity

39

Content Diversity[1]

One of the key ways to ensure a healthy public service ecology is to maintain a rich and heterogeneous provision of programming. British television, thanks to its public service tradition, is well known for the wide variety of genres that have helped to provide a diversity of cultural expression. These genres enable public service broadcasters to engage with a range of subject matters, both familiar and new, and to entertain, challenge and expose audiences to different experiences. Some of those genres – such as big entertainment, quiz shows, reality and comedy – fulfil entertainment values and are in good health,[2] while others are in crisis, due to rising costs, a highly competitive pay TV market and the scaling down of commitments following changes to the quota regime in the 2003 Communications Act.

Genres that have been traditionally associated with public service broadcasting – such as education, natural history, science, arts, current affairs, children's and religion – have now been in steady decline for over a decade. Public service channels produce by far the highest levels of original content in these genres and, despite the introduction of tax relief for certain areas including high-end drama, live-action children's programming and animation,[3] spending across all genres on first-run original programmes fell by 15% between 2008 and 2014.[4]

A shift to on-demand viewing in recent years has further segmented our viewing habits. Although the vast majority of our viewing continues to be live, some genres are increasingly viewed on catch-up services. Big entertainment shows and sports events often account for the highest proportion of live viewing, compared to drama series, which have the highest proportion of on-demand viewing.[5] These trends are significant as they point to the increasing complexity of maintaining public service mixed genre provision given an increasing reliance on 'big data', consumer preferences and taste algorithms that may serve to limit the diversity and visibility of a broad range of genres.

In particular, creating a programme in a more fragmented television landscape that reaches a 'mass' audience and that contributes to a shared cultural life represents a considerable challenge. Today, that responsibility increasingly lies with the 'big entertainment' shows that have traditionally occupied primetime weekend evening slots, and, together with drama, are the most popular genre with the highest audience share at 17%. These shows are costly – a 14-week run of BBC One's *Strictly Come Dancing* or ITV's *X Factor* costs in excess of £20 million, as they often involve a long production cycle.[6] Nonetheless as talent shows generate several hours of programming each week, their cost per hour remains lower than that of drama. But they are hugely important to public service channels, who are the most successful innovators of entertainment genres and biggest producers of television entertainment formats, with the ability to commission 'more new titles every year than any other TV system in the world',[7] because of the ratings and profile they generate. Other genres, such as children's, arts and current affairs for example, are in a far more fragile condition for a variety of reasons.

News and Current Affairs

Television news has for over half a century been one of the most valued and popular public service genres. It remains the key platform through which ordinary citizens access news with two-third of adults turning to television compared to 41% who go online.[8] The traditional narrative is that through the provision of impartial and accurate information across a range of domestic and international topics, television news has sought to develop informed citizenship and to promote active participation in democratic processes. Current affairs complements these noble objectives and, through research and in depth analysis, aims to investigate events of interest to the public and to monitor the affairs of powerful elites. Yet, both genres are now in crisis albeit in different ways.

The crisis is not one of falling levels of output. Buoyed by the growth of a multichannel environment, total news and current affairs output has actually increased across the PSB networks since 2009. The picture is uneven of course: while hours have increased on BBC One and Channel 5, they have fallen on BBC Two, ITV and Channel 4.[9] It is partly a crisis of investment – spending declined by 14% between

2008–2014[10] while at the same time newsrooms are producing more hours of material with fewer staff.

In terms of television news, however, the crucial development is that audiences are simply starting to switch off. While the average viewer watched 119 hours a year in 2010, this had fallen to 108 hours by the end of 2014, a decline of some 10% in four years.[11] The change is particularly intense on the public service channels: Thinkbox, using BARB figures, reports an 18% decline since 2003 in audiences for TV news bulletins on the public service channels in contrast to only a 7% decline across the whole multichannel environment.[12] Some broadcasters are feeling the effect more than others – viewing figures for the *ITV Evening News*, for example, had dropped from 3.4 million viewers on weekday evenings in 2010 to 2.5 million by late 2015[13] while the BBC's share of a declining field has actually increased in recent years.

The crisis is particularly acute when it comes to younger audiences who are far more developed in their consumption of online news. Thinkbox reports that 16–34 year olds now watch an average of six minutes a day of TV news on the main channels, down from 13 minutes a day in 2003. 16–24 year olds watch even less TV news with their annual consumption down 29% since 2008 with the average young adult watching only just over two minutes a day.[14] This reflects an attitudinal as well as a technological shift. As Luke Hyams, formerly of short-form video specialists Maker Studios, told us at our launch event:

Generation Z are growing up in a time when these ever present social media platforms are free, unfiltered and enable anyone with a smartphone to become a broadcaster. The voices and opinions that they interact with on these services are devoid of perceived journalistic bias or an agenda and tell it like it is with a raw, unedited delivery that most young people have come to accept as the norm.

Where does this leave public service broadcasters like the BBC? The independent position and distinct voice of the corporation seems to be lost upon swathes of this generation, who look upon the BBC as just another voice of authority in an increasingly crowded media landscape.[15]

We therefore agree with the authors of a recent report on the future of television news that the crisis needs to be immediately addressed. 'Television news is still a widely used and important source of news, and will remain so for many people for years to come, but if television news providers do not react to the decline in traditional viewing and the rise of online video ... they risk irrelevance.'[16]

None of this means that television bulletins have lost their influence as a source of news and television is still, according to Ofcom, 'by far the most-used platform for news.'[17] Jeremy Tunstall argues that the BBC remains the UK's news agenda setter[18] and, while social media and online video are central to any future vision of the news, the role of existing PSB news providers is still crucial in shaping how we talk about matters of public interest. For example, in Northern Ireland, the media academic Ken Griffin told us that the BBC and UTV remain 'the main source of objective news and current affairs coverage' and are able to offer an alternative to the country's print media which, he argues, 'consistently exhibit political bias'.[19] Similarly, in Wales, Sian Powell and Caitriona Noonan of Cardiff University argue that there is only a limited range of news sources about devolved politics in Wales and so the need for effective public service broadcasting is 'central to the future of a well-informed citizenry and a publicly accountable government in Wales'.[20]

Indeed, social media have not replaced the ability of the major news bulletins to set the tone for ongoing national debates around major political issues like elections and economic matters. It is, we believe, a sign of the increasing politicisation of the whole media landscape (and therefore a reminder of the need for the BBC and other broadcast organisations to be meaningfully independent in editorial matters) that serious complaints were made, for example, about the BBC's coverage of the independence referendum in Scotland in 2014 and about its approach to austerity where, as one researcher concluded, its bulletins were 'almost completely dominated by stockbrokers, investment bankers, hedge fund managers and other City voices'.[21] Impartiality is a worthwhile objective as long as it is not used to police the divisions that burst to the surface at times of major political conflict.

Public service news media must meet especially demanding standards of impartiality when dealing with topics where there are significant differences of opinion (although, of course, they should not seek to avoid topics that are deemed to be 'controversial'). Impartiality is not secured merely by allocating similar amounts of time to 'pro' and 'anti' voices. Many issues that matter for the public, or for specific sections of the public, are complex and there should be no expectation either that there are only two positions to be covered or that the 'Westminster consensus' is necessarily the most appropriate starting place. On the other hand, neither does impartiality refer to the affordance of equal airtime to 'sense' and 'nonsense'. According to Professor Steve Jones, the BBC's coverage of climate change, for example, has at times given unwarranted attention to a small number of climate change 'deniers': 'Attempts to give a place to anyone, however unqualified, who claims interest can make for false balance: to free publicity to marginal opinions and not to impartiality but its opposite'.[22] Impartial coverage requires both an engagement with a *range* of informed positions and a commitment to drawing on credible evidence as opposed to unsubstantiated claims.

The nature of the 'crisis' in current affairs is rather different. There was a steep decline in current affairs provision in the 1980s and 1990s[23] followed by a 35% fall in output between 1992 and 2002.[24] Yet, in recent years, far from falling off a cliff, the average consumption of current affairs appears to be increasing with a 52% rise in viewing time since 2003 across all channels (albeit with a slightly smaller rise of 23% on the main PSB channels).[25] Ofcom figures also show a 10% rise in hours produced across the schedule between 2009–2014 with BBC Two and Channel 5 showing increases of nearly 60%. The situation is not quite so rosy when it comes to peak-time current affairs where a majority of the overall increase is accounted for by the BBC's digital news channel and where both BBC One and ITV show less than one hour a week of current affairs.[26]

The problem, therefore, is not about the total number of hours transmitted but with the very delicate position that current affairs occupies in a ratings-driven environment. Despite the public's appetite for high quality investigations and analysis, current affairs programmes remain expensive to produce and do not attract the largest audiences. That they still continue to feature in prime-time schedules is largely to do with the obligations placed on public service broadcasters by regulators. According to one anonymous producer quoted in a 2013 report on the future of current affairs, 'if there were no regulation, current affairs would disappear overnight. It would legitimise the race for ratings'; another argued that 'broadcasters' commitment to current affairs is dubious and is slipping fast. They are doing our stuff, but grudgingly, because they have to. There is relentless pressure to soften what we do'.[27]

There appears to be a quite different atmosphere – and of course a very different financial landscape – from the 1970s when programmes such as ITV's *World in Action, This Week* and *Weekend World* reached collectively 20 million viewers a week and had huge resources thrown at them.[28] The small increase in peak-time current affairs output since 2009 has been matched by a 14% fall in spending and there is anecdotal evidence, according to Steven Barnett, 'that there is now more emphasis on the personal, the human interest and on celebrity issues than in the late 1990s'.[29] 'Infotainment' is gradually replacing output that used to focus on international stories and costly investigations. Channel 4's *Dispatches* and the BBC's *Panorama* remain the cornerstones of this latter genre but they are becoming increasingly reliant on 'safer' topics such as consumer or lifestyle stories.[30] In light of these shifts, we want to reiterate our commitment to the democratic importance of 'accountability journalism'.[31] We believe that not only should the quotas remain (and in the case of ITV, as we have already argued, be increased) but that there needs to be a revival, monitored by Ofcom, of the 'hard-hitting' investigative strands that have produced some of the most celebrated output of British television like *World in Action*'s programmes on thalidomide in the 1970s.

We believe that the lives and concerns of all citizens, but especially young people and ethnic and other minorities, are too often underserved by the journalism of existing public service providers. Young people, for example, often don't see their world and their concerns covered in a comprehensive and relevant manner. This alienates them and pushes them towards more energetic newcomers such as Vice Media who operate outside of the formal public service compact. The dominant culture of journalism fails to reach these and other minorities and too often seeks to manufacture an unsatisfactory consensus by over-representing the centre ground. At a time of growing disillusionment with traditional parliamentary politics and, especially, in the light of increased devolutionary pressures, we believe that news providers need to adopt not simply a more technologically sophisticated grasp of digital media but a model of journalism that is less wedded to the production of consensus politics and more concerned with articulating differences. Television, as Richard Hoggart reminds us in relation to the Pilkington Inquiry, 'should not hesitate to reflect 'The quarrel of this society with itself', even though politicians may not like the result'.[32] We believe that this is the case today just as much as it was in 1962.

Drama

Drama, including soaps, is one of the most popular genres associated with the remit of public service. The genre's popularity, with an average audience share of 17% in 2015, is matched by its high costs. As one of the most expensive genres, a typical, prime-time homegrown drama costs between £500,000 and £1 million per hour.[33] While public service channels continue to be the highest investors in the genre, Ofcom's 2015 review of public service broadcasting reported a 31% fall in investment in original drama since 2008.[34] Although audience satisfaction with drama is stable,[35] BARB figures show that the average time spent watching drama series and soaps on the main channels fell by 50% between 2003 and 2015.[36]

This does not appear to signal a lack of interest in drama itself as falling levels of investment by PSBs has been, at least in part, offset by a huge increase in co-productions and pay TV platforms offering globally appealing US content. Streaming services such as Netflix and Amazon appeal to younger demographics, and the subscription take-up has been exponential, with almost a quarter of UK households subscribed to Netflix by the end of 2015.[37] They are changing our viewing habits too, with 'binge viewing' becoming an increasingly popular way of engaging with quality, complex drama. The domination of US content is also clear with a doubling of American scripted shows, from 200 to an estimated 409, with content produced for streaming media experiencing the largest jump.[38] Netflix has recently promised to spend $5 billion on programming and to produce 600 hours of new content in 2016 alone.[39]

While this increase is not directly linked to the fall in drama spending in the UK, it clearly makes the market more competitive, a situation welcomed by the BBC's head of drama, Polly Hill, who points to the need for PSBs to take risks in developing fresh ideas and engaging scripted content.[40] Yet, while this might be an extra push to increase the already high standards of UK drama, not all public service channels are able or willing to take up this challenge, with ITV, a channel traditionally associated with high-end dramas such as *Downton Abbey* or *Mr Selfridge,* recording an alarming 65% drop in drama investment since 2008.[41]

The pressure to produce popular, high-budget drama has also led to an increasing dependence on US investment – reflected in a growing reliance on UK/US coproductions. The BBC's recent adaptation of John Le Carre's novel *The Night Manager* was coproduced with US TV channel AMC while Andrew Davies' adaptation of Tolstoy's *War and Peace* for BBC One was coproduced with the Weinstein Company. Netflix has also invested in UK specific content, namely the British drama *The Crown,* filmed at Elstree Studios, but also the third season of Charlie Brooker's *Black Mirror,* with Channel 4 losing the right to show the season's first run as a result.[42] This is not limited only to high-end drama and the US market alone. Michael Winterbottom's quirky *The Trip* moved from the BBC to Sky Atlantic for its third season, reflecting the BBC's inability to 'compete with the financial resources which Sky Atlantic was able to commit'.[43]

While there are clear advantages and benefits to such collaborations, notwithstanding increased investment opportunities and increased international recognition for British talent and content, US/UK co-productions tend to cultivate a specific subgenre of 'period and fantasy world dramas'[44] or novel adaptations, which do not necessarily lead either to risk-taking or, for that matter, making 'British stories for British audiences' but to content with a broadly international appeal. This is an issue that has been repeatedly addressed in relation to the need for television to reflect the full diversity of life in the UK as we discussed in the previous chapter and that was raised at several of our events.[45] Furthermore, much like sports, commercial pressures have, in the past, resulted in audiences losing out, with quality scripted shows like *Mad Men*, *The Wire* and *The Sopranos* migrating behind paywalls, out of reach of their loyal viewers who had previously watched them on public service channels.

Notes

1 Edited extract from Chapter 10 of the Puttnam Report, http://futureoftv.org.uk/wp-content/uploads/2016/06/FOTV-Report-Online-SP.pdf.

2 Ofcom, *Public Service Broadcasting in the Internet Age: Ofcom's Third Review of Public Service Broadcasting* (London: Ofcom, July 2015). See also Jeremy Tunstall, *BBC and Television Genres in Jeopardy* (New York: Peter Lang, 2015).

3 Mark Sweney, 'Children's TV Gets Tax Break for Live-Action Productions in Autumn Statement', *The Guardian*, 3 December 2014, www.theguardian.com/media/2014/dec/03/childrens-tv-tax-break-autumn-statement-george-osborne.

4 Ofcom, *PSB in the Internet Age*, 7.

5 Oliver & Olhbaum and Oxera, *BBC Television, Radio And Online Services: An Assessment Of Market Impact And Distinctiveness*, (London: O&O, 2016) 56.

6 According to Richard Holloway the shows take up anything between 43–46 weeks to produce. In Tunstall, *BBC and Television Genres in Jeopardy*, 293.

7 Oliver & Olhbaum and Oxera, *BBC Television, Radio And Online Services*, 62.

8 Ofcom, *News Consumption in the UK: research report* (London: Ofcom, December 2015), http://stakeholders.ofcom.org.uk/binaries/research/tv-research/news/2015/News_consumption_in_the_UK_2015_report.pdf.

9 Ofcom, *PSB Annual Report 2015: Output and spend index* (London: Ofcom, July 2015), http://stakeholders.ofcom.org.uk/binaries/broadcast/reviews-investigations/psb-review/psb2015/PSB_2015_Output_and_Spend.pdf, 23.

10 Ofcom, 'Public Service Broadcasting in the internet age' [paper presented to the Inquiry, September 29, 2015].

11 Ofcom, *News Consumption*.

12 Thinkbox, *TV Viewing Across the UK* (London: Thinkbox, 2016).

13 Rasmus Kleis Nielsen and Richard Sambrook, 'What is Happening to Television News?' (Oxford: Reuters Institute for the Study of Journalism 2016), 9.

14 Ofcom, *PSB in the Internet Age*, 11.

15 Luke Hyams, 'Do Young People Care About Public Service Broadcasting?', *Huffington Post*, 4 December 2015, www.huffingtonpost.co.uk/luke-hyams/do-young-people-care-about-psb_b_8706282.html.

16 Nielsen and Sambrook, 'What is Happening to Television News?', 3.

17 Ofcom, *News Consumption*, 2.

18 Tunstall, *BBC and Television Genres in Jeopardy*, 143.

19 Ken Griffin, submission to the Inquiry, http://futureoftv.org.uk/wp-content/uploads/2016/05/KenGriffinNI.pdf.

20 Sian Powell and Caitriona Noonan, submission to the Inquiry, http://futureoftv.org.uk/wp-content/uploads/2016/05/Sian-Powell-and-Caitriona-Noonan.pdf.

21 Mike Berry, 'Hard Evidence: How biased is the BBC?' *New Statesman*, 23 August 2013, www.newstatesman.com/broadcast/2013/08/hard-evidence-how-biased-bbc.

22 Steve Jones, 'BBC Trust review of impartiality and accuracy of the BBC's coverage of science', *BBC,* July 2011, http://downloads.bbc.co.uk/bbctrust/assets/files/pdf/our_work/science_impartiality/science_impartiality.pdf, 16.

23 Steven Barnett, *The Rise and Fall of Television Journalism* (London: Bloomsbury, 2011).

24 David Bergg, 'Taking a Horse to Water? Delivering Public Service Broadcasting in a Digital Universe', in *From Public Service Broadcasting to Public Service Communications,* ed. Jamie Cowling and Damien Tambini (London: IPPR, 2002), 12.

25 Thinkbox, *TV Viewing in the UK,* 2016.

26 Ofcom, *PSB Annual Report 2015: Output and spend index,* 30–31.

27 Quoted in Jacquie Hughes, *An Uncertain Future: The Threat to Current Affairs* (London: International Broadcasting Trust, 2013), 11.

28 Jeremy Tunstall, submission to the Inquiry. http://futureoftv.org.uk/wp-content/uploads/2016/03/Jeremy-Tunstall.pdf.

29 Quoted in Hughes, *An Uncertain Future,* 12.

30 Tunstall, *BBC and Television Genres in Jeopardy,* 184.

31 Hughes, *An Uncertain Future,* 4.

32 Richard Hoggart, *A Measured Life* (London: Transaction Publishers, 1994), 66.

33 Tunstall, *BBC and Television Genres in Jeopardy,* 91.

34 Ofcom, 'Public Service Broadcasting in the internet age' [paper presented to the Inquiry, September 29, 2015].

35 Ofcom, *PSB in the Internet Age.*

36 Thinkbox, [paper presented to the Inquiry, April 19, 2016].

37 Jasper Jackson, 'Netflix Races Ahead of Amazon and Sky with 5m UK Households', *The Guardian,* 22 March 2016, www.theguardian.com/media/2016/mar/22/netflix-amazon-sky-uk-subscribers-streaming.

38 Josef Adalian, 'There Were Over 400 Scripted TV Shows on the Air in 2015', *Vulture,* 16 December 2015, www.vulture.com/2015/12/scripted-tv-shows-2015.html#.

39 Nathan McAlone, 'It Would Take 25 Days to Binge-Watch all the New Netflix Original Content Coming out this Year', *Business Insider,* 6 January 2016, http://uk.businessinsider.com/new-netflix-original-content-in-2016-would-take-25-days-to-watch?r=US&IR=T.

40 Hannah Furness, 'BBC has Nothing to Fear from Netflix or Amazon, Head of Drama says', *Daily Telegraph,* 28 December 2015. www.telegraph.co.uk/news/bbc/12067689/BBC-has-nothing-to-fear-from-Netflix-or-Amazon-head-of-drama-says.html.

41 Ofcom, *PSB in the Internet Age,* 12. It should be noted that 2008 marked a highpoint of investment prior to the crash that saw ITV deliberately shift to factual commissioning in the face of a significant drop in advertising revenue.

42 John Plunkett, 'Netflix Deals Channel 4 Knockout Blow over Charlie Brooker's Black Mirror', *The Guardian,* 29 March 2016. www.theguardian.com/media/2016/mar/29/netflix-channel-4-charlie-brooker-black-mirror.

43 Adam Sherwin, 'The Trip: Steve Coogan and Rob Brydon take Hit TV Show to Sky', *The Independent,* 15 February 2016 www.independent.co.uk/news/the-trip-steve-coogan-and-rob-brydon-take-hit-tv-show-to-sky.

44 Oliver & Ohlbaum and Oxera, *BBC Television, Radio And Online Services: An Assessment Of Market Impact And Distinctiveness,* 52.

45 For example, 'Are You Being Heard? Representing Britain on TV' held at Goldsmiths, 22 March 2016; 'Future for Public Service Television – Inquiry Event for Wales', 6 April 2016; 'Does Television Represent Us?' Held at the Black-E, Liverpool, 4 May 2016. http://futureoftv.org.uk/public-events/.

40

Children and Public Service Broadcasting

Sonia Livingstone and Claire Local[1]

[...] Children are defined as persons aged from 0–17 years old in the UN Convention on the Rights of the Child, ratified by the UK.[2] Children represent one fifth of the UK population (and the entirety of the future population), yet in discussions about broadcasting, they are often overlooked as a group with specific needs and are easily lost under umbrella terms ('audience', 'public', 'viewers', 'households', 'population'). Given broadcasters' practical struggle to appeal to teenagers, they appear tempted to define 'children' as under 12. Problematically, the BBC and other public service broadcasters (PSBs) have substantially cut provision for teenagers and, increasingly, for younger groups.[3] But children's needs from infancy through to adolescence should be recognised and provided for, as children develop intellectually, emotionally and socially.

Much has been said on the future of public service content, the growth of multiple platforms, new market and regulatory pressures, and changing audience preferences and practices, among other widely debated topics.[4] However, little attention has been paid to the role that public service television plays in educating, entertaining and broadening the horizons of children in the UK. This paper focuses on how public service television[5] can better serve a child audience that spends on average at least 35 hours per week consuming broadcast, on-demand and online content.[6]

We divide the domain of children's content by defining 'broadcasting', 'television' and 'public service' as shown in Table 40.1. While our focus is on the two cells that encompass 'public service television', we argue that it is crucial to grasp the relationships between public service television and commercial content on the one hand, and between public service television and other public service content on the other.

Table 40.1

Forms of Children's Content.

Children's content		Public service content	Commercial content
Television	Television on a TV set *Live and recorded viewing on a TV set*	e.g. CBBC programmes	e.g. CITV or the Disney Channel
	Television not on a TV set *Live viewing on a network connected device, such as a laptop, tablet, smartphone, games console etc.*	e.g. CBBC programmes via iPlayer live	e.g. CITV online live
	Other television *On-demand viewing on a network connected TV or other device, such as a laptop, tablet, smartphone, games console etc.*	e.g. CBBC programmes viewed via iPlayer or YouTube	e.g. CITV or Disney programmes viewed via CITV online or YouTube
Other content	*Other content* *Includes audiovisual (e.g. games, film) and print, web, music and other content*	e.g. information and games on www.bbc.co.uk/cbbc, NASA Kids' Club site, Wiki_for_Kids, KidzSearch	e.g. CITV or Disney web content, Miniclip, MovieStarPlanet

Children Still View Public Service Television on a Television Set

Care is needed regarding popular claims about children's changing media practices as they are easily overstated and often under-evidenced.[7] Despite pessimistic predictions about children and TV, children are not deserting broadcast television in general, or public service television in particular. According to Ofcom's 2015 *Children and Parents: Media Use and Attitudes Report*, 'the amount of time 8–11s and 12–15s spend online has more than doubled since 2005, with 12–15s now spending more time online than watching TV' – where 'watching TV' is defined as watching TV exclusively *on a TV set*.[8] But for children aged 3–11, viewing on a TV set exceeds internet use. For 12–15 year olds, although internet use now exceeds television viewing on a TV set, they nonetheless watch as much or more television on a TV set as do 3–11 year old children.[9] Indeed, 96% of children aged 5–15 use a TV set to watch television and the majority (87%) of viewing of broadcast TV among 4–15 year olds is of live television.[10] This matters because of the social situation that such viewing is typically associated with.[11] [...] It also matters because of issues of social and digital inequality and inclusion. The assumption that all children are able to access content via the internet neglects the minority who lack internet connectivity at home: Ofcom estimates one in ten 8–11 year olds and one in twenty 12–15 year olds are without internet at home or elsewhere.[12] In short, children's public service broadcasting on television continues to serve a valuable and valued function in UK society.

Is Children's Television Viewing Really in Decline?

The above data are insufficient for claims and predictions about children's changing media practices. While it may seem obvious that children are increasingly watching television on devices other than a TV set – usually a tablet or laptop – to the best of our knowledge this trend has not been measured. We know 96% of children aged 5–15 view television content on a TV set, and 45% of the same age range view television on other devices.[13] We do not know how much time children spend watching TV on devices other than a TV set, nor how many of the hours spent 'using the internet' include viewing television content. Thus we do not know the balance between time spent watching TV on a TV set and on an internet-enabled device, nor the balance between time spent on TV content and other online content.[14]

[...] Discussion of PSB in the current media landscape must distinguish television content from television viewing devices and measure both, by age group. Are children moving away from live TV to on-demand services? Are they replacing TV content with other activities such as (non-TV content on) YouTube or online games? Without answers to such questions, we cannot say how much time in total children spend consuming PSB services offline or online, or evaluate how valuable (or not) PSB provision may be to children. It would be premature to determine future provision of public service television viewed by children without clear answers.

Yet this has already happened. BBC Three, the BBC's 'youth' channel whose target audience includes 16 and 17 year olds, now exists only as an online service[15] and no PSB in the UK offers systematic programming across the full age range of children (0–17).[16] This contravenes Article 17 of the UN Convention on the Rights of the Child, ratified by the UK, which stipulates that, concerning media, children should have access to a variety of information and material.[17] This lack of provision is often framed as broadcasters responding to children's preferences for accessing content, but it cannot be distinguished from the alternative, that children are responding to broadcasters' reduction in provision for them.

The Case for Online Provision of Children's Public Service Television

[...] The BBC is the only PSB to have an online platform dedicated to children's content that neither collects their personal data nor carries commercial sponsorship or advertising.[18] ITV and Channel Five have online platforms targeted at children (CITV and Milkshake TV) but both include adverts.[19] Channel 4 does not have an online platform dedicated to children nor a dedicated space on its online website, All 4. For children, therefore, options for viewing non-commercial public service television content are limited.[20] This matters both because of the adverse effects of exposure to advertising[21] and because commercial broadcasters tend to omit a range of content of value to children.[22]

Meanwhile commercial services increasingly target the child audience (or 'market'). Either providers of children's online content collect and exploit children's personal data or the boundary between advertising and programming is increasingly blurred.[23] Paid advertising on digital platforms is subject to guidelines, but 'commercial' content is not. This is problematic insofar as Youtube becomes increasingly popular with children, where they can watch vloggers, 'unboxing' videos (where presenters discuss new products they have bought), and other 'endorsement' videos.[24] [...]

In addition to legitimate concerns surrounding children's increasing exposure to commercial content, any decline in public service provision risks the loss of positive opportunities for children to engage with quality content that informs, inspires and entertains them.

The Case for Online Provision for Children of Other Public Service Content

It is increasingly difficult (and inadvisable, in terms of children's experiences of content) to evaluate the contribution of television content separately from the proliferation of other forms of online content and services available to children online – think of web content, games, quizzes, parental guidance, links to further content options (both television and other), online communities, and so forth.

It seems obvious that online content of all kinds can, and do, enhance the experience of television content, including public service content.[25] It also seems obvious that the choice to spend time on television content – broadcast or not, public service or not – along with the benefits to be gained will be shaped significantly by the wider online environment in which television content is positioned and viewed. PSBs and other providers have long been working on exactly this assumption. Our point here is that a range of other public service content must be considered when evaluating the situation for public service television. But here we face a further evidence gap that impedes effective decision-making. We know remarkably little about what content children engage with on the internet. The main systematic measurement of online content use relies on Ofcom's[26] reporting of comScore data of the 'top 50 web entities accessed by children aged 6–14 from desktop and laptop computers'. This is problematic for our present purposes as it excludes tablets, smartphones and other devices. The use of the concept 'web entity' is also problematic, as this includes entities which are not updated with content such as Microsoft, or websites for downloading apps. Still, it shows that children's top twenty web entities accessed in 2015 were, in rank order: 1. Google; 2. MSN; 3. BBC; 4. YouTube.com; 5. Facebook and Messenger; 6. Yahoo; 7. Amazon; 8. Wikipedia.org; 9. Windows Live; 10. Roblox.com; 11. Mode Tend Parenting; 12. O.UK; 13. eBay sites; 14. Disney Entertainment; 15. Microsoft; 16. Steam (App); 17. Safesearch.net; 18. Origin; 19. Animaljam.com; 20. Adobe.com.

Even if one considers all 50 sites, it is immediately apparent that children are accessing considerable amounts of commercial content, much of it designed for a general (adult) audience. It is also clear that these data tell us little about children's choices of television or other content, public service or commercial content, child-appropriate content or other. Nor is it clear where such data are to come from. Children's online activities constitute a major part of their media experiences, but there is little information about the content involved or the consequences of engaging with it available in the public domain to inform policy. Such data as are collected rarely evaluate content and use against child-specific criteria of value or benefit.[27] [...]

The Case for Enhancing the 'Discoverability' of Children's Public Service Content

Insofar as there is good quality content – television and other – available for children online, how are children (or their parents) to discover it? Discoverability poses a new and pressing challenge for public service content providers – and for the children and their families who could benefit from such content. [...]

This in turn poses a major challenge to the scalability and sustainability of public service providers, especially those that are small, niche or catering to minority groups.

PSBs have traditionally played a valuable role in exposing their audience to mixed diet schedules, thereby encouraging viewers to watch programmes on subject matter that they may not seek out unprompted but may yet enjoy. A concern with children locating content through search engines or YouTube is that these 'mainstream' as many people as possible towards highly ranked sites (or to other sites like those the child has already visited[28]).

Safety considerations also lead parents to restrict their children's freedom to search the internet widely,[29] as well as favouring the 'walled gardens' built for children online by both public and, especially, commercial providers. Our risk-averse society worries about – rather than welcomes – support for children's freedom to search in creative ways online, discovering new and surprising content and exploring at will according to interest.

What can be done? We are intrigued at the investment of the German government in Ein Netz für Kinder, a search engine designed for children to increase the discoverability of high quality content for children online.[30] We also note the efforts of Google to produce KidzSearch, and possibly other initiatives exist. We are not aware of independent evaluations that show how many children these reach, whether they are effective, or whether they help in the discoverability of public service content by children. [...]

Notes

1 Sonia Livingstone, OBE, is Professor of Social Psychology in the Department of Media and Communications at the London School of Economics. Claire Local is Content and Media Policy Advisor at Ofcom. This an edited extract from their submission to the Puttnam Inquiry, http://futureoftv.org.uk/wp-content/uploads/2016/01/Sonia-Livingstone-and-Claire-Local.pdf. An expanded version of the original submission has also been published: Sonia Livingstone and Claire Local, 'Measurement Matters: Difficulties in Defining and Measuring Children's Television Viewing in a Changing Media Landscape', *Media Information Australia*, 2017, DOI: 10.1177/1329878X17693932.

2 See www.ohchr.org/en/professionalinterest/pages/crc.aspx.

3 See Sonia Livingstone, *On the Future of Children's Television – A Matter of Crisis?* (Oxford: Reuters Institute, 2008). http://eprints.lse.ac.uk/27102/.

4 Peter Lunt and Sonia Livingstone, *Media Regulation: Governance and the Interests of Citizens and Consumers* (London: Sage, 2012).

5 The notion of public service television is that specified by the remit of this inquiry, but we note here and in our main text that this conflates public service broadcasters with public service content in ways that can obscure.

6 See Ofcom, *Children and Parents: Media Use and Attitudes Report* (London: Ofcom, 2015), www.ofcom.org.uk/__data/assets/pdf_file/0024/78513/childrens_parents_nov2015.pdf, 64.

7 For example, see Stuart Dredge, 'Traditional TV Viewing for Teens and Tweens is Dead. Not Dying. Dead'. *The Guardian*, 15 April 2015, www.theguardian.com/technology/2015/apr/15/traditional-tv-viewing-teens-tweens-awesomenesstv-meg-deangelis.

8 See Ofcom, *Children and Parents*, 22.

9 For 3–4 year olds, TV viewing (14.5 hours per week) is more than double the internet use (6.8 hours per week), and it is considerably greater for all age groups except 12–15 year olds (TV viewing 15.5 hours per week, internet use 19.8 hours per week).

10 See Ofcom, *Children and Parents*, 215.

11 We base these claims on a sizeable body of qualitative research observing and evaluating family interactions around shared viewing. See, for example, Dafna Lemish, *Children and Television* (London: Wiley-Blackwell, 2006), also Patti M. Valkenburg, *Children's Responses to the Screen: A Media Psychological*

Approach (Mahwah, NJ: Lawrence Erlbaum Associates, 2004). Ofcom data supports this claim insofar as it shows the majority of children's 'live viewing' (and time-shifted viewing) occurs with adults.

12 See Ofcom, *Children and Parents*, 23.

13 See See Ofcom, *Children and Parents*, 50: Figure 21.

14 Indeed, in the aforementioned figures, watching TV on an online device is presumably counted not under 'TV' but under 'internet', confusing further claims that children have shifted from TV to the internet. Relatedly, Ofcom reports a large increase in the popularity of YouTube amongst children in recent years, and even that among 12–15s, more prefer watching YouTube to TV (29% v. 25%)(Ofcom, 2015, 27). But viewing YouTube can include watching television programmes, so it is not easy to claim that children are shifting their attention from television to the internet (though certainly they are shifting from the TV set to viewing on other devices). Further, in its claim that 'hours of total viewing of TV are in decline among all children' (213), Ofcom refers to data from BARB, which appears to include viewing live or time-shifted on a TV set (or on other devices connected to a TV set) but not to time spent viewing television on other devices (such as iPlayer on a tablet), as BARB does not measure this yet. In Ofcom's Digital Day research (2014) with some 359 children, 52% of 'watching' (for 11–15s) and 64% (for 6–11s) was of 'live' TV, the rest being viewed via streaming, on-demand, recorded and short clips. This suggests that viewing TV on other devices – seemingly not counted in either Ofcom or BARB statistics – is considerable and growing, but still a minority of TV viewing. See Ofcom, *PSB Annual Report 2015 TV Viewing Annex* (London: Ofcom, July 2015), http://stakeholders.ofcom.org.uk/binaries/broadcast/reviews-investigations/psb-review/psb2015/PSB_2015_TV_Viewing.pdf.

15 See Damian Kavanagh, 'The Future is Here. BBC Three is Moving Online. It'll be Great. Promise', *BBC Blog*, 26 November 2015, www.bbc.co.uk/blogs/aboutthebbc/entries/c15f39ce-8978-4c6c-9c8b-d03178d3fcc6.

16 Even the BBC, which currently has the most comprehensive provision for children amongst UK PSBs, neglects children in the age bracket 13–15. The BBC target audience for CBeebies is children aged 0–6, for CBBC it is 6–12, and for BBC Three, 16–24, thereby leaving a significant gap between 12 and 16. Older children will soon have even less provision as BBC Three moves online and will cease to broadcast live.

17 Article 17 of the UN Convention on the Rights of the Child recognises 'the important function performed by the mass media and shall ensure that the child has access to information and material from a diversity of national and international sources, especially those aimed at the promotion of his or her social, spiritual and moral well-being and physical and mental health. To this end, States Parties shall: (a) Encourage the mass media to disseminate information and material of social and cultural benefit to the child and in accordance with the spirit of article 29 [development to the child's full potential]; ... (d) Encourage the mass media to have particular regard to the linguistic needs of the child who belongs to a minority group or who is indigenous; (e) Encourage the development of appropriate guidelines for the protection of the child from information and material injurious to his or her well-being, bearing in mind the provisions of articles 13 [freedom of expression] and 18 [parental responsibilities].' The state is responsible for ensuring that the provisions of the Convention are fully enacted.

18 No account is needed to watch its content and, unlike the other PSBs, the BBC distinguishes between accounts for 'under 16s' and those who are '16 or over'. When watching content on CBBC iPlayer, all recommendations are child-friendly and include a variety of programmes and genres within children's content.

19 In the case of ITV, there are click-through adverts (which cannot be skipped) screened during on-demand content. The Channel Five platform is sponsored by FisherPrice and although there are no screened adverts, there are adverts at the top of each page, many of which might not be considered age-appropriate (such as, on the same page as Peppa Pig, the UK Government's Drink Driving Campaign

with a link to a video containing disturbing content of a road traffic collision – last accessed December 16, 2015).

20 For detailed information on this point, including the claim based on Ofcom data that 97% of money spent on children's productions is spent by the BBC, see House of Lords, *The Select Committee on Communications, Inquiry on BBC Charter Renewal* (HoL: Evidence Session No. 6, 27, October 2015).

21 Reg Bailey (2011), Letting Children be Children. Report of an Independent Review of the Commercialisation and Sexualisation of Childhood. London: Department for Education. Sandra Calvert (2008), Children as Consumers: Advertising and Marketing. *The Future of Children* 18(1) 205–243. Valkenburg (2004).

22 See Ofcom (2007), House of Lords (2015) and Maire Messenger Davies and Helen Thornham, *Academic Literature Review: On the Future of Children's Television Programming* (London: Ofcom, 2007).

23 In the US, the Federal Trade Commission is investigating numerous complaints about YouTube Kids over 1) inappropriate content; and 2) advertising. See Sarah Perez, 'YouTube Kids Faces Further FTC Complaints Related to Junk Food Ads Targeting Young Children', *Techcrunch*, 24 November 2015, http://techcrunch.com/2015/11/24/youtube-kids-faces-further-ftc-complaints-related-to-junk-food-ads-targeting-young-children/.

24 See Ofcom, *Children and Parents*, 27.

25 See, for example *Common Sense Media*, www.commonsensemedia.org/website-lists.

26 See Ofcom, *Children and Parents* Annex, 224. Most recent data available, as reported here, are from May 2015.

27 Having observed the reluctance of policy makers to define 'what good looks like' for children online, the first author has attempted to gather guidelines for providers so that online content for children could be created that is high quality, diverse and imaginative, meeting the expectation of this inquiry that it 'informs & inspires, entertains & educates, connects & challenges'. See Sonia Livingstone, 'What Does Good Content look like? Developing Great Online Content for kids', in *Children's Media Yearbook 2014*, ed. Lynn Whitaker (Milton Keynes: The Children's Media Foundation, 2014).

28 See Eli Pariser, *The Filter Bubble: What the Internet Is Hiding From You* (London: Penguin, 2011).

29 See Sonia Livingstone, Leslie Haddon, Anke Görzig, *Children, Risk and Safety Online: Research and Policy Challenges in Comparative Perspective* (Bristol: The Policy Press, 2012).

30 See http://enfk.de/ – Ein Netz für Kinder is a programme of the German Federal Government of Culture and Media. It aims to encourage the production of high-quality, nationwide content in the areas of information, education and entertainment for children aged 6–12. The programme budget amounts to a maximum of 1 million euro annually. Ein Netz für Kinder aims to increase the number, quality, as well as discoverability, of high-quality content for children on the internet. The initiative encourages children to use the internet as a playful way of learning and to develop their creative potential to express themselves in the digital world.

41

Public Service Television and Sports Rights

Paul Smith and Tom Evens[1]

Sport has long been a vital part of the range of different programme genres provided by UK public service broadcasters (PSBs). In fact, the very existence of the UK's sporting calendar owes much to the growth of public service broadcasting during the twentieth century. As described by the broadcasting historian, Paddy Scannell:

Consider the FA Cup Final, the Grand National or Wimbledon. All these existed before broadcasting, but whereas previously they existed only for their particular sporting publics they became, through radio and television, something more. Millions now heard or saw them who had little direct interest in the sports themselves. The events became, and have remained, punctual moments in a shared national life. Broadcasting created, in effect, a new national calendar of public events.[2]

If anything, the ability of PSBs to bring the nation together with live coverage of major sporting events is even more valuable today. In an era of multi-channel digital television and increasingly fragmented audiences, live television coverage of major sporting events remains one of the few forms of programming able to bring the nation together for a shared viewing experience. In 2013, for instance, when Andy Murray became the first British winner of the men's singles title at Wimbledon for 77 years, he was watched by a (BBC) television audience of over 17 million. Perhaps even more impressively, over 90 per cent of the UK's population watched (at least some of) the BBC's coverage of the 2012 London Olympic Games, with audiences for the opening and closing ceremony each exceeding 25 million. However, the access of viewers to live television coverage of events like these in such huge numbers is dependent on their continued availability via the BBC, and/or other commercially funded PSBs.

The Twin Threat to PSB Sports Coverage

A combination of the escalating costs of sports rights and a squeeze on its own finances means that there is a very real danger that sport (and particularly live sport) will become an increasingly marginal feature of the BBC's (and other PSBs) output.

Driven largely by the growth of pay-TV since the 1990s, the increased value of the rights to popular sports and competitions, such as Premier League football (see Table 41.1), means that without regulatory intervention (see below) live coverage (or even highlights coverage) is increasingly beyond the budget of PSBs. Since its inception in 1992, not a single live Premier League football match has been broadcast live by a UK PSB. Instead, PSB coverage has been restricted to highlights coverage, and even here there has been a significant increase in the value of the rights, from £104 million paid by the BBC (seasons 2004–2005 until 2006–2007) to £204 million agreed by the Corporation in 2015 (seasons 2016–2017 to 2018–2019).

While most extreme in the case of Premier League football, other sports have also seen significant increases in the value of their rights over the last couple of decades, perhaps most notably the Olympic Games (see Table 41.2) and English cricket, which saw a trebling of the value of its rights – from £15million to £50 million – when it moved from free-to-air PSB coverage to pay-TV.[3]

Alongside rights inflation, the BBC's capacity to secure sports rights has also been undermined by recent cuts to its own funding. Following the 2010 licence fee settlement, the BBC cut its sports rights budget by 15 per cent and committed itself to limit

Table 41.1

The Value of (UK) Live Premier League Football Rights.

Years	Value (£ millions)
1992–1997	191
1997–2001	670
2001–2004	1,200
2004–2007	1,024
2007–2010	1,706
2010–2013	1,773
2013–2016	3,018
2016–2019	5,136

Source: BBC (2015)

Table 41.2

The Value of Europe-Wide Olympic Games TV rights (summer and winter).

Years	Value (US$ millions)
1998–2000	422.1
2002–2004	514.0
2006–2008	578.4
2010–2012	848

Source: IOC (2015)

spending on sports rights to an average of 9p in every licence fee pound.[4] Furthermore, the announcement in the 2015 Budget that the BBC is to take on from the government the £600 million-plus annual cost of providing free TV licences for people aged over 75, has resulted in further reductions in spending on sports rights, with an additional annual saving of £35 million targeted by the Corporation.[5]

The impact of the BBC's shrinking sports rights budget is already evident. In February 2015, it was announced that the BBC had lost the live rights to the Open Golf Championship to Sky, bringing to an end 61 years of live coverage of the event on free-to-air television. In a similar vein, in December 2015, the Corporation announced that it had decided to terminate ahead of schedule its contract with Formula One (originally due to end in 2018). To avoid a similar fate with other sports, the BBC has looked to share the cost of rights with other PSBs where once it was able to command exclusive coverage. Most notably, in July 2015, the BBC and ITV announced a joint six year deal to offer live coverage of Six Nations Rugby, with ITV offering all England, Ireland and Italy home matches and the BBC covering Wales and Scotland home matches. This strategy may well enable live coverage of at least some key sporting events to remain on free-to-air television, but it cannot disguise a significant dilution in the capacity of the BBC to achieve its key public service objectives.

The Public Value of BBC Sport

For the BBC, sports coverage provides an important means to achieve some of its key 'public purposes'. Specifically, the BBC has emphasised the importance it attaches to continuing to offer a broad mix of UK and international sports coverage that includes: major events that bring communities and nations together; minority sports that bring communities of interest together and broaden cultural horizons; and sports serving audiences that are otherwise under-served by the BBC, such as young men, lower-income and ethnic minority audiences.[6]

Alongside its already pragmatic attempts at alliances with other PSBs, the BBC also should look to maximise the public value of its sports coverage by continuing to provide extensive coverage of minority and or growing sports, which are often available at a relatively affordable cost. For example, the BBC has recently agreed deals: to provide live coverage of snooker's three biggest tournaments until 2019; to launch innovative new coverage of the increasingly popular, particularly amongst younger sports fans, mixed martial arts competition, Ultimate Fighting Championship (UFC), via BBC Three; and, continues to build on its popular coverage of women's international football.

The BBC and other PSBs should also highlight the benefits of the universally available free-to-air coverage they can provide. Some major sports organisations, such as the AELTC (Wimbledon tennis) have long appreciated the value of such coverage for the long-term popularity of (and commercial sponsorship opportunities available for) their sport and have opted to remain available via PSBs. Other sporting organisations, such as the ECB (English cricket) have experienced the disadvantages of moving to pay-TV. In 2005, Channel 4's coverage of Ashes cricket reached a peak audience of 8.2 million. Four years later, following the sale of the exclusive TV rights to Sky, the audience peaked at 1.9m and, in 2013, just 1.3 million. Last summer, when England clinched victory in the First Test Match of the series, the TV audience was just 474,000, only marginally more than a repeat of *Columbo* being aired at the same time on ITV3! Cricket may well be earning far more from the sale of its rights to pay-TV, but it is less and less part of the national consciousness.

Regulation: Protecting the 'Crown Jewels' of Sport and PSB

Against the background of escalating rights costs and reduced funding for PSBs, the position of the BBC (and, albeit to a slightly lesser extent, other PSBs) in the UK sports rights market is more dependent than ever on the continued existence (and effective enforcement) of listed events legislation, which effectively guarantees that certain key national sporting events (the so-called 'crown jewels' of sport) remain available on free-to-air television. [...]

The listed events policy remains a vital safeguard for the preservation of major sporting events and competitions on public service television. For example, in June

2015, the IOC announced that it had agreed a Pan-European deal with Discovery, the owner of the pay-TV broadcaster, Eurosport, for the exclusive rights to the Olympic Games, between 2018 and 2024 (although only for 2022 onwards in the UK). This meant that the BBC had lost control of the rights to broadcast the Olympic Games. However, listed events legislation has ensured the sub-licensing of rights for free-to-air coverage in the UK, which was agreed between Discovery and the BBC in 2016, as part of an exchange deal, which also included the sub-licensing (from the BBC to Discovery) of pay-TV rights for 2018 and 2020.

Just as, if not more significantly, Sky agreed an exclusive deal for live coverage of Formula One racing between 2019 and 2025. As part of the deal, Sky has proposed to broadcast the British Grand Prix (as well as two other races) free-to-air via its planned new channel, *Sky Sports Mix*, intended to showcase Sky Sports programming to potential new subscribers. While the British Grand Prix is not a listed event, as the law stands (the Broadcasting Act 1996, as amended by the Television Broadcasting Regulations 2000 and the Communications Act 2003) it may be possible for a pay-TV broadcaster, such as Sky or BT, to broadcast a listed event by adopting a similar approach. This is because the existing legislation only requires an event be available via a 'qualified service', which is defined as available free-to-air to 95 per cent of the population. In such a scenario, the letter of the law would not be breached, but the spirit of legislation intended to ensure easily accessible coverage of national sporting events and a shared viewing experience almost certainly would be. For example, BT recently employed this type of approach as part of its exclusive live UK coverage of UEFA Champions League football. According to reports, BT's commitment to offer a number of high profile matches, including some of those involving English teams, on a free-to-air basis, via its *BT Showcase* channel, was an important factor in convincing UEFA to agree to an exclusive pay-TV deal. However, the matches broadcast via *BT Showcase* have, to the frustration of UEFA's sponsors, attracted far fewer viewers than the free-to-air coverage previously offered via PSB (ITV). Taken together, these developments highlight the need for the tightening of the listed events legislation so as to restrict live coverage of listed events to designated PSB channels, either by making this requirement a clearer part of the legislation and/or by amending the existing regulation to include a more detailed audience requirement (e.g. a minimum average peak time audience rating).

The growth of pay-TV has provided benefits for both viewers and sports organisations, but this does not lessen the case for listed events legislation. The argument for such legislation is based on its potential to promote (and/or preserve) 'cultural citizenship' in two key ways. First, listed events legislation may be justified on grounds of equity. For instance, Ofcom has highlighted the rising cost of pay-TV subscriptions for UK viewers[7] and, given the spiralling cost of recent rights deals, these costs are only set to increase. For example, in 2016, Sky announced that the price of its Sky Sports package was to increase by £2.75 a month to £27.50, meaning that the cost of a year's subscription to Sky Sports will be more than double the cost of an annual television licence. The continued (and growing) exclusion of low income groups from access to sporting events broadcast exclusively on pay-TV is exacerbated by the UK government's reluctance to fully implement changes to listed legislation as recommended by the Davies Review.[8]

Secondly, one of the main benefits of ensuring that major sporting events are broadcast on free-to-air television is the generation of what economists refer to as 'positive network externalities'. In simple terms, an individual not only enjoys the event and the 'conversational network' through viewing, their participation also adds value to the network for everyone. This concept is highly significant to the debate on the future of PSB, and listed events legislation in particular, because it can be seen to apply to the difficult to quantify, but no less real, shared benefits that can result from the coverage of major sporting events on universally available free-to-air television – think London 2012 and the 'feel good factor'.

The opposition of many sports organisations to the listing of their sports is based on the belief that they are best placed to judge how to further the interests of their own sport, and in particular how to balance the potentially increased revenue to be gained via pay-TV with the benefits (not least commercial via increased sponsorship revenue) of greater exposure through free-to-air broadcasting. Even though the example of English cricket suggests that this may not always be the case, the key argument in support of listed events legislation is not that policy makers and regulators know better than individual sports organisations how to promote the best interests of a particular sport. Rather, it is, as noted above, that the wider public interest in the form of cultural citizenship is served by the availability of particular sporting events on free-to-air PSB television. For sports organisations whose events are protected for free-to-air coverage, the existence of listed events legislation may well be a source of frustration, but it is not particularly unusual in democratic societies for certain property rights to be subject to state regulation in the public interest. Planning laws mean that those who live in heritage properties cannot do with them exactly what they want. To promote cultural citizenship and to preserve public service broadcasting, the same is true for sports organisations and listed events.

Notes

1 Dr Paul Smith is Senior Lecturer in Media and Communications at the Leicester Media School, De Montfort University. Dr Tom Evens is Senior Researcher at the Centre for Media & ICT (iMinds-MICT) at Ghent University, Belgium. This is an edited extract of their submission to the Inquiry, http://futureoftv. org.uk/wp-content/uploads/2016/04/Smith-and-Evens.pdf.

2 Paddy Scannell, 'Public Service Broadcasting and Modern Public Life', in *Culture and Power: A Media, Culture and Society Reader*, eds. Paddy Scannell, Philip Schlesinger and Colin Sparks (London: Sage, 1992), 322–23.

3 Tom Evens, Petros Iosifidis and Paul Smith, *The Political Economy of Television Sports Rights* (Basingstoke: Palgrave Macmillan, 2013), 116.

4 BBC, *BBC Strategy Review: Supporting Analysis* (London: BBC, 2010), 32, http://downloads.bbc.co.uk/ bbctrust/assets/files/pdf/review_report_research/strategic_review/supporting_analysis.pdf.

5 Barbara Slater, 'Formula 1', *BBC Blog*, 21 December 2015, www.bbc.co.uk/blogs/aboutthebbc/entries/ d65fc069-ca52-41dd-a7be-674ca1614e14.

6 BBC, *The BBC Response to the Government's Free-to-Air Events Consultation*, July 2009, http://downloads. bbc.co.uk/bbctrust/assets/files/pdf/review_report_research/listed_events/bbc_response.pdf; also BBC Trust, *The BBC's Processes for the Management of Sports Rights, Review by MTM London*, January 2011, http://downloads.bbc.co.uk/bbctrust/assets/files/pdf/review_report_research/vfm/sports_rights.pdf.

7 Ofcom, *Cost and Value of Communications Services in the UK* (London: Ofcom, 2014), http:// stakeholders.ofcom.org.uk/binaries/research/consumer-experience/tce-13/cost_value_final.pdf

8 See *Review of Free-to air Listed Events*, November 2009. http://webarchive.nationalarchives.gov.uk/+/http:/ www.culture.gov.uk/images/consultations/independentpanelreport-to-SoS-Free-to-air-Nov2009.pdf.

42

Securing the Future for Arts Broadcasting

Caitriona Noonan and Amy Genders[1]

[...] In 2014 Tony Hall, director general of the BBC, announced that the Corporation would place the arts centre-stage across all BBC platforms.[2] While this offers some good news for both those working in that area of programming and audiences with an interest in the subject of arts, the wider trend is one of decline.

Research commissioned by Ofcom categorises arts television as a genre 'at risk' of disappearing as relatively small audiences are unable to offset increased production costs.[3] A decline is also evident in Ofcom's own research which finds that in the five years to 2011 spending on arts programming by the five main terrestrial broadcasters fell by 39%.[4] Regular strands, which arguably are the lifeblood of any genre, have been cancelled, for example, *The Review Show* (BBC 1994–2014), or moved to niche subscription channels, for example *The South Bank Show* (ITV 1978– 2010; Sky Arts 2012 –). Meanwhile, ITV and Channel 5 broadcast little regular arts content, and Channel 4's peak-time arts output fell from 30 hours in 2009 to just 19 hours in 2014.[5] There is a clear downward trend in the visibility of arts content within the schedules, particularly during peak-time.

This decline is the confluence of a number of factors. Decreases in commissioning and production budgets mean fewer resources for producers. Within specialist factual genres such as arts, this can have a limiting effect on the coverage of the subject, access to expertise, and the aesthetics of the final programme. Furthermore, our research directly highlights that even within the PSBs traditionally aligned to serving niche audiences, in a more competitive, multi-channel environment the commercial necessity of appealing to a mass audience has become the norm. This has had a direct impact on the tone, subject-choice and scheduling of arts leading to accusations of marginalisation and 'dumbing down'.

Two further unique elements underscore the need for intervention within the arts genre specifically. Unlike UK-originated children's content, which has had some success in accessing international markets, arts television has historically been a national construct and rarely sells beyond national markets. It often struggles to find an international audience at a time when many PSBs and independent producers are looking to expand their revenue streams through overseas content and format sales. Furthermore, the genre's decline in peak-time makes arts content less attractive to independent production companies and over the past decade there has been a marked decline in the number of production companies specialising in this content. Given the structural and commercial changes in the television sector, both of these limitations suggest that the downward trajectory of arts content on British public service broadcasting is unlikely to be reversed without a deliberate strategy to save it.

Why is Arts Broadcasting Vital?

One of the founding principles of PSB is that broadcasters should engage with the totality of life in Britain including its cultural life and artistic community. For a large portion of the population, television is their primary way to engage with the arts across the UK. If this genre disappears from free to air channels it denies access to the whole population to the range of arts and culture available. It also renders a whole area of society effectively invisible at a time when the Warwick Commission argues that 'too few of the population have access to as rich a culturally expressive life as might otherwise be open to them'.[6] Therefore, marginalising art and culture on television further marginalises these spheres in everyday life.

British arts and culture is a globally successful sector communicating British creativity and ideas worldwide. It is a significant employer, with cultural and creative sectors constituting 'the fastest growing industry in the UK'.[7] However, it also contributes to local wellbeing and features in a variety of national policy agendas including health, education, urban regeneration and social inequality. Television offers it further reach and visibility that has both economic and cultural value. Content which is critical and engages with debates on the sustainability of the arts contributes to a more vibrant sector. In return the arts sector makes important contributions to the television sector and the wider creative industries in terms of creative innovation and talent. This is a delicate ecosystem and so the health of the creative industries depends on the health of the arts and cultural sectors and vice versa.

Despite claims that we have moved into a 'post-broadcast age', the findings of reports such as that by the Warwick Commission suggest that 'TV remains a key feature of most people's everyday cultural life.'[8] In order for broadcasters to fulfill their remit for public service, a mixed ecology of programmes is needed ranging from drama to sports, news to comedy. Arts, of course, needs to be part of this mix. A strong arts proposition will serve to offer audiences high quality, informative programmes which speak to the diversity of creativity in the UK and beyond. Everyone should have access to the arts regardless of economic or social background and a healthy public broadcasting system will reflect these values. [...]

We believe there are both economic and cultural rationales for greater provision of content for and with young people. Undoubtedly, young people have different consumption habits (e.g. preference for online consumption rather than linear schedules). Yet arts programming has the potential to offer novel forms of engagement and opportunities for creative expression and further investment should be made into developing innovative content creation and distribution strategies that reflect this. Engaging with young people in this way will also encourage them to see the arts as a viable career aspiration thereby strengthening the sectors.

We also believe there is a need for greater diversity in arts broadcasting in terms of subject matter and form, and in the diversity of those working in this genre behind and in front of the camera. Our research found that many within the field regard current arts provision as too narrow in its focus and often reluctant to take creative risks. The arts are one of the most vibrant and diverse areas of public life and it is important to have programming which reflects this. Public service broadcasters must provide space to take creative risks and should strengthen their distinctiveness through investment in programming that is creative in both style and content. We also advocate an ongoing commitment within all PSBs to diversity through paid training opportunities extending access to this professional space beyond those from more privileged backgrounds. [...]

Notes

1 Dr Caitriona Noonan is lecturer in Media and Communication in the School of Journalism, Media and Cultural Studies (JOMEC), Cardiff University. Amy Genders is a PhD candidate at University of South Wales. This is an edited extract from their submission to the Puttnam Inquiry, http://futureoftv.org.uk/wp-content/uploads/2016/04/Caitriona-Noonan-and-Amy-Genders.pdf.
2 BBC. 'Tony Hall announces greatest commitment to arts for a generation', 25 March 2014. www.bbc.co.uk/mediacentre/latestnews/2014/bbc-arts-release.

3 Mediatique. *PSB Review: Investment in TV Genres,* 1 December 2014, http://stakeholders.ofcom.org.uk/binaries/broadcast/reviews-investigations/psb-review/psb3/Investment_in_TV_Genres.pdf

4 Ofcom. *Public Service Broadcasting Annual Report 2012* (London: Ofcom, June 2012), www.ofcom.org.uk/__data/assets/pdf_file/0027/72666/section-a.pdf.

5 Ofcom, *PSB Annual Report 2015: Output and Spend annex* (London: Ofcom, July 2015), 26. http://stakeholders.ofcom.org.uk/binaries/broadcast/reviews-investigations/psb-review/psb2015/PSB_2015_Output_and_Spend.pdf.

6 Warwick Commission, *Enriching Britain: Culture, Creativity and Growth* (Coventry: University of Warwick, 2015), 32. www2.warwick.ac.uk/research/warwickcommission/futureculture/finalreport/

7 Warwick Commission, *Enriching Britain,* 20.

8 Ibid., 33.

43

Public Service Television and Civic Engagement

Daniel Jackson[1]

In most appraisals of democracy today the news media figures prominently. This is for good reason: it is the main channel of communication between elected representatives and citizens; and (self-appointed) watchdog of the powerful. The performance of the news media with respect to civic engagement is thus much debated and often maligned. [...]

While news organisations are sometimes reluctant to accept the responsibility that comes with such power, it is implicit in the core principles of journalistic philosophy, whereby attempts to constrain or censor the news media are seen as threats to democracy itself.[2] But these normative roles also are surrounded by many tensions that surround the ability of our news media to perform their democratic functions. Borrowing from Bennett and Entman,[3] I'll discuss four of these tensions.

Tension 1: Diversity versus Commonality

The media landscape continues to expand rapidly. Media fragmentation and segmentation have expanded the genres of what can be termed 'political'. There is also undoubtedly more news and journalism circulating in the public sphere than ever before, which should be considered a good thing.

However, segmentation and fragmentation do bring potential dangers as well. Firstly, in a commercially dominated system that is driven by the demands of advertisers, audiences can be segmented by technological access and spending power, not cultural or civic needs.[4] The resulting risk is that the market disregards some citizens who are less desirable to advertisers. As Gandy[5] explains, the targeting of ever more specialised and smaller groups serves to undercut a common public culture. In this sense, segmentation can be implicitly anti-civic and anti-collectivist.

Secondly, changes in the way we engage with media (increasingly mobile, networked, web-based), together with the affordances of these devices and plat-forms (e.g. algorithms, data-driven, user-led 'pull mediums') are all pointing in the direction of increased personalisation of our media consumption, including news. This has numerous consequences. Two that I would like to highlight here are that for the interested citizen, there has never been more information avail-able to learn about political issues, but conversely, at the same time it has never been easier to *avoid* political fare either. Secondly, as research in online news con-sumption is beginning to show, increasing personalisation in media consumption can lead to ideological homogeneity (also knows as a 'filter bubble'), where we consume news that fits within our ideological biases, and can filter out that which doesn't.[6]

The challenge for PSBs is to maintain a sense of shared identity in their offerings, so as to foster a culture that still values civic life. It should also offer moments where audiences can (inadvertently or through choice) be challenged by political views that may contrast with their own. This means that PSBs must offer a range of ideological viewpoints from across the political spectrum.

Tension 2: The Information Necessary for Citizens to Participate Effectively in Democratic Life, versus the Entertainment-Driven Focus of an Increasingly Commercial-Oriented Media

Here, I will spare readers from the somewhat staid arguments about dumbing down,[7] but instead warn of some other dangers of the increasing corporate and commercial bias of our news media, which emanate from the organisation and structure of the media itself. As profit-seeking entities, commercial media organisations are reliant on advertising as the primary source of their income. As political economists have noted, this dependence can come at the expense of editorial independence.[8]

Whilst many journalists and editors might scoff at such suggestions of advertiser influence, there is growing evidence of other subtle ways in which the relationship between journalism and promotional industries (advertising, marketing and PR) are changing. For instance, a number of recent studies have documented the growing influence of public relations material in the news, raising questions of editorial inde-pendence.[9] Similarly, news organisations – in the search for new income streams – are increasingly working collaboratively with brands through 'branded content' and 'native advertising' initiatives, which blur the lines between news and advertising.

Whilst the response of news organisations to such accusations is often one of defiance, there is no doubt they are still very real threats – to editorial independence, to the normative concept of a fourth estate and in my view, to democracy. I will explain why, with respect to the next tension.

Tension 3: The Need of the Media to Treat People as Citizens on the One Hand and as Consumer Publics on the Other

If we consider the media environment as a whole, there can be little doubt that we are overwhelmingly addressed as consumers rather than citizens. The circulation of goods, the material and symbolic meanings of commodities, and the dominant position of advertising in its many forms make civic culture look diminutive in comparison to consumer culture.

News and journalism are not immune from this process. The consumer model of news is now well established in the UK.[10] It is precisely because of news organisations' treatment of the audience as consumer and not citizen that some of the processes described above are able to take place.

According to McChesney, the consequences for democracy of a consumer-centric news media system are serious, as they carry a huge implicit political bias: 'Consumerism, class inequality and individualism tend to be taken as natural and even benevolent, whereas political activity, civic values and anti-market activities are marginalised.'[11] The news media are thus central in the definition of culture in terms of consumerism and not citizenship. For McChesney, the combination of neoliberal media policies and corporate media culture tends to promote a deep and profound de-politicisation of society, evidence of which can be seen across the western world, and the USA in particular. In the UK – to the extent that it is not with us already – we should not think we are immune to such developments, especially given recent developments in media policy.

Tension 4: Broadcasters' Relationship with the Press

UK news broadcasters are mandated to be impartial, accurate and fair. As such, they provide a counterbalance to a highly partisan press. But this is a delicate balance. Studies consistently show that UK broadcasters are susceptible to following the news agendas of the press. This might not be so problematic if our press were a) not so overwhelmingly right wing and b) concentrated in so few hands. In the 2015 UK general election, we saw a super-charged Tory press, aligned with the agenda of the Conservative Party, that was remarkably successful at setting the news agenda of the terrestrial broadcasters.[12] Just as worrying was the 2014 survey[13] that found that the UK public holds a number of (quite grave) misapprehensions about many key public policy issues, such as immigration, welfare and crime. Such a collective failure is something our news media, including PSBs, should be ashamed of.

We know what to expect now from the UK press. Therefore it is imperative that public service broadcasters offer us news that is distinctive, independent, and as free as possible from the biases implicit in commercial news and broadcasting.

Conclusion

Underwriting all of these tensions are the market forces of a largely commercial media landscape. Compared to the US system, British broadcasting has traditionally remained relatively protected from the worst excesses of the market, but this is not inevitable or permanent, especially given the current political landscape.

The BBC is also not immune from these tensions. Whilst its news operations have seen relatively fewer newsroom cuts compared to the commercial sector, the BBC arguably acts too much like a commercial broadcaster at times, and news output is not always as distinctive or independent as it could or should be. But the question here is whether the BBC's funding model is *driving* this type of news, or whether there are other factors, such as journalistic culture and corporation strategy. I would argue the latter. [...]

Notes

1 Dr Daniel Jackson is Associate Professor of Media and Communication at Bournemouth University. This is an edited extract from his submission to the Puttnam Inquiry, http://futureoftv.org.uk/wp-content/uploads/2015/12/Dan-Jackson.pdf.

2 Daniel Jackson, 'Citizens, Consumers and the Demands of Market-Driven News', in *Voter as Consumer: Imagining the Contemporary Electorate*, eds. Richard Scullion and Darren Lilleker (Newcastle: Cambridge Scholars Publishing, 2007).

3 Lance Bennett and Robert M. Entman *Mediated Politics: Communication in the Future of Democracy*, (Cambridge: Cambridge University Press, 2000).

4 Robert W. McChesney, 'Corporate Media, Global Capitalism', in *Media Organization and Production*, ed. Simon Cottle (London: Sage, 2000).

5 Oscar H. Gandy, 'Dividing Practices: Segmentation and Targeting in the Emerging Public Sphere', in *Mediated Politics: Communication in the Future of Democracy*, eds. W. Lance Bennett and Robert M. Entman (Cambridge: Cambridge University Press, 2000).

6 Eli Periser, *The Filter Bubble* (London: Penguin, 2011).

7 Though you can read some of my thoughts on this and related subjects in Daniel Jackson, 'Citizens, Consumers and the Demands of Market-Driven News', in *Voter as Consumer: Imagining the Contemporary Electorate*, eds. Richard Scullion and Darren Lilleker (Newcastle: Cambridge Scholars Publishing, 2007) and Daniel Jackson, 'Time to get Serious? Process News and British Politics', in *Retelling Journalism: Conveying Stories in a Digital World,* eds. Marcel Broersma and Chris Peters (Leuven-Paris-Walpole, MA: Peeters, 2014).

8 For example, Robert Hackett, 'News Media and Civic Equality: Watch Dogs, Mad Dogs, or Lap Dogs?', in *Democratic Equality: What Went Wrong?* ed. Edward Broadbent (Toronto: University of Toronto Press, 2001). See also Edward Herman and Noam Chomsky, *Manufacturing Consent: The Political Economy of the Mass Media* (New York: Pantheon, 1988).

9 See Kevin Moloney, Daniel Jackson and David McQueen, 'News Journalism and Public Relations: A Dangerous Relationship', in *Journalism: New Challenges*, eds. Stuart Allan and Karen Fowler-Watt. (London: Routledge, 2013); and Daniel Jackson and Kevin Moloney, 'Inside Churnalism: PR, Journalism and Power Relationships in Flux', *Journalism Studies* (2015): 763–80.

10 See Jackson, 'Citizens, Consumers'.

11 McChesney, 'Corporate Media, Global Capitalism', 36.

12 Steven Barnett 'Four Reasons why a Partisan Press Helped win it for the Tories', in *UK Election Analysis 2015: Media, Voters and the Campaign,* eds. Daniel Jackson and Einar Thorsen (Bournemouth: The Centre for the Study of Journalism, Culture and Community, 2015); David Deacon, John Downey, James Stanyer and Dominic Wring, 'News Media Performance in the 2015 General Election Campaign', in *UK Election Analysis 2015: Media, Voters and the Campaign,* eds. Daniel Jackson and Einar Thorsen (Bournemouth: The Centre for the Study of Journalism, Culture and Community, 2015).

13 Jonathan Paige, 'British Public Wrong about Nearly Everything, Survey Shows', *The Independent*, 9 July 2013 www.independent.co.uk/news/uk/home-news/british-public-wrong-about-nearly-everything-survey-shows-8697821.html.

44

Tunnel Vision: The Tendency for BBC Economic and Business News to Follow Elite Opinion and Exclude Other Credible Perspectives

Gary James Merrill[1]

Introduction

This chapter focuses on a very specific element of television's role in promoting a more creative and robust public culture, namely the economic and business journalism produced by the BBC. Despite increased competition from online news organisations, BBC News holds three of the top five positions in the overall rankings.[2] The BBC is also rated as a highly trustworthy source of economic and business news[3] and, since the financial crisis, there has been a substantial increase in audience interest.[4] Economic and business news equips the public with essential knowledge and understanding to make informed decisions – as consumers, workers, tax payers and citizens – and is a vital element of the BBC's journalistic output. Unfortunately, only around a fifth of viewers believe that the Corporation gives a 'fair and balanced picture' of the economic environment.[5] For these reasons, it is important to take a critical look at the impartiality of the BBC's journalism in this sphere.

The Left-Wing Problem

The central role that public service broadcasting plays in British democracy was clearly demonstrated during the 2015 UK general election campaign. Among the countless interviews, numerous speeches and the endless analysis, one quote from BBC TV's coverage stands out because it exemplifies a formidable challenge facing the Corporation. It came from then Ukip leader Nigel Farage who, during a heated interaction on BBC 1's *Question Time*, said:

There just seems to be a total lack of comprehension on this panel and, indeed, amongst this audience, which is a remarkable audience even by the *left-wing standards of the BBC.*[6] (emphasis added)

Farage's contention that BBC habitually tilts left is not unique. Indeed, this belief is oft-repeated[7] with columnists in the Conservative-supporting press particularly strident in their views. The *Daily Mail's* Richard Littlejohn, for example, even saw evidence of this tendency in the BBC's 2014 'austerity-laden' Christmas TV schedule.[8] Public opinion also sways in the same direction: according to an Opinium/*Observer* poll in 2013, 41 percent of respondents said the BBC displays some bias, and of these twice as many people thought it favours the left rather than the right.[9] Overall, only 37 percent placed the BBC as neutral which, for an organisation that has a statutory duty to impartiality, should be a concern for regulators, journalists, programme makers and the general public alike. The challenge is even greater in the context of economic and business news: the Corporation's own research[10] has revealed that audiences do not fully understand the coverage; they would like it to 'relate more to their own circumstances'; and 'only 22 percent believe it gives a fair and balanced picture.'

The 'Anti-Business' Fallacy

Impartiality is a legal requirement of the BBC Charter and a core editorial value[11] and applies equally to all output. It is impossible to be perfectly impartial in every news item, of course, and so BBC journalists are obliged to show 'due impartiality', which

requires us to be fair and open minded when examining the evidence and weighing all the material facts, as well as being objective and even handed in our approach to a subject. *It does not require the representation of every argument or facet of every argument on every occasion or an equal division of time for each view.*[12] (emphasis added).

Achieving due impartiality clearly requires high-level editorial decisions that take into account the reporting of an issue over time. The BBC has a devolved editorial structure by which authority is given to programme and unit editors who follow the principles of the Corporation's code of conduct but take responsibility for their team's output.[13] Hence, although the BBC has clear guidelines on impartiality at an institutional level, how these are implemented is largely the decision of editors.

The BBC did not have a business editor until 2001 when it appointed Jeff Randall to the role. In his five-year tenure, Randall remodelled the Corporation's business coverage on the premise that the BBC he had joined: 'was culturally and structurally biased against business.'[14] This tendency was so pronounced, said Randall in another interview, that: 'on the whole, they [the BBC] treated business as if it was a criminal activity.'[15] Randall's perception that the Corporation was fiercely 'anti-business' added fuel to the perennial belief that the BBC was 'left wing.' In 2007, such views contributed to the commissioning of a major study by the BBC Trust into the impartiality of the Corporation's business reporting.

If the BBC's journalism were 'anti-business', as Randall *et al* maintain, one would expect BBC news to generally favour organisations and viewpoints traditionally associated with the left. However, the author of a 2007 study, Alan Budd, found no 'evidence of systemic (anti-business) bias'[16] but he did discover a neglect of news from the perspectives of investors and workers. Indeed, the author noted indifference to the coverage of labour issues:

Around 29 million people work for a living in the UK and spend a large proportion of their waking hours in the workplace. However, little of this important part of UK life is reflected in the BBC's business coverage.[17]

The belief that the BBC is 'anti-business' or indeed, 'left-wing' in general, is often informed by naked opinion and prejudice. However, as suggested by the Budd Report, the systematic and methodical analysis of news points to economic and business journalism of quite a different hue.

Whereas Budd noted an underrepresentation of organised labour, a study of BBC Radio 4's *Today* programme's coverage of the UK government's bank rescue plan in 2008 revealed the exclusion of other arguments from the left.[18] The author assessed the extent to which competing solutions to the banking crisis were given exposure and credence. If the BBC had been true to the accusations of left-wing tendencies, then nationalisation, a flagship of left-wing thought, would have dominated. However, the option that took centre stage in the programme's coverage, injecting public money or the 'bank bailout', was supported by virtually all key sources, mostly from the City of London. In contrast, nationalisation was barely mentioned or quickly dismissed.

Opinion beyond Westminster and the City

It is important to note that Budd's assessments were made against a definition of impartiality from a previous BBC publication, the Neil Report, which requires the

Corporation's journalism to be 'fair and open minded in reflecting all significant strands of opinion, and exploring the range and conflict of views'.[19]

What constitutes: 'significant strands of opinion' is clearly open to debate. The leading political parties' viewpoints are naturally included, but because no party, at the time of writing, currently offers a coherent and comprehensive alternative economic narrative, one could argue that views from beyond the broad parliamentary consensus are insignificant by definition. However, in the light of capitalism's evident fragility – the financial crisis and the subsequent global recession – one might expect the BBC, as a public service broadcaster, to give greater exposure and credence to alternative discourses.

Alternative economic discourses undoubtedly exist, and high-profile critics include Nobel Prize winners Joseph Stiglitz and Paul Krugman, and the international financier, George Soros.[20] Numerous writers, journalists and academics have also questioned core neoliberal assumptions and asked whether markets serve the public interest well. Other authors have focussed on the brutality of an economic system that prioritises unfettered profit maximisation over social concerns.[21] In addition, there are countless NGOs, trade unions, grassroots pressure groups and other organisations that have their own specific issue with the modern market economy. Indeed, despite the major political parties' convergence in policy, there is evidence that the world's publics are not convinced. In 2009, a major international survey[22] found widespread disillusionment with capitalism. In only two countries did more than 20 per cent of people think it was working well, and a higher proportion thought it 'fatally flawed'. Globally, there was also significant support for more government regulation of business and a fairer distribution of wealth.[23]

Despite this widespread – and often very credible – scepticism towards capitalism, academic research suggests that BBC journalists focus their attentions on opinions from Westminster and the City of London. The inevitable consequence of this tunnel vision is that viewpoints from outside a narrow corridor of power are largely excluded. In the case of the 'bank bailout', nationalisation was advocated by some leading economists but it had little support among the British financial and political elite. As a result, BBC journalists felt no obligation to include this option in debates.[24] The same phenomenon was also evident in a comparative analysis of the reporting of economics and business by four British news providers.[25] In total, the author analysed some 1,625 articles published by the BBC News website, the *Guardian/Observer*, the Telegraph Group, and the *Times/Sunday Times*. For each of three issues – economic globalisation, private finance in public services, and supermarket power – all four news organisations tended to gravitate to the business and political elite for information and opinion and, with the possible exception of the *Guardian/Observer*, generally excluded the views of individuals and organisations critical of neoliberalism and big business. Most significantly, the analysis revealed that the political content of the BBC's reporting had far more in common with newspapers traditionally associated with the *right* than the left.[26] [...]

Toward a More Inclusive Journalism

Despite the studies outlined above, the economic and business news of the BBC has received relatively little interest from researchers. Consequently, the belief that the BBC is 'left-wing' in this context has never been robustly tested, and so it persists. There is scant evidence to suggest that this perception might be well-founded and research to date suggests that the *converse* is more likely. Indeed, the case for the antithesis was strengthened by a candid admission from the former economics (and business) editor of the BBC, Robert Peston, who said the Corporation tends to follow the agenda of the *Daily Telegraph* and the *Daily Mail*. 'If I'm honest', he said, '... the BBC's routinely so anxious about being accused of being left-wing, it quite often veers in what you might call a very pro-establishment, [a] rather right-wing direction, so that it's not accused of that.'[27]

Given the importance of impartiality in BBC economic and business journalism to public debate, more research is clearly needed. The Budd Report was a useful start but it was published before a period of deep introspection for many practitioners that was prompted by the financial crisis.[28] Furthermore, there was no comparative element and so Budd could not benchmark the BBC's output against other news providers that have distinct political positions. Hence, this paper recommends the commissioning of a post-crisis study which assesses the extent to which the BBC and other news organisations award exposure and credence to ideas from outside of the Westminster–City of London axis. It is vital that such an inquiry is carried out with the support of BBC journalists and other editorial staff, and it is hoped that the findings of the subsequent report would inform the evolution of the BBC's economic and business output.

Notes

1 Gary Merrill is a Senior Lecturer in the Department of Media, Culture and Language at the University of Roehampton. This is an edited extract from Gary Merrill's submission to the Puttnam Inquiry, http://futureoftv.org.uk/wp-content/uploads/2016/02/Gary-Merrill.pdf.

2 Ofcom, 'News Consumption in the UK' (London: Ofcom, 2014), http://stakeholders.ofcom.org.uk/market-data-research/other/tv-research/news-2014/.

3   BBC Trust, 'Seminar on Impartiality and Economic Reporting', 6 November 2012, http://downloads.bbc.
    co.uk/bbctrust/assets/files/pdf/our_work/economics_seminar/impartiality_economic_reporting.pdf.

4   Pre-Financial Crisis most people kept up with financial issues 'about once a month, and 40 percent
    never, or rarely. Post-Crisis, one third are doing so daily and more than three quarters are doing
    so at least weekly', quoted in Diane Coyle, 'Comment on the BBC Trust seminar on impartiality
    and economic reporting', 6 November 2012, www.bbc.co.uk/blogs/bbctrust/entries/80a3ecff-
    58c9-4686-8d5c-abd182fecad1.

5   BBC Trust, 'Seminar on Impartiality'.

6   Nigel Farage, *Question Time*, BBC One, 16 April 2015.

7   For example, Boris Johnson, 'The Statist, Defeatist and Biased BBC is on the Wrong Wavelength', *Daily
    Telegraph*, 14 May 2012; Alan Budd, 'Report of the Independent Panel for the BBC Trust on Impartiality
    of BBC Business Coverage', April 2007, www.bbc.co.uk/bbctrust/our_work/other/business_news.
    shtml; BBC Journalism Group 'The BBC Trust Impartiality Review, Business Coverage', submission
    to the Panel, 23 January 2007, http://downloads.bbc.co.uk/bbctrust/assets/files/pdf/review_report_
    research/impartiality_business/f1_journalism_submission.pdf Budd, 'Report of the Independent
    Panel', 14.

8   Richard Littlejohn, 'So the BBC Denies it has a Left-Wing Bias? Just Take a Look at this Austerity-Laden
    Christmas Schedule ...', *Daily Mail*, 9 December 2014.

9   Toby Helm, 'More People Think BBC has Bias to Left than Bias to Right', *The Guardian*, 2 November
    2013, www.theguardian.com/media/2013/nov/03/bbc-left-right-poll.

10 BBC Trust, 'Seminar on Impartiality.

11 Ronald Neil, 'The BBC's Journalism after Hutton: The Report of the Neil Report Team' June 2004,
    downloads.bbc.co.uk/aboutthebbc/insidethebbc/howwework/reports/pdf/neil_report.pdf.

12 BBC, 'Impartiality and Diversity of Opinion', 2005, http://downloads.bbc.co.uk/guidelines/
    editorialguidelines/Legacy_Guidelines/2005-section04-impartiality.pdf, 27.

13 Neil, 'The BBC's Journalism'.

14 Jeff Randall, quoted in Lisa Kelly and Raymond, 'Business on Television: Continuity, Change, and Risk
    in the Development of Television's "Business Entertainment Format"', *Television and New Media*, 12, 3
    (May 2011), 232.

15 Jeff Randall, quoted in Confederation of British Industry, 'Written Evidence: BBC Trust Impartiality
    Review, Business Coverage', 2007, http://downloads.bbc.co.uk/bbctrust/assets/files/pdf/review_
    report_research/impartiality_business/e_written_evidence.pdf.

16 Budd, 'Report of the Independent Panel ', 14.

17 Budd, 'Report of the Independent Panel ', 19.

18 Mike Berry, 'The *Today Programme* and the Banking Crisis', *Journalism* 14, 2 (2012) 253–70.

19 Neil, quoted in Budd, 'Report of the Independent Panel ', 6.

20 See for example, Joseph Stiglitz, J (2002) *Globalisation and Its Discontents*, London: Penguin, 2002);
    Joseph Stiglitz, *The Roaring Nineties* (London: Penguin, 2003); Joseph Stiglitz, *Making Globalization
    Work: The Next Steps to Global Justice* (London: Penguin, 2007); Paul Krugman, *The Return of Depression
    Economics and the Crisis of 2008* (London: Penguin, 2008); Georg Soros, *The Crisis of Global Capitalism*
    (London: Little, Brown, 1998).

21 For example, Noam Chomsky, *Profit Over People: Neoliberalism and Global Order* (London and
    Toronto: Seven Stories Press, 1998); Naomi Klein, *No Logo* (London: Flamingo, 2000); Naomi Klein,
    The Shock Doctrine: The Rise of Disaster Capitalism (London: Penguin, 2008); George Monbiot, *Captive
    State: the Corporate Takeover of Britain* (London: Pan Books/Macmillan, 2001); John Pilger, *The New
    Rulers of the World* (London: Verso, 2002), Joel Bakan, *The Corporation: Pathological Pursuit of Profit*

and Power (London: Constable & Robinson, 2004); Thomas Piketty, *Capital in the Twenty-First Century* (Cambridge, MA: Harvard University Press, 2014).

22 The researchers questioned 29,000 people in 27 countries.

23 BBC World Service (2009) 'Wide Dissatisfaction with Capitalism— Twenty Years after Fall of Berlin Wall', opinion poll, produced in conjunction with PIPA and Globescan, 9 November 2009, www.globescan.com/news_archives/bbc2009_berlin_wall/.

24 Berry, 'The *Today Programme*'.

25 Gary Merrill, 'Convergence and Divergence: a Study of British Economic and Business Journalism' (PhD Diss., Goldsmiths, University of London, 2015).

26 Ibid.

27 Quoted in Jack Sommers, 'Robert Peston Claims BBC is "Obsessed" with Covering Same Stories as Daily Mail', *Huffington Post*, 10 June 2015, www.huffingtonpost.co.uk/2014/06/06/robert-peston-says-bbc-leftwing-bias-bollocks_n_5458619.html?utm_hp_ref=uk.

28 It is generally agreed by academics and practitioners alike that the news media as a whole failed to warn of possible negative outcomes prior to the financial crisis of 2007/8. See Merrill, 'Convergence and Divergence', 9–10 and 191–195 for a summary of the arguments.

Part Seven

Recommendations and Afterword

45

How to Strengthen Public Service Television

Chris Tryhorn[1]

Recommendation 1: The UK's Broadcasting Ecology Must be Maintained

The word 'ecology' is often used to describe the mix of broadcasting provision in the UK, and it seems an apt term. It conjures up the way in which different broadcasters and producers feed off each other, in terms of ideas and nurturing talent, and how a number of idiosyncratic entities live alongside each other, competing and co-existing. These organisations play a role in supporting the wider creative industries and encourage grassroots creativity too. The notion of an ecology has for some time been gaining ground in the arts world, where there are increasing attempts to map the subtle interdependencies and overlaps of interest between the commercial, publicly subsidised and amateur spheres.[2]

The point about most ecologies is that they permit delicate balances to be maintained; pull on one thread and things can unravel. We should be very careful about upsetting those balances in the television world. It is not unimaginative or somehow reactionary to argue for the preservation in broad terms of the *status quo* when the *status quo* has a proven track record of working.

In such a context, and given the lack of clarity from the government, it is important to make the case that public funding for the BBC should be maintained and Channel 4 must not be privatised. Such radical upheaval at either broadcaster could have untold consequences, and the uncertainty needs to be cleared up.

BBC and Channel 4 have a virtuous effect on the rest of the TV market, by setting standards and thereby improving the overall quality of output. ITV's aspiration to make high quality UK-originated drama is bolstered by creative competition with the publicly funded BBC, for example. Would Sky be as committed to arts programming were it not for the existence of the BBC?

The quality of the BBC and Channel 4's output, and their ability to be freed, at least to some extent, of commercial constraints stems directly from the way they are set up as organisations: their public ownership and, in the BBC's case, the direct relationship with viewers that public funding entails. Commercial ownership – or in the BBC's case, commercial funding – would inevitably dilute their commitments to public service output, whatever regulatory constraints were formally imposed.

Thinking about the BBC and Channel 4 together can help to define their respective purposes. It makes sense to treat the two questions in conjunction for a more holistic sense of the broadcasting ecology and the role of public service television within it.

Recommendation 2: The BBC Needs to Have its Funding Futureproofed

The BBC should remain publicly funded, as argued above. Crucially, though, this funding needs to be futureproofed to take account of changes in technology and consumption. It needs to be funded honestly and not continually raided by opportunistic governments using it to pay for media infrastructure projects or politically motivated schemes.

The TV licence fee is a peculiar throwback to a bygone era of broadcast monopoly and spectrum scarcity. But as a means of funding today's BBC it seems preferable to the alternatives, including the most suggested option, voluntary subscription, which would probably lead to the BBC becoming a premium service for more affluent citizens.[3] The universality of the fee guarantees the BBC scale and allows it to aspire to reach everyone in the UK. The government has narrowed the options down to a reformed licence fee, a media levy, or a hybrid licence fee and subscription model.[4] The first two options are preferable.[5]

It is time to ditch the anachronism of a 'TV licence fee' and adopt the platform-neutral term 'BBC licence fee', creating at the same time an opportunity to make the funding more honest. The BBC should no longer have to bear the costs of projects that the government ought to be funding.[6] The level of its funding should be set sustainably to bring an end to continual cost-cutting and debilitating uncertainty.

Recommendation 3: Channel 4's Status as a Publicly Owned Publisher Broadcaster Needs to be Set in Stone

Channel 4 has enormously enriched British television since it was launched in 1982. Viewed now, in the multichannel age, when ITV is arguably no longer the public service broadcaster it was, its importance seems greater than ever. It is vital that the UK retains at least two broadcasting organisations that are unambiguously committed to public service, as the BBC and Channel 4 are. If it were left just to one, ie the BBC, the competition in terms of quality that exists across the television marketplace would be diminished.

Channel 4's business model – publicly owned, commercially funded, a 'publisher-broadcaster' that produces none of its programmes – is idiosyncratic, but none the worse for that. It takes its public remit seriously, but privatisation would threaten that remit. Even if a sale tied the buyer to certain regulatory requirements, it would necessarily change Channel 4. No one is likely to buy it without wanting to make a profit; and regulatory requirements can always be gamed.

Such an existential threat will always lurk in the background, without more fundamental backing for the Channel 4 model. The government should set out unambiguously that Channel 4 will remain in public hands and that it sees it as a critical part of the UK's broadcasting ecology.

Action may also be needed to support Channel 4's ability to make public service programmes. Ofcom has put forward interesting suggestions about changing the regulatory model as the advantage of EPG prominence diminishes.[7]

Recommendation 4: We Need to Decide Whether we Really Want ITV and Channel 5 to do Public Service Television – and Get a Fair Deal from That

One could be forgiven for not realising that ITV and Channel 5 are public service broadcasters.[8] It is not something either broadcaster makes much of.[9] Programmes outside news that are obvious examples of public service television do not always spring readily to mind.

In the past, when ITV was the UK's only commercial broadcaster, the viewers switched on and the advertising revenues rolled in, so public service output was no threat to the business model, and in fact was a crucial part of the licence award process. Now, amid vast competition, getting 'eyeballs' is ITV's priority, and the main ITV channel's position at channel 3 on the EPG has been a key advantage.

The question for the public here is: who is getting the better deal, the broadcasters or the public? Are ITV and Channel 5 doing sufficiently good public service programmes to deserve their high EPG slots? The problem is that the regulatory requirements are made in vague terms: a better audit needs to be made. We need to decide whether they really are fulfilling their remit as public service broadcasters, and whether they should up their game as such.

This is an important question as the next generation of TV interfaces is likely to see the traditional EPG model transformed beyond recognition. It would definitely be in ITV's interests to gain prominence on future interfaces as their business model depends on being the UK's biggest commercial broadcaster, with the advertising benefits such scale brings. In such a scenario, should they be asked to do more to justify that prominence – and should we hold them to account more for the quality of their output? Without regulation, they would have to rely on the status and appeal of their shows to justify such prominence.

Recommendation 5: Public Service Television Must Be Defined Better, Sold Better – And Be Better

Outside broadcasting and political circles, how many people really know what public service television is? Ofcom's research found serious gaps in public understanding. If we view it as such a public good, should it not be sold to the public better?

First, of course, we must be clear about what it is. We should perhaps start by defining public service television less generally; should we be calling ITV and Channel 5 'public service broadcasters' when this is far from a sufficient description of them? Maybe we need to think more in terms of public service programmes, as the public increasingly does.[10] [...] The task of 'selling' public service television need not be overwhelmingly difficult. All organisations that lie somewhere between the purely profit-seeking and the state have to justify receiving any kind of public subsidy.[11] There are ways of explaining why not everything worthwhile need be popular. For example, there is a value in something that you do not currently watch but might choose to watch one day, and a value in something you have no intention of ever watching but are glad that other people can watch. These kinds of ideas are not complicated and can bolster the validity of content that appears to have limited appeal.[12]

But nothing will boost the case for public service television like the programmes themselves. A notable finding of Ofcom's research was the public's tendency to be more concerned with whether content was good or bad rather than designated public service or not.[13] There are definitely areas where broadcasters could do better: a personal wish list would include more investigative journalism, more challenging documentaries, a revival of regional (not local) programming, and better long-form drama to emulate the heights reached by recent American shows. This kind of programming is not cheap and it may be necessary for more 'popular' programmes to lose out as a result. Real quality of output, however, will provide the best argument in favour of public service television in the 21st century.

Notes

1 Chris Tryhorn is a freelance journalist and researcher and one of the key contributors to the Puttnam Report. He has worked on a number of projects relating to the media and culture sectors. He was a media reporter at *The Guardian* between 2003 and 2010. This is an edited extract of his submission to the Inquiry, http://futureoftv.org.uk/submissions/public-service-television-must-be-defined-better-sold-better-and-be-better/.

2 See for example John Holden's *The Ecology of Culture* (Arts & Humanities Research Council, 2015), for which the author was the researcher.

3 The government has excluded funding by advertising or through general taxation. *BBC Charter Review: Public Consultation* (London: Department for Culture, Media & Sport, 2015), 50.

4 *BBC Charter Review: Public Consultation*, 51–52.

5 The problem with a partial subscription is that it could be the 'thin end of the wedge' allowing for a full subscription model at a later date, and by definition would exclude those unable to pay from whatever services were placed behind the paywall.

6 Recent examples include the costs of digital switchover, local TV, superfast broadband, and free licences for the over-75s.

7 *Public Service Broadcasting in the Internet Age: Ofcom's Third Review of Public Service Broadcasting* (London: Ofcom, 2015), section 6.23.

8 Ofcom-commissioned research reported that 'Participants of all ages expressed some surprise that Channel 5 had public service obligations.' *An Investigation into Changing Audience Needs in a Connected World* (London: Ipsos MORI for Ofcom, 2014), 48.

9 ITV's 2014 annual report appears to contain no reference to 'public service'. The company does, however, make much of the importance of its 'high quality content' and makes passing reference to adhering to its regulatory requirements.

10 See the findings in Ipsos MORI, *An Investigation into Changing Audience Needs*, 41.

11 Arts organisations that have lost Arts Council funding and have been subject to pressure on outreach and public engagement have more experience of this than broadcasters.

12 There is a significant literature to draw on here, and indeed the notion of 'public value' was adopted by BBC in the last charter review process, with the BBC Trust using 'public value tests' to assess the merits of new services.

13 Ipsos MORI, *An Investigation into Changing Audience Needs*, 41.

46

Recommendations of the Puttnam Report[1]

General

The UK's public service television system is a vital political, economic and cultural resource and should be viewed as an *ecology* that needs careful protection and coordination. Public service media should not be viewed as synonymous with market failure and therefore should not be regulated simply in relation to the impact of their content and services on the wider media market. Principles of independence, universality, citizenship, quality and diversity need to be embedded into the regulation and funding of an emerging digital media landscape.

1. In return for public service broadcasters meeting the obligations of their licences, their content should be guaranteed prominence on electronic programme guides, smart TVs and on the interfaces of on-demand players as they emerge.
2. Retransmission fees should be paid by pay-TV platforms to public service television operators to address the current undervaluation of public service content by these distributors.
3. Ofcom should supplement its occasional reviews of public service broadcasting with a regular qualitative audit of public service content in order to ensure that audiences are being served with high-quality and diverse programming. This should include detailed data on the representation and employment of minority groups and a comprehensive account of the changing consumption patterns of younger audiences.
4. Ofcom should continue to monitor the independent production sector and take action, where necessary, if consolidation continues to increase and if diversity of supply is affected.

The BBC

We support the inclusion of diversity as a specific public purpose for the BBC but strongly reject the abolition of the purpose focusing on the delivery of emerging communications technologies and services to the public. We believe the BBC should be encouraged to pursue networked innovation, to embrace the internet and to develop a range of content and services for the online world.

The BBC should continue to provide mixed programming and cater to all audiences as well as competing with other broadcasters to produce high quality programmes. The BBC needs to demonstrate further commitments to creative ambition and to address shortfalls in specific areas, for examples its services to Black, Asian, and minority ethnic (BAME) audiences, its relationships with audiences in the devolved nations, its institutional commitment to impartiality and its willingness to embrace new types of collaborative partnerships.

5. The government should replace the licence fee as soon as is practically possible with a more progressive funding mechanism such as a tiered platform-neutral household fee, a supplement to Council Tax or funding via general taxation with appropriate parliamentary safeguards. We do not believe that advertising or subscription are appropriate to the aspiration that BBC content and services should be free at the point of use.
6. The government should hand over decision-making concerning the funding of the BBC to an independent advisory body that works on fixed settlement periods.
7. The BBC should be reconstituted as a statutory body, as with Channel 4, thus abolishing its royal charter or – at the very minimum – providing statutory underpinning to a continuing royal charter.
8. Appointments to the BBC's new unitary board should be entirely independent from government. We recommend that the process should be overseen by a new independent appointments body and based on a series of tests drawn up by the former commissioner for public appointments, Sir David Normington. There should be an opportunity for BBC staff to take part in the selection of at least some board members while representative voices from the devolved nations must be involved in selecting the members for Scotland, Wales and Northern Ireland.
9. If Ofcom is handed the responsibility of regulating the BBC, it must be given the resources and the structures to regulate the BBC independently of both government and its commercial rivals.

Channel 4

Channel 4 occupies a critical place in the public service ecology – supporting the independent production sector and producing content aimed specifically at diverse audiences.

10. Channel 4 should not be privatised – neither in full or in part – and we believe that the government should clarify its view on Channel 4's future as soon as possible.
11. Channel 4 should significantly increase its provision for older children and young adults and restore some of the arts programming that has been in decline in recent years.
12. Channel 4 should continue to innovate and experiment across different platforms and it should aim to arrest the fall in the number of independent suppliers that it works with.

ITV and Channel 5

We believe both ITV and Channel 5 should remain part of the public service television ecology but that they have been contributing less to it than they might have.

13. We recommend that ITV and Channel 5 continue to receive the privileges afforded to other public service broadcasters but we believe that their commitment to public service needs to be strengthened.
14. Ofcom should be asked to conduct a major review of how best ITV can contribute to the PSM ecology for the next decade and beyond, including explicit commitments for programming and investment, alongside a fresh look at the range of regulatory support that can be offered.
15. ITV should be asked to take on a more ambitious role in regional TV and in current affairs. Measures to be considered might include increasing the minimum amount of regional current affairs from 15 to 30 minutes a week and an increase in network current affairs output to the equivalent of 90 minutes a week.
16. Channel 5's voluntary commitment to children's programming should from now on be embedded in its licence, with specific commitments to UK-originated children's content, in return for the channel continuing to receive the benefits of its public service status.

A New Fund for Public Service Content

We recognise that there are important new sources of public service content coming from commercial operators such as Sky and Discovery as well as subscription video-on-demand services like Netflix and Amazon. We note, however, that this output is dependent on the extent to which it serves a larger commercial purpose and is not part of any regulatory obligation. We also note the importance of traditional public service television providers in creating an environment in which commercial operators are able to thrive through their investment in training and high quality content, which boosts the 'brand' of television in the UK.

We wish to highlight the growing contribution to a digital media ecology of a broad range of cultural institutions – including museums, performing arts institutions and community organisations – who are producing video content in areas such as science and the arts.

17. In order to increase the levels, quality and security of this provision, we propose to set up a new fund for public service content. This would consist of a series of digital innovation grants – the DIG – that would be open to cultural institutions and small organisations that are not already engaged in commercial operations.
18. DIG funding would not be limited to linear video content but to other forms of digital content that have demonstrable public service objectives and purposes. We would expect applicants to partner with existing public service broadcasters and platform owners in order to promote their content.
19. The DIG would be funded by the proceeds of a levy on the revenues of the largest digital intermediaries and internet service providers and would be disbursed by a new independent public media trust.

Diversity

There is clear evidence of dissatisfaction with the performance of public service television from ethnic, regional, national and faith-based minorities and it is vital that PST operators address these issues if they are to retain any legitimacy with these audiences.

There is also evidence that the television workforce is not representative of the wider UK population and that there is a systematic under-representation of, for example, ethnic minorities and those from poorer backgrounds at top levels of the industry.

We welcome the various 'diversity strategies' adopted by all broadcasters, but these have not achieved the desired change either in representation or employment. We believe that there are systemic failures that account for an enduring lack of diversity on- and off-screen and therefore that more systemic solutions are required alongside the setting of targets and provision of training schemes.

20. The 2010 Equality Act should be amended so that public service television commissioning and editorial policy would be covered by public service equality duties.
21. A renewed commitment to diversity must be accompanied by sufficient funds. We agree with the proposal by Lenny Henry that the BBC (and in our view other public service broadcasters) should ringfence funding – taking its cue from the BBC's funding of its nations and regions output – that is specifically aimed at BAME productions (though this could apply to other minority groups in the future).

Nations and Regions

The public service television system has failed to reflect the changing constitutional shape of the UK such that audiences in Scotland, Wales, Northern Ireland and the English regions are being under-served. We believe that the English regions have failed to benefit from the existing 'nations and regions' strategies of the main public service broadcasters and that the broadcasters have a responsibility to shift both production and infrastructure to areas that are currently marginalised.

We welcome the increase in 'out of London' production as well as recent commitments from public service broadcasters to step up their investment in the devolved nations. We are concerned, however, that their proposals will fail to challenge the underlying centralisation of the UK television ecology.

We propose a 'devolved' approach to public service television that ultimately aims at sharing responsibility for broadcasting matters between the UK parliament and the devolved nations.

22. Commissioning structures and funding need to better reflect devolutionary pressures and budgets for spending in the devolved nations should be wholly controlled by commissioners in those nations.

23. We firmly believe that it is time for a 'Scottish Six' – and indeed a 'Welsh Six' and a 'Northern Irish Six.'

24. The government (or governments in the future) should both protect and enhance funding aimed at minority language services that play such a crucial role in maintaining cultural diversity and identity. The government needs to identify stable sources of funding for S4C other than the BBC in its review of the channel in 2017.

25. The BBC should be allowed to revisit its local television proposal and strike up meaningful partnerships with a range of commercial and not-for-profit news organisations in order to galvanise television at the local level.

Genres and Content Diversity

We note that there has been a decline in investment in some of the genres traditionally associated with public service television: arts, current affairs and children's programming. Other genres, for example drama and sports, have been negatively affected by rising costs and competition from heavily capitalised commercial rivals.

Regulators and broadcasters need to work together to consider how best to address these pressures in order to maintain a diverse public service ecology. There should be no automatic assumption that a particular genre is no longer 'affordable.'

We believe that the creation of the Digital Innovation Grants (DIG) fund will create significant opportunities for a broader range of public service organisations to contribute to reversing the decline in, for example, arts, history, science, religious and children's programming.

We note the decline in viewing of television news across the main public service broadcasters, especially among younger audiences. We believe that this is partly due to wider changes in consumption patterns but also that new sources of news are providing an energetic and robust challenge to television bulletins that are sometimes seen as 'staid' and unrepresentative.

We note that there has been a steady migration of live sports from free to air channels to pay TV and that the vast majority of sports coverage is now to be found on pay TV channels. Public service broadcasters are increasingly unable to compete with companies like Sky and BT in rights to the most popular sports. While some 46% of all investment in first-run original programming in the UK is devoted to sports, only a small proportion of the audience is able fully to benefit from this.

26. At a time of increasing disengagement with mainstream political parties, public service news content ought to adopt a model of journalism that is less wedded

to the production of consensus politics and more concerned with articulating differences.

27. We have earlier recommended that Channel 4 significantly increases its provision for older children and young adults, while Channel 5 should have its commitment to children's programming embedded in its licence. The BBC must also be required to maintain its engagement with younger audiences and to reverse its recent cuts in this area.

28. We support the efforts of the European Broadcasting Union to protect audiences' access to major sporting events and believe that the government needs to protect the number of 'listed events' available to UK audiences on a free-to-air basis.

Talent Development and Training

Employment in the television industry is growing but it is a sector that, due to some significant barriers to entry, does not yet reflect the demographic make-up of the UK. There is an urgent need for a more consolidated approach to maximising entry-level opportunities and increasing investment in training and professional development at all levels of the industry.

29. Creative Skillset, as the key industry body that is charged with developing skills and talent, should coordinate a sector-wide response to challenges concerning entrance into and training within the television industry.

30. The government should meet urgently with industry bodies and broadcasters to consider how best to make the forthcoming apprenticeship levy work effectively for the television industry.

Note

1 The recommendations reproduced here are taken directly from the final report of the Puttnam Inquiry and reflect some of the pressing issues of the time, not least the impending publication of the new BBC Charter and Framework Agreement (eventually published in December 2016) and discussions inside government about whether to privatise Channel 4. The full report is at: http://futureoftv.org.uk/wp-content/uploads/2016/06/FOTV-Report-Online-SP.pdf.

Afterword

Vana Goblot and Natasha Cox[1]

The Inquiry into the Future of Public Service Television examined the role and purpose of public service television during what were (and remain) turbulent times for the industry. From new players and platforms settling in and expanding our television experience in the UK, to genuine government threats to the historical continuum of public service institutions, the current period is one in which upholding and strengthening PSB's key principles – such as universality, quality and diversity – seems to us to be more important than ever before.

Over the course of a short but intense period in 2015–2016, our focus as members of the core team of the Inquiry was to bring together television industry professionals, civil society groups, academics and campaigners in order to facilitate discussions and findings about the best way forward to ensure a robust and diverse public service ecology. We gathered together more than 50 submissions (and even penned one of them[2]), organised fourteen events,[3] and held regular meetings in order to consider the much-needed guidance by our two advisory bodies.[4] Across the Inquiry's own social media platforms – Twitter, Facebook and a YouTube channel – conversations ranged from audience representation and diversity to what a future Communications Act might look like. All of this proved essential in formulating ideas for the final report as our task was to consider how best to embed public service principles in a diverging and unsettling digital world. Judging by the response we have received to the report, we would like to think that, by and large, we met this brief. However, while were busy working through a broad and dissonant range of positions and arguments, we were also aware that we were only skirting around some bigger questions facing the television industry. While we hope that this volume goes a step further in updating research originally produced for the Inquiry with important contributions by industry professionals and international scholars, we believe there are two issues that remain underdeveloped: the significance of the independent production sector and the contribution of non-traditional sources of public service content.

Public Service Values, Freelancers and the Independent Sector

Public service television indisputably offers a range of qualities and values that remain essential for a healthy democracy. It has also been at the forefront of technological innovations like internet-distributed television: take, for example, the BBC's online news, red-button technology and the iPlayer. But if the PSBs are to continue being media pioneers, they might want to reconfigure and reimagine what the independent sector can offer to an evolving public service media environment.

We believe that PSBs play a central part in supporting the independent television production sector. But more needs to be done in an environment of increased competition, especially one fuelled by the consolidation and dominance of 'super-indies' and their focus on international sales, both of which are, according to Natasha Cox, 'stunting creative freedoms'.[5] The launch of BBC Studios in 2016 could be seen as an important step to readjust this market imbalance. However, we are yet to witness how BBC Studios' newfound ability to directly compete with the independent sector will contribute either to the strengthening of the BBC's public service mission or to satisfying the government's appetite for more competition in the media market. Indeed, it is too early to be sure whether BBC Studios will offer more creative freedom to independent producers or simply maintain the power of commissioners inside the Corporation.

Furthermore, for a healthy PSB ecology, it is important to carefully consider the impact of this big brand's entry into the market on smaller independent production companies. In order for the future of public service television to be more open and pluralistic, we may need to develop new relationships between broadcasters and all levels of the independent sector – not just the BBC and 'super-indies' but crucially, smaller units that are more organically tied to communities and interest groups – in the interest of delivering a broader range of public service content.

While our Inquiry attended to some of these concerns, the desired focus may have been overshadowed by other more pressing issues facing public service organisations. Threats to diminish the BBC's scope and serious deliberations to privatise the publicly owned Channel 4 required robust and focused responses. As Lord Puttnam observed at the Inquiry's launch event,[6] the BBC has been 'a permanently endangered, permanently threatened organisation' and public service organisations can only prosper 'in an atmosphere of confidence'.[7] Indeed, independent television production can only thrive if it is supported by confident, 'healthy 21st century broadcasters'.[8]

With these threats seemingly averted for the time being, British television continues to demonstrate why it continues to be one of the leading creative industries: producing multi-award winning programmes – from Channel 4's *Educating Yorkshire* and *Gogglebox* to the BBC's *Storyville* strand with Nick Fraser at the helm as commissioning editor. At the heart of this success is an independent production sector that attracts programme makers from different backgrounds and that brings in a greater diversity of perspectives, services and choice. If we want television to provide a more honest and accurate portrayal of life outside what many see as the privileged bubbles that dominate UK society, the BBC needs to nurture relationships as a curator across its platforms, to see the value and variety in working with freelancers in order to breathe fresh air into the Corporation.

Hidden in Plain Sight? New Sources of Public Service Content

The Puttnam Report revealed the resilience of the viewing figures of public service channels. Despite the exponential rise of online streaming services, the five main PSB channels still account for over half of total TV viewing.[9] While there is a steadily widening gap between old and young viewers, even the latter are reported to watch over two hours of TV content daily.[10] With UK subscription to VOD services such as Netflix and Amazon Prime on the rise,[11] traditional PSB broadcasters are responding to the connected world in order to keep pace with audiences' changing consumption habits. For example, BBC Three – now exclusively online – commissions something 'that has public service value in the short form space', according to its controller, Damian Kavanagh, and values how audiences respond on Twitter and Facebook as much as ratings.

Yet Vice, a media giant, synonymous with cutting-edge, provocative content that originally eschewed traditional platforms went the other way and, in October 2016, launched a UK TV channel, Viceland. Undeterred by gloomy reports of how the new initiative was progressing,[12] it soon began discussing plans to launch a TV studio in order to develop scripted programmes.[13] At the time of writing, Vice remains a pioneering and somewhat isolated example of this reverse move. Yet, even so, it highlights how important it is to keep a fresh view of the dynamics between platforms and content and the importance that partnerships play in the mix, as attested by Vice and the Guardian joining forces in December 2016 in order to develop new media formats across different genres and platforms.[14]

Indeed, our report sought to highlight that public service is an ecology situated in a multiplatform, connected world, and that its content therefore can no longer be seen as limited to the traditional public service broadcasting providers.[15] We were also frequently presented with arguments that some of the content produced by non-public service broadcasters is made to engage with viewers who feel they are underserved by traditional PSBs. According to Shane Smith, the CEO of Vice, 'elusive' young, diverse audiences do have an appetite for intellectually stimulating content. In his 2016 MacTaggart Lecture, he argued that challenging topics such as climate change and LGBTQ issues are all of interest to 16–24 year olds, yet they are very rarely covered by established television companies.[16] Similarly, short-form expert Luke Hyams, sections of whose talk at the Inquiry's launch is reproduced in this volume, insists that it is not only the choice of topics but also the tone and positioning of content that is crucial. He spoke powerfully of what he saw as the failure of due impartiality in the BBC's coverage of the 2011 London Riots to address issues of particular relevance to younger audiences and identified the broader issue of institutional political bias as one of the key reasons why sections of Generation Z have started to abandon traditional broadcasters.

Our Inquiry made several attempts to further probe this perceived gap in provision and to engage with the online platforms and social media that extend and enrich public service provision, but we faced some difficult issues. These were less about the reluctance of traditional public service providers to engage with these questions or the result of a clichéd 'commercial' versus 'publicly owned' binary. Instead, they involved rather more abstract and intangible problems: how do we involve non-PSB content providers when 'they' are not defining their own work as 'public service content'? And how do we persuade this highly individualised, unsystematised and unregulated sector of YouTubers and multiplatform independent companies even to begin to consider potential continuities between their socially engaged content and more traditional public service approaches? As much as social media can deliver the experience of communality and conversation, involving this disparate community in a dialogue about the future was by no means straightforward while the formal response from self-styled 'technology platforms' such as Facebook and Google was that of political neutrality and polite distance from the topic.

If we learnt anything in the time between the launch of our report and the publication of this volume, it is that this experience of detachment has become increasingly problematic with the emergence of fake news on the one hand and the enormous influence of these companies on the other. We are yet to witness the ideological transformation of these technology giants into responsible and fully-fledged media organisations, let alone institutions with formal public service responsibilities. What we know for sure is that we urgently need an open debate on the importance of public service principles beyond established platforms and institutions. Perhaps this is a topic for a future independent Inquiry.

Notes

1 Vana Goblot was the research associate and Natasha Cox was the media manager for A Future for Public Service Television Inquiry.
2 See Natasha Cox, submission to the Inquiry, http://futureoftv.org.uk/wp-content/uploads/2016/01/Natasha-Cox.pdf.
3 Our partners included the British Academy, BAFTA, the Guardian Events and the Hansard Society and Goldsmiths, University of London. For more information about our public events, go to http://futureoftv.org.uk/public-events/.
4 See more information and members of our two supporting bodies, including Advisory Committee http://futureoftv.org.uk/people/advisory-committee/ and Broadcast Panel, http://futureoftv.org.uk/people/broadcast-panel/.
5 Natasha Cox, submission to the inquiry.
6 The transcript of the event and press coverage is available at http://futureoftv.org.uk/events/do-we-still-need-public-service-television/.
7 Ibid.
8 Ibid.
9 Ofcom, *PSB Annual Report* (London: Ofcom, 2017), 2.
10 Ibid.
11 BARB, 'Subscription VOD Households', *BARB,* 17 January 2017, www.barb.co.uk/tv-landscape-reports/tracker-svod/.
12 For example, see Mark Sweeney, 'Viceland UK Scores Zero Ratings on Some Nights after Sky TV Launch', *The Guardian,* 4 October 2016, www.theguardian.com/media/2016/oct/04/viceland-uk-ratings-sky-tv.
13 Manori Ravindran, 'Vice to Launch Scripted Studio', *Broadcast,* 20 June 2017. www.broadcastnow.co.uk/news/vice-to-launch-scripted-studio/5119263.article.
14 GNM press office, 'Vice and The Guardian Announce New Partnership', *The Guardian,* 8 December 2016, www.theguardian.com/gnm-press-office/2016/dec/08/vice-and-the-guardian-announce-new-partnership.
15 See Lord Puttnam, *A Future for Public Service Television: Content and Platforms in a Digital World* (London: Goldsmiths, University of London, 2016), Chapter 7.
16 Shane Smith, The McTaggart Lecture, 25 August 2016. www.youtube.com/watch?v=ZSdkxXGQAPA.

Contributors (Editors and Commissioned Authors)

Tess Alps is the Chair of Thinkbox, the body owned by the UK broadcasters, whose role is to help advertisers get the best out of today's diverse, multi-platform TV. She set up the company in 2006 and was its first CEO. Immediately prior to this she was the Chairman of the PHD Group in the UK. Tess is also a Council member of the Advertising Standards Authority and a member of the corporate board of the Royal Academy of Arts. She is a Fellow of the RTS, and a member of BAFTA. In 2007 she won the Outstanding Achievement award from Women in Film and Television and in 2013 she was voted Media Industry Leader of the Decade.

Patrick Barwise (www.patrickbarwise.com) is Emeritus Professor of Management and Marketing at London Business School. He joined LBS in 1976 after an early career at IBM and has published widely on management, marketing and media. He is also the former Chairman of Which?, Europe's largest consumer organisation and an experienced expert witness in international commercial, tax and competition cases.

James Bennett is Professor of Television and Digital Culture at Royal Holloway, University of London. He is co-director of the Centre for the History of Television Culture & Production and currently Principal Investigator on the ADAPT social media television history project.

Georgina Born is Professor of Music and Anthropology at Oxford University. Her work combines ethnographic and theoretical writings on media, music and cultural production. She directs the research programme 'Music, Digitisation, Mediation', funded by the European Research Council. Her books include *Rationalizing Culture* (1995), *Western Music and Its Others* (2000), *Uncertain Vision: Birt, Dyke and the Reinvention of the BBC* (2004), *Music, Sound and Space* (2013) and *Interdisciplinarity* (2013). She is Honorary Professor of Anthropology at University College London, and in 2014 she was elected a Fellow of the British Academy, where she chairs the Culture, Media and Performance group.

Natasha Cox is a documentary and television producer working in London. She has produced documentaries for the BBC and Channel 4 as well as contributing to the award winning film, *(Still) The Enemy Within* released in 2014. Natasha has written about the media for Open Democracy and Our Beeb and is a regular writer for Through the Cracks. Natasha was the Media Manager for *A Future for Public Service Television* Inquiry.

Gunn Enli is Professor of Media Studies at the University of Oslo, Norway. She is the co-author of the *Routledge Companion to Social Media and Politics* (2016) and *The Media Welfare State* (2015) and author of *Mediated Authenticity* (2015). She has published widely on political campaigning, social media and media policy.

Des Freedman is Professor of Media and Communications at Goldsmiths, University of London. He is the author of *The Contradictions of Media Power* (2014) and *The Politics of Media Policy* (2008) and co-author (with James Curran and Natalie Fenton) of *Misunderstanding the Internet* (2nd edn, 2016). He is a former Chair of the Media Reform Coalition and was project lead for the Inquiry into a Future for Public Service Television.

Vana Goblot lectures at Goldsmiths, University of London. She was a research associate on A Future for Public Service Television Inquiry. Her research interests include the history and production cultures of television industry and issues of quality and cultural value. Her PhD, which she completed in 2013, examines public service values in the digital, multiplatform age, using the case study of BBC Four. Vana worked previously as a translator and journalist in the UK, and a broadcast journalist in the former Yugoslavia.

David Hendy is Professor of Media and Cultural History at the University of Sussex, and has been commissioned to write an authorised single-volume history of the BBC for its Centenary in 2022. His previous books include *Life on Air: a History of Radio Four* (2007), *Public Service Broadcasting* (2013), *Noise: a Human History of Sound and Listening* (2013), and *Radio in the Global Age* (2000). He has also written and presented several series for BBC Radio 3 and BBC Radio 4, including *Rewiring the Mind* (2010), *Noise: a Human History* (2013), and *Power of Three* (2016).

Jennifer Holt is Associate Professor of Film and Media Studies at the University of California, Santa Barbara. She is the author of *Empires of Entertainment* (2011) and co-editor of *Distribution Revolution* (2014); *Connected Viewing: Selling, Streaming &*

Sharing Media in the Digital Age (2013); and *Media Industries: History, Theory, Method* (2009). Her work has appeared in journals and anthologies including *Cinema Journal, Journal of Information Policy, Moving Data*, and *Signal Traffic: Critical Studies of Media Infrastructures*. She is a founding member of the *Media Industries* journal editorial collective.

Amanda D. Lotz is Professor in the Department of Communication Studies at the University of Michigan and Fellow at the Peabody Media Center. She is the author, coauthor, or editor of eight books that explore television and media industries including *The Television Will Be Revolutionized* (2014) and *Portals: A Treatise on Internet-Distributed Television* (2017).

Sarita Malik is Professor of Media, Culture and Communications at Brunel University London. Her research explores issues of social change, inequality, communities, and cultural representation. Sarita is currently leading 'Creative Interruptions', a large international research project funded by the UK Arts and Humanities Research Council. The project examines how the arts, media and creativity are used to challenge marginalisation. Sarita previously worked in broadcast journalism, research development and arts management.

Matthew Powers is Assistant Professor in the Department of Communication at the University of Washington in Seattle. His research has been published in *Journal of Communication, International Journal of Press/Politics*, and *Journalism: Theory, Practice & Criticism*, among others. He holds a Ph.D. in Media, Culture and Communication from New York University.

Lord Puttnam was the chair of the *A Future for Public Service Television* Inquiry. He is an independent producer of award-winning films including *The Mission, The Killing Fields, Local Hero, Chariots of Fire* and *Midnight Express*. Lord Puttnam was Deputy Chairman of Channel 4 Television (2006–2012), Vice President and Chair of Trustees at BAFTA from whom, in 2006, he received a Fellowship. From 2012–2017, he was the UK Prime Minister's Trade and Cultural Envoy to Vietnam, Laos, Cambodia and Burma. He is the chair of Atticus Education, an online education company based in Ireland and President of the Film Distributors' Association. He is also chair of Nord Anglia International School, Dublin; Life President of the National Film & Television School; a UNICEF Ambassador; Adjunct Professor of Film Studies and Digital Humanities at

University College Cork; and an international Ambassador for WWF. He is the recipient of over 50 Honorary Degrees, Diplomas and Fellowships from the UK and overseas.

Trine Syvertsen is Professor of media studies at the University of Oslo. She is author of *Media Resistance: Dislike, Protest, Abstention* (2017) and co-author of *The Media Welfare State* (2014). She has published extensively on television, media history and policy.

Jon Thoday is the joint-founder and Managing Director of Avalon, a multi-award winning talent management, live promotion and television production group. Since founding the company, Jon has produced numerous ground-breaking television shows including Emmy and Peabody award winner *Last Week Tonight with John Oliver* (HBO), multi-award winning and Emmy-nominated *Catastrophe* (Channel 4/Amazon Prime), RTS and Rose d'Or winning *Not Going Out* (BBC1's longest running sitcom currently on air), multi-BAFTA award winning *Harry Hill's TV Burp* (ITV1), *Russell Howard's Good News* (BBC2), *Fantasy Football League* (BBC/ITV), *The Frank Skinner Show* (ITV) and *Workaholics* (Comedy Central USA).

Mark Thompson became president and chief executive officer of The New York Times Company in November, 2012. He has directed the Company's strategy and presided over an expansion of its digital and global operations. Previously, he served as Director-General of the BBC. He joined the BBC in 1979. He left for two years in 2002 to become CEO of Channel 4 Television in the U.K. before returning in 2004 as Director-General. His book, *Enough Said: What's Gone Wrong with the Language of Politics?* was published in the UK and US in September 2016. Mark Thompson was educated at Stonyhurst College and Merton College, Oxford.

Illustrations

Index